W9-ADV-694

Analytic Theory of
Partial Differential Equations

Analytic Theory of Partial Differential Equations

David L. Colton

University of Delaware

π

Pitman Advanced Publishing Program

Boston · London · Melbourne

PITMAN PUBLISHING LIMITED
39 Parker Street, London WC2B 5PB

PITMAN PUBLISHING INC
1020 Plain Street, Marshfield, Massachusetts

Associated Companies
Pitman Publishing Pty Ltd., Melbourne
Pitman Publishing New Zealand Ltd., Wellington
Copp Clark Pitman, Toronto

AMS Subject Classifications: (main) 35A20, 35B50, 35B60, 35R25, 35R30
(subsidiary) 35J18, 35K10, 30G20, 35A35

Library of Congress Cataloging in Publication Data

Colton, David L 1943–
Analytic theory of partial differential equations.

(Monographs and studies in mathematics; 8)
Bibliography: p. 225
Includes index.
1. Differential equations, Partial—Numerical solutions. I. Title. II. Series.
QA374.C648 515.3′53 80-14112
ISBN 0 273 08462 3

Printed in Great Britain by Pitman Press, Bath.

Contents

Preface

This book is concerned with the development of the analytic, or function theoretic, behavior of solutions to partial differential equations. This area of mathematics has had a long and rich history and certain topics in the field, such as maximum principles and the Cauchy–Kowalewski theorem, are by now part of the standard menu of any introductory course in partial differential equations. In recent years a wealth of new results has been obtained and reported on in a variety of research monographs (Bergman [12]; Bers [16]; Colton [27], [28]; Gilbert [70], [71], [72]; Krzywoblocki [107]; Tutschke [168]; Vekua [171], [172]; Wendland [175]), symposium proceedings (Meister, Weck and Wendland [125]; Ruscheweyh [154]), special 'anniversary' volumes (Gilbert and Weinacht [77]), and textbooks (Bergman and Schiffer [15]; Bers, John and Schechter [17]; Garabedian [61]; Haack and Wendland [80]; Protter and Weinberger [146]). However, for the most part these texts concentrate only on certain aspects of the general theory and as a result the graduate student or professional mathematician may be hard-pressed to arrive at an overview of the entire area. This problem is compounded by the fact that many important new results have not yet found their way into any book at all. It is therefore our purpose to provide an introductory treatment of the general area of the analytic theory of partial differential equations that will be accessible to anyone having a basic knowledge of analytic function theory and partial differential equations (e.g. Ahlfors [3] and Epstein [51]), as well as a passing familiarity with the elements of functional analysis (the material in Epstein [51] is more than sufficient). In our attempt to survey as broad an area as possible, while at the same time keeping an introductory flavor, we have reluctantly been forced to exclude certain topics which we have considered to be of a more specialized nature. However, having read the present book, the reader can easily fill in these gaps by consulting one or more of the above-cited references.

The plan of the book is as follows. The first chapter is concerned with maximum principles for elliptic and parabolic equations, as well as an elementary discussion of Phragmén–Lindelof principles. As an example of the use of maximum principles we give an application to a problem appearing in nuclear reactor theory. The first two parts of Chapter 2 are a continuation of Chapter 1 and are devoted to deriving a Harnack's inequality for elliptic equations and its application in the derivation of Liouville's theorem for solutions of elliptic equations defined in the entire plane. The last section of Chapter 2 discusses one of the simplest problems in scattering theory: that of the scattering of a plane acoustic wave by a spherically stratified medium of compact support. In addition to providing the background for later sections concerned with more complicated problems in scattering theory, this section provides us with an opportunity to introduce in a simple manner the concept of integral operator methods for solving partial differential equations.

Chapters 3 and 4 deal with the analyticity of solutions to partial differential equations and the problem of the unique continuation of these solutions. After a brief discussion of the content of the Cauchy–Kowalewski theorem, we proceed to establish Holmgren's uniqueness theorem and the equivalence of the unique continuation and Runge approximation properties for elliptic equations. This is followed by a proof of the analyticity of solutions to semilinear elliptic equations in two independent variables and an example of an equation that far from having only analytic solutions has no solutions whatsoever. We then turn in Chapter 4 to the problem of the analytic continuation of solutions to elliptic and parabolic equations and establish reflection principles for equations of this type in two independent variables. Included in our discussion is the concept of Huygen's principle for reflection, the failure in general of establishing generalized reflection principles for partial differential equations in more than two independent variables, and a Schwarz reflection principle for solutions of the Helmholtz equation that vanish on a portion of a sphere. Chapter 4 is concluded by a proof of Gilbert's envelope method, its application to the location of singularities of the axially symmetric potential equation, and the analytic continuation of analytic solutions to the heat equation. This last topic arises in connection with the inverse Stefan problem and is in the spirit of earlier work by Garabedian in the use of inverse methods to solve free boundary problems (cf. Garabedian [65]).

Chapter 5 is concerned with the development of Runge approximation theorems for elliptic and parabolic equations in interior domains and the Helmholtz equation in exterior domains. The proofs of these theorems are based in one way or another on the results on the analytic behavior of solutions established in Chapters 3 and 4. Included in our discussion is a treatment of integral operators for elliptic equations in two independent variables (based on the complex Riemann function), the Bergman–Vekua theorem, the Bergman kernel function, and polynomial solutions of the heat equation. The results of this chapter have applications to the numerical solution of boundary value problems for partial differential equations, and references are given for examples of this. Of particular importance in this context is that since the approximating functions are in fact solutions of the governing differential equation, the dimensionality of the problem is reduced by one.

Chapter 6 is an elementary introduction to the theory of pseudoanalytic or generalized analytic functions as developed by Bers [16] and Vekua [171]. After deriving certain results on a class of singular integrals, the similarity principle is obtained and used to prove Carleman's theorem, Liouville's theorem and a maximum modulus principle for elliptic systems in two unknown functions defined in the plane. The last section of this chapter is devoted to an alternate approach for treating such systems based on the use of integral operators, and by this method we obtain a version of Cauchy's integral representation and a reflection principle for solutions of the elliptic system under consideration. The discussion in this chapter is purposely kept elementary, and for a more detailed development and generalizations to elliptic systems in more than two unknown functions the reader is referred to Bers [16], Gilbert [72], Haack and Wendland [80], Tutschke [168], Vekua [171] and Wendland [175].

The last two chapters of this book make use of our previously derived results to study two of the classical inverse problems of mathematical physics, the backwards heat equation and the inverse scattering problem. Both of these are basically problems in the analytic continuation of solutions to partial differential equations, but where now the solutions are not known exactly but only in some approximate sense. In the case of the backwards heat equation the domain of regularity of the solution is given, whereas in the case of the inverse scattering problem this is in general unknown and hence the problem becomes considerably more complicated. We begin in Chapter 7 by studying the backwards heat equation using the methods of logarithmic convexity and quasireversibility. Included here is a discussion of pseudoparabolic equations in one-space variable based on the use of the Riemann function and the fundamental solution. Following our treatment of the backwards heat equation we turn in Chapter 8 to the inverse scattering problem. Here we first establish a relationship between the indicator diagram of an entire function related to the far field pattern and the location of the singularities of the scattered wave. This is followed by a short treatment of the problem of determining the speed of sound in a spherically stratified medium by using the Born approximation and the Backus–Gilbert method for improperly posed moment problems. Finally, we give an introductory treatment of potential theory making use of the concept of a dual system and use these results along with conformal mapping techniques to solve the inverse scattering problem for both a 'hard' and 'soft' cylinder given a knowledge of the far field pattern at low frequencies.

This book evolved from a one-year course given at the University of Delaware during the academic year 1978–79. A one-semester course can be based on the first four chapters, or alternatively on Chapters 3 through 6. Another possibility is to base a seminar on the analytic continuation of solutions to partial differential equations (both 'exact' and 'approximate' continuation) on Chapters 3, 4, 7 and 8. By combining selected chapters from the present volume along with others taken from a book dealing with Hilbert space methods in partial differential equations (e.g. Showalter [159]) it is also possible to fashion a basic one-year graduate course on partial differential equations that gives a balanced view of both function theoretic and Hilbert space methods in studying partial differential equations. For the ease of the instructor we have placed a short collection of exercises at the end of each chapter.

The author gratefully acknowledges financial support from the Air Force Office of Scientific Research under grant AFOSR 76-2879 and the National Science Foundation under grant MCS 78-02452. A particular note of thanks is given to Alison Chandler for her careful typing of the manuscript.

David Colton
Newark, Delaware
September 1979

List of symbols

Sets

$\partial\Omega$	boundary of Ω
$\bar{\Omega}$	closure of Ω
$\overline{PQ}$	the line segment joining the points P and Q
D^*	the reflection of the set $D \subset R^2$ with respect to the x axis
$\rho(\mathbf{A})$	resolvent set of operator $\mathbf{A}$
$\sigma(\mathbf{A})$	spectrum of operator $\mathbf{A}$

Scalars

$\min(a, b)$	the smaller of a and b
$\max(a, b)$	the larger of a and b
$\max\limits_D u$	maximum of function u over the set D
$\min\limits_D u$	minimum of function u over the set D
$\underline{\lim}$	limit inferior
$\overline{\lim}$	limit superior
$\sup\limits_D f$	supremum of function f over the set D
$\inf\limits_D f$	infimum of function f over the set D
$\|P\|$	distance of point P from the origin
$\bar{z}$	the complex conjugate of z
$\|m\|$	$m_1 + \cdots + m_n$ where $m = (m_1, \ldots, m_n)$
$r(\mathbf{A})$	spectral radius of operator $\mathbf{A}$

Linear spaces

R^n	Euclidean n space
C^n	space of n complex variables
$(.,.)$	inner product

$\langle .\,, . \rangle$	bilinear form
$\|\cdot\|$	norm
$\langle X, Y \rangle$	dual system
$f \perp A$	f is orthogonal to all elements of the set A

Functions

exp	exponential
J_n	Bessel function of order n
$j_{n_{(1)}}$	spherical Bessel function of order n
$H_{n_{(1)}}$	Hankel function of first kind of order n
h_n	spherical Hankel function of first kind of order n
H_n	Hermite polynomial of degree n
Γ	gamma function
S_{nm}	spherical harmonic
P_n^l	associated Legendre polynomial
P_n	Legendre polynomial
O-notation	$u = O(v)$ if there exists a positive constant K such that $\|u\| \leq Kv$
$S \ll \tilde{S}$	the series $\tilde{S}$ dominates the series S

Function spaces

$C(D)$	set of continuous functions defined on D
$C^n(D)$	set of n times continuously differentiable functions defined on D
$C_0^n(D)$	set of n times continuously differentiable functions defined on D and having compact support in D
$C^{n+\epsilon}(D)$	set of n times continuously differentiable functions defined on D where nth derivative is Hölder continuous with exponent ϵ
$L_2(D)$	set of functions which are square integrable over D in the sense of Lebesgue

Operators

Δ_n	Laplacian in R^n
grad u, ∇u	gradient of u
Re	real part
Im	imaginary part

1

Maximum principles

In this chapter we shall show that, analogous to analytic functions of a complex variable, solutions of elliptic and parabolic partial differential inequalities achieve their maximum (and minimum) on the boundary of their domain of definition. These results, due in their final form to Hopf [89] and Nirenberg [138], are now considered to be among the 'classical' results in the theory of partial differential equations, and can be found in numerous textbooks on the subject. We assume that the reader is already acquainted with the elementary proofs of the maximum principle for the Laplace and heat equations. An excellent source, which has strongly influenced our own presentation, is Protter and Weinberger [146], and the reader is urged to consult this book for further developments and applications. For the sake of simplicity and ease of visualization we shall present proofs only for equations defined in the plane, and simply state the results for the case of higher dimensions. We also assume that all solutions are smooth enough such that the derivatives appearing in the differential equation exist and are continuous. The plan of this chapter is briefly as follows. We first consider maximum principles for elliptic equations and prove Hopf's so-called 'first and second lemmas' (Theorems 1.1 and 1.2, respectively). We then prove the analogous results for parabolic equations as given by Friedman [57] and Nirenberg [137]. These theorems on maximum principles for partial differential equations are then followed by a short discussion of the Phragmén–Lindelof principle for elliptic and parabolic equations, and the chapter is concluded by an example of the use of maximum principles in the study of a certain problem arising in nuclear reactor theory.

1.1 Elliptic equations

We consider the differential inequality

$$L[u] \equiv a(x, y)\frac{\partial^2 u}{\partial x^2} + b(x, y)\frac{\partial^2 u}{\partial x\, \partial y} + c(x, y)\frac{\partial^2 u}{\partial y^2} + d(x, y)\frac{\partial u}{\partial x}$$

$$+ e(x, y)\frac{\partial u}{\partial y} + h(x, y)u \geqslant 0 \tag{1.1}$$

where the coefficients are assumed to be uniformly bounded functions of x and y in a domain D, and L is assumed to be uniformly elliptic. The uniform ellipticity of (1.1) means that there exists a positive constant γ such that for x, y in D and real numbers ξ, η we have

$$a(x, y)\xi^2 + 2b(x, y)\xi\eta + c(x, y)\eta^2 \geqslant \gamma(\xi^2 + \eta^2). \tag{1.2}$$

Theorem 1.1 *Let u satisfy* (1.1) *in D where $h(x, y) \leqslant 0$ for $(x, y) \in D$. Then u cannot achieve its nonnegative maximum in D unless u is a constant.*

Note: If D is bounded and u is continuous in $\bar{D}$ then, since a continuous function on a compact set achieves its maximum values on that set, the theorem states that if the maximum of u is nonnegative, it must occur on the boundary ∂D of D. If $h \equiv 0$ then, since any constant satisfies the inequality (1.1), we can drop the assumption that the maximum is nonnegative. Finally, if we replace u by $-u$ we obtain corresponding results when functions satisfying $L[u] \leqslant 0$ can achieve a (nonpositive) minimum.

Proof We shall assume that u achieves its maximum in D and show that u must be a constant. Let $M \geqslant 0$ be the maximum value of u in D. Suppose u is not identically a constant. Then there exists an open disk Ω_1 contained in D such that on $\partial\Omega_1$ there is a point (ξ, η) where $u = M$, but $u < M$ in Ω_1. Let Ω be an open disk such that $\bar{\Omega} \subset \Omega_1$ except at the point (ξ, η) where $\partial\Omega$ is tangent to $\partial\Omega_1$. Let the origin be chosen at the center of Ω, the radius of Ω be R, and Ω_2 an open disk with center at (ξ, η) of radius less than R and such that $\bar{\Omega}_2 \subset D$ (see Fig. 1.1). We now observe that $\partial\Omega_2$ can be divided into two components, one of which is the closed arc $\sigma_1 = \partial\Omega_2 \cap \bar{\Omega}$ and the other $\sigma_2 = \partial\Omega_2 \backslash \sigma_1$. On σ_1 we have that there exists a positive number ε such that

$$u \leqslant M - \varepsilon \quad \text{on} \quad \sigma_1. \tag{1.3}$$

Now consider the comparison function defined by

$$U = e^{-\alpha r^2} - e^{-\alpha R^2} \tag{1.4}$$

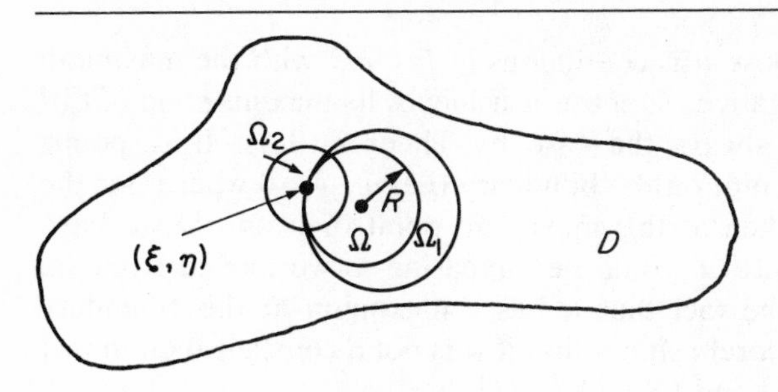

Fig. 1.1

where $r=(x^2+y^2)^{1/2}$ and α is a positive parameter to be chosen later. Note that for $\alpha>0$ we have

$$\begin{aligned} &U>0 \quad \text{in} \quad \Omega \\ &U=0 \quad \text{on} \quad \partial\Omega \\ &U<0 \quad \text{outside} \quad \Omega \end{aligned} \tag{1.5}$$

and from (1.1), (1.2) we have

$$\begin{aligned} L[U]&=\{4\alpha^2(ax^2+2bxy+cy^2)-2\alpha(a+c+dx+ey)+h\}e^{-\alpha r^2}-he^{-\alpha R^2} \\ &>0 \end{aligned} \tag{1.6}$$

for $(x, y)\in\bar{\Omega}_2$ and α sufficiently large. Furthermore, from (1.3), (1.5) we have that there exists a positive constant δ such that

$$u+\delta U<M \quad \text{on} \quad \partial\Omega_2. \tag{1.7}$$

Note that from (1.1), (1.6) we have

$$L[u+\delta U]>0 \quad \text{in} \quad \bar{\Omega}_2. \tag{1.8}$$

At the point (ξ, η) we have $u+\delta U=M\geqslant 0$, and hence from (1.7) we can conclude that $u+\delta U$ must achieve a nonnegative maximum at some point $(\xi', \eta')\in\Omega_2$. From the calculus of several variables we have that at (ξ', η') the function $v=u+\delta U$ satisfies

$$\partial v/\partial\xi=\partial v/\partial\eta=0, \qquad \partial^2 v/\partial\xi^2\leqslant 0, \qquad \partial^2 v/\partial\eta^2\leqslant 0 \tag{1.9}$$

for any coordinates ξ, η obtained from x, y by a linear transformation. However, at (ξ', η') we can make a change of variables such that at (ξ', η') the principal part of (1.1) becomes the Laplacian, and hence from (1.8) we conclude that $u+\delta U$ does *not* achieve a maximum at (ξ', η'), a contradiction. Hence u must be identically constant, contrary to assumption, and this proves Theorem 1.1.

Now suppose $L[u]\geqslant 0$ in a domain D with smooth boundary ∂D.

Assume $h \leq 0$ and suppose u is continuous in $D \cup \partial D$ with the maximum of u in D being nonnegative. Suppose u achieves its maximum on ∂D (if D is bounded this is always the case by Theorem 1.1). If $\vec{\mu}$ points outward from ∂D at a point on the boundary (i.e. $\vec{\mu} \cdot \vec{\nu} > 0$ where $\vec{\nu}$ is the unit outward normal) then at this maximum point of u on ∂D we have $\partial u/\partial \mu \geq 0$, since otherwise u would be increasing as we moved from ∂D into D, contradicting the fact that u has a maximum at this boundary point. The following theorem shows that if u is not a constant then in fact $\partial u/\partial \mu > 0$ (cf. Hopf [91] and Oleĭnik [141]).

Theorem 1.2 *Let u satisfy (1.1) in D where $h(x, y) \leq 0$ for (x, y) in D. Suppose that $u \leq M$ in D, $M \geq 0$, and $u(\xi, \eta) = M$ where $(\xi, \eta) \in \partial D$. Suppose (ξ, η) lies on the boundary of an open disk $\Omega \subset D$. Then if u is continuous in $D \cup \{(\xi, \eta)\}$ and the outward directional derivative $\partial u/\partial \mu$ exists at (ξ, η), we have*

$$\partial u/\partial \mu > 0 \quad \text{at} \quad (\xi, \eta)$$

unless $u \equiv M$.

Proof Let Ω be an open disk of radius R such that $\bar{\Omega}$ is contained in $D \cup \{(\xi, \eta)\}$ and Ω is tangent to ∂D at (ξ, η). Without loss of generality assume that Ω is centered at the origin. Let Ω_1 be an open disk with center at (ξ, η) and radius $R/2$ (see Fig. 1.2). Let U be defined as in (1.4), where α is such that $L[U] > 0$ in Ω_1. By Theorem 1.1 we have that if $u \not\equiv M$ then $u < M$ in $\bar{\Omega}$ except at the point (ξ, η). Since $U = 0$ on $\partial\Omega$ we can find $\delta > 0$ such that $u + \delta U \leq M$ on $\partial\Omega_1 \cap \bar{\Omega}$. Hence $u + \delta U \leq M$ in the shaded region of Fig. 1.2. Since $L[u + \delta U] > 0$ in this region, by arguments as in Theorem 1.1, we have that the maximum of $u + \delta U$ occurs at (ξ, η) and is equal to M. Hence at (ξ, η)

$$(\partial/\partial \mu)(u + \delta U) = \partial u/\partial \mu + \delta(\partial U/\partial \mu) \geq 0. \tag{1.10}$$

But at (ξ, η) we have that the unit normal $\vec{\nu}$ points in the direction of the

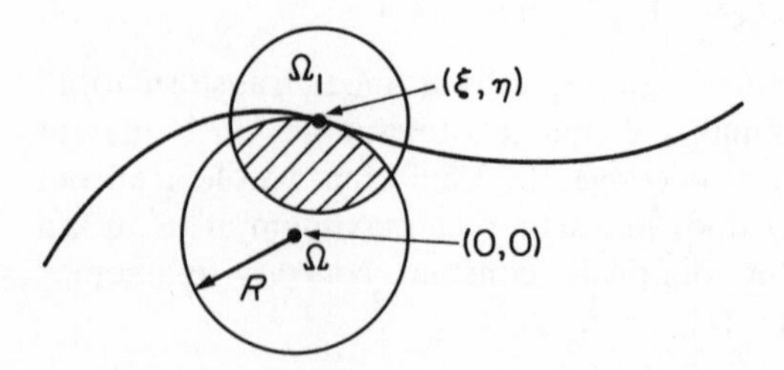

Fig. 1.2

radius vector and hence

$$\partial U/\partial \mu = \vec{\mu} \cdot \vec{\nu}(\partial U/\partial r) = -2\alpha R e^{-\alpha R^2} \vec{\mu} \cdot \vec{\nu} < 0 \tag{1.11}$$

since $\vec{\mu} \cdot \vec{r} > 0$. Hence from (1.10) we have $\partial u/\partial \mu > 0$.

Note: If $h \equiv 0$ then the condition $M \geq 0$ can be removed from the hypothesis of Theorem 1.2. Furthermore, if we replace u by $-u$ we can obtain minimum principles associated with $L[u] \leq 0$, where if $h \not\equiv 0$ we must require that the minimum is nonpositive.

The proofs of Theorems 1.1 and 1.2 can be easily extended to higher dimensions, and for the sake of completeness we give these extensions in the theorems below (cf. Protter and Weinberger [146]).

Theorem 1.3 *Let u satisfy the inequality*

$$L[u] \equiv \sum_{i,j=1}^{n} a_{ij}(\mathbf{x}) \frac{\partial^2 u}{\partial x_i \, \partial x_j} + \sum_{i=1}^{n} b_i(\mathbf{x}) \frac{\partial u}{\partial x_i} + h(\mathbf{x}) u \geq 0$$

in a domain D in which the coefficients of L are uniformly bounded, $h \leq 0$, and L is uniformly elliptic, i.e. there exists a positive constant γ such that for all n-tuples of real numbers $(\xi_i, \ldots, \xi_n)$ and $\mathbf{x} \in D$,

$$\sum_{i,j=1}^{n} a_{ij}(\mathbf{x}) \xi_i \xi_j \geq \gamma \sum_{i=1}^{n} \xi_i^2.$$

If u attains a nonnegative maximum M at a point in D, then $u \equiv M$.

Theorem 1.4 *Let μ satisfy the inequality $L[u] \geq 0$ where L is the operator of Theorem 1.3. Suppose that $u \leq M$ in D, $u = M$ at a boundary point P, and that $M \geq 0$. Assume that P lies on the boundary of a ball in D. If u is continuous in $D \cup P$, any outward directional derivative of u at P is positive unless $u \equiv M$ in D.*

1.2 Parabolic equations

We now want to prove the analogue of Theorem 1.1 for the parabolic inequality

$$L[u] \equiv a(x, t) \frac{\partial^2 u}{\partial x^2} + b(x, t) \frac{\partial u}{\partial x} + c(x, t) u - \frac{\partial u}{\partial t} \geq 0 \tag{1.12}$$

where the coefficients are again assumed to be uniformly bounded functions of x and t in a domain D, and L is uniformly parabolic in D, i.e.

there exists a positive constant γ such that

$$a(x, t) \geq \gamma \tag{1.13}$$

for (x, t) in D. As will be seen, the maximum principle for (1.12) is quite different from that for (1.1). The extension of the maximum principle for (1.12) to higher dimensions is again straightforward. Following Protter and Weinberger [146] we break the proof of the maximum principle into three lemmas.

Lemma 1.1 *Let Ω be an open disk such that $\bar{\Omega} \subset D$. Suppose u satisfies* (1.12) *in D where $c(x, t) \leq 0$ in D and u achieves its maximum value $M \geq 0$ at a point (ξ, τ) on $\partial\Omega$. Suppose $u < M$ in Ω. Then the tangent to Ω at (ξ, τ) is parallel to the x-axis.*

Proof Without loss of generality assume the center of Ω is at the origin and let R be the radius of Ω. We shall assume (ξ, τ) is not at the top or bottom of Ω (i.e. $\xi \neq 0$) and reach a contradiction. Without loss of generality we can assume (ξ, τ) is the only boundary point where $u = M$, since otherwise we can replace Ω by a smaller disk Ω' such that $\Omega' \subset \Omega$ and Ω' and Ω are tangent at (ξ, τ).

Let Ω_1 be a disk with center at (ξ, τ) and radius $R_1 < |\xi|$ such that $\Omega_1 \subset D$. Let σ_1 be the closed arc of Ω_1 contained in $\bar{\Omega}$ and let $\sigma_2 = \partial\Omega_1 \backslash \sigma_1$ (see Fig. 1.3).

On σ_1 we have that there exists a positive constant ε such that

$$u \leq M - \varepsilon \quad \text{on} \quad \sigma_1. \tag{1.14}$$

Since the maximum of u in D is M we have

$$u \leq M \quad \text{on} \quad \sigma_2. \tag{1.15}$$

Now consider the comparison function

$$U = e^{-\alpha r^2} - e^{-\alpha R^2} \tag{1.16}$$

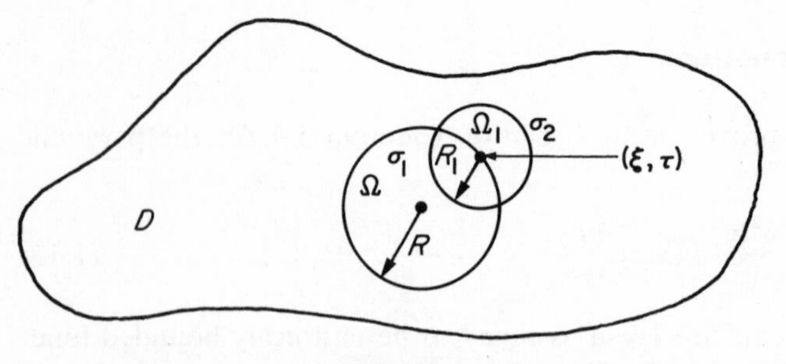

Fig. 1.3

where $r=(x^2+t^2)^{1/2}$ and α is a positive parameter. Then, as in Theorem 1.1, we have

$$\begin{aligned} U&>0 \quad \text{in} \quad \Omega \\ U&=0 \quad \text{on} \quad \partial\Omega \\ U&<0 \quad \text{outside} \quad \Omega \end{aligned} \tag{1.17}$$

and for α sufficiently large

$$L[U]=\{4\alpha^2ax^2-2\alpha a-2\alpha bx+2\alpha t+c\}e^{-\alpha r^2}-ce^{-\alpha R^2}>0 \tag{1.18}$$

for (x, t) in $\bar{\Omega}_1$ since $|x|\geqslant|\xi|-R_1>0$ in $\bar{\Omega}_1$. Again, as in Theorem 1.1, there exists a positive constant δ such that

$$u+\delta U<M \quad \text{on} \quad \partial\Omega_1 \tag{1.19}$$

and

$$L[u+\delta U]>0 \quad \text{in} \quad \bar{\Omega}_1. \tag{1.20}$$

Since $u+\delta U=M\geqslant 0$ at (ξ, τ) we conclude that the nonnegative maximum of $u+\delta U$ in Ω_1 must occur at an interior point and this leads to a contradiction since at this point $\partial u/\partial x=\partial u/\partial t=0$ and $\partial^2u/\partial x^2\leqslant 0$.

Lemma 1.2 *Let u satisfy (1.12) where $c(x, t)\leqslant 0$ in D. Suppose $u<M$ at some point (ξ, τ) of D and that $u\leqslant M$ in D, where $M\geqslant 0$. If l is any horizontal line segment in D which contains (ξ, τ), then $u<M$ on l.*

Proof The proof is by contradiction. Suppose $u=M$ at some point (ξ', τ) of l. Without loss of generality we assume that $\xi'<\xi$ and $u<M$ for $\xi'<x\leqslant\xi$. Let d_0' be the distance of the line segment $\xi'\leqslant x\leqslant\xi$, $t=\tau$, to ∂D and $d_0=\min(\xi-\xi', d_0')$. We now define the function $d(x)$ for $\xi'<x<\xi'+d_0$ to be the distance from (x, τ) to the nearest point in D where $u=M$. Since $u(\xi', \tau)=M$ we have $d(x)\leqslant x-\xi'$. Lemma 1.1 implies that the nearest point is directly above or below (x, τ), i.e. $u(x, \tau+d(x))=M$ or $u(x, \tau-d(x))=M$. Hence, for any $\delta>0$ (see Fig. 1.4)

$$d(x+\delta)\leqslant(d(x)^2+\delta^2)^{1/2}<d(x)+\delta^2/2d(x). \tag{1.21}$$

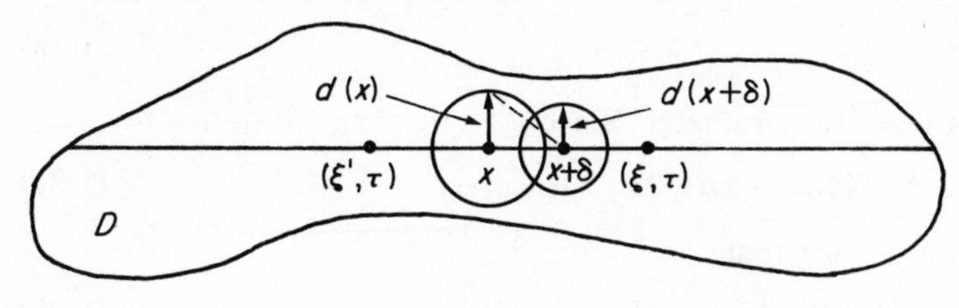

Fig. 1.4

Similarly, for $0<\delta<d(x)$

$$d(x+\delta)>(d(x)^2-\delta^2)^{1/2}. \tag{1.22}$$

Now keep δ such that $0<\delta<d(x)$. Subdivide the interval $(x, x+\delta)$ into n equal parts and apply (1.21), (1.22) to conclude that for $j=0, 1, \ldots, n-1$

$$\begin{aligned} d(x+((j+1)/n)\delta)-d(x+(j/n)\delta) &\leqslant \frac{\delta^2}{2n^2 d(x+(j/n)\delta)} \\ &\leqslant \frac{\delta^2}{2n^2(d(x)^2-\delta^2)^{1/2}} \end{aligned} \tag{1.23}$$

and hence, summing from $j=0$ to $n-1$,

$$d(x+\delta)-d(x)\leqslant\frac{\delta^2}{2n(d(x)^2-\delta^2)^{1/2}} \tag{1.24}$$

for any integer n. Letting $n\to\infty$ we have

$$d(x+\delta)\leqslant d(x) \tag{1.25}$$

for $0<\delta<d(x)$, i.e. $d(x)$ is a nonincreasing function of x. Since $d(x)\leqslant x-\xi'$, which is arbitrarily small for x sufficiently close to ξ', we have $d(x)\equiv 0$ for $\xi'<x<\xi'+d_0$, i.e. $u(x, t)\equiv M$ on this interval, a contradiction. Hence $u<M$ on l.

Note that Lemma 1.2 says that if $u=M\geqslant 0$ at a point of D, where $M=\max u$, then $u\equiv M$ along the longest horizontal segment containing this point and lying in D.

Lemma 1.3 *Let u satisfy* (1.12) *where $c(x, t)\leqslant 0$ in D. Suppose $u<M$ in a portion of D lying in the strip $\tau_0<t<\tau_1$ for some fixed numbers τ_0 and τ_1. Assume $M\geqslant 0$. Then $u<M$ on the portion of the line $t=\tau_1$ contained in D.*

Proof The proof is again by contradiction. Suppose there exists a point on the line $t=\tau_1$ where $u=M$. Without loss of generality assume that this point is the origin (in particular $\tau_1=0$). Let Ω be an open disk centered at the origin and small enough such that the lower half of Ω is contained in that part of D where $t>\tau_0$. Now consider the comparison function

$$U=e^{-x^2-\alpha t}-1 \tag{1.26}$$

where α is a positive parameter. Since

$$L[U]=e^{-x^2-\alpha t}\{4ax^2-2a-2bx+\alpha+c\}-c \tag{1.27}$$

we can choose α such that

$$L[U]>0 \quad \text{in} \quad \Omega \quad \text{for} \quad t<0. \tag{1.28}$$

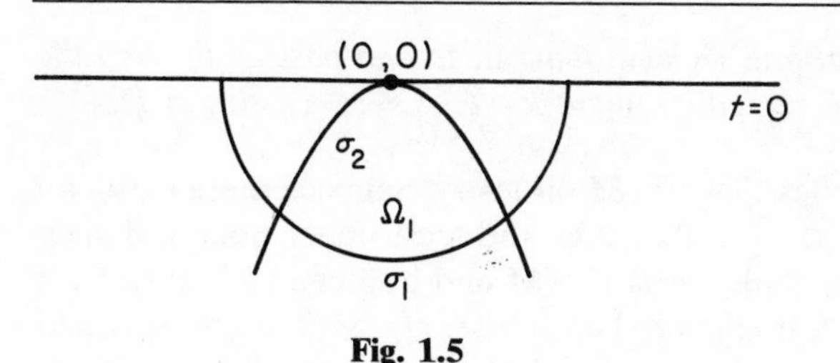

Fig. 1.5

The parabola

$$x^2+\alpha t=0 \tag{1.29}$$

is tangent to the line $t=0$ at the origin. Let σ_1 be the (closed) portion of $\partial\Omega$ which is below this parabola and σ_2 the part of the parabola lying inside Ω. Let Ω_1 be the region bounded by σ_1 and σ_2 (see Fig. 1.5). Since $u<M$ on σ_1 there exists an $\varepsilon>0$ such that

$$u\leqslant M-\varepsilon \quad \text{on} \quad \sigma_1. \tag{1.30}$$

Now consider the function $u+\delta U$ where δ is a positive parameter. Since $U=0$ on σ_2 we can choose δ such that:

(1) $L[u+\delta U]>0$ in Ω_1;
(2) $u+\delta U<M$ on σ_1;
(3) $u+\delta U\leqslant M$ on σ_2.

Condition (1) implies that $u+\delta U$ cannot achieve its (nonnegative) maximum in Ω_1. Therefore the maximum of $u+\delta U$ in $\bar{\Omega}_1$ is M, occurring at the origin. Hence

$$(\partial/\partial t)(u+\delta U)\geqslant 0 \tag{1.31}$$

at the origin. Furthermore,

$$(\partial U/\partial t)=-\alpha<0 \tag{1.32}$$

at the origin, and hence

$$\partial u/\partial t\geqslant-\delta(\partial U/\partial t)>0 \tag{1.33}$$

at the origin. But since the maximum of u on $t=0$ occurs at the origin,

$$\partial u/\partial x=0, \qquad \partial^2 u/\partial x^2\leqslant 0 \tag{1.34}$$

at this point. But (1.33), (1.34) and the fact that $c(x, t)\leqslant 0$, $M\geqslant 0$, contradict the fact that $L[u]\geqslant 0$ at the origin.

***Theorem* 1.5** *Let u satisfy* (1.12) *in D where $c(x, t)\leqslant 0$ for (x, t) in D. If the nonnegative maximum M of u is attained at any point (ξ, τ_1) of D, then*

$u \equiv M$ on each horizontal line segment lying in D and passing through the point (ξ, τ_0) provided the vertical segment $x = \xi$, $\tau_0 \leq t \leq \tau_1$, lies in D.

Proof Lemma 1.2 implies that $u = M$ on $t = \tau_1$. Suppose there exists a t such that $u(\xi, t) < M$, $t_0 < t < \tau_1$. Let τ be the least upper bound of such values of t. Then, by continuity, $u(\xi, \tau) = M$ and by Lemmas 1.2 and 1.3 we have $u(\xi, \tau) < M$, a contradiction. Hence $u(\xi, \tau) = M$ for $\tau_0 \leq t \leq \tau_1$ and hence by Lemma 1.2, $u \equiv M$ on each of the horizontal line segments stated in the theorem.

Note: As with Theorem 1.1 we can remove the assumption that $M \geq 0$ if $c \equiv 0$. Again as in Theorem 1.1, we can obtain results on where solutions of $L[u] \leq 0$ can achieve a (nonpositive) minimum if we replace u by $-u$. Note, however, that Theorem 1.3 is quite different from Theorem 1.1. In particular, it is possible for a solution of (1.12) to achieve its maximum at a point in D without being identically a constant. To see this assume that $c = 0$ and consider the initial-boundary value problem

$$\begin{aligned} a(x, t)(\partial^2 u/\partial x^2) + b(x, t)(\partial u/\partial x) = \partial u/\partial t \qquad & 0 < x < 1, \qquad t > 0 \\ u(0, t) = 0, \qquad & t \geq 0 \\ u(1, t) = f(t), \qquad & t \geq 0 \\ u(x, 0) = 0, \qquad & 0 \leq x \leq 1 \end{aligned} \tag{1.35}$$

where $f(t) = 0$ for $0 \leq t \leq \tau$ and $f(t) < 0$ for $t < \tau$. Then from Theorem 1.5 and the above remarks we can conclude that $u \equiv 0$ for $0 \leq x \leq 1$, $0 \leq t \leq \tau$, and $u < 0$ for $0 < x \leq 1$, $t > \tau$. (Here we have assumed that $f(t)$ is smooth enough such that a solution to (1.35) exists.) Similar examples can be constructed when $c \neq 0$. Indeed, Theorem 1.5 and Lemma 1.2 can be combined to identify the entire region in which a solution which attains an interior nonnegative maximum must be constant: if u achieves its nonnegative maximum M at a point (ξ, τ) in D and (ξ', τ') is a point of D which can be connected to (ξ, τ) by a path in D consisting only of horizontal and 'upward' pointing vertical segments, then $u(\xi', \tau') = M$. To see this simply 'work backwards' from (ξ, τ) to (ξ', τ') applying Lemma 1.2 or Theorem 1.5 as appropriate. Finally, we observe that if c is not nonpositive the change of variables $v = u \exp(-\lambda t)$ replaces the equation $L[u] \geq 0$ by $(L - \lambda)[v] \geq 0$, and if λ is sufficiently large a maximum principle applies to v.

We now turn to the analogue of Theorem 1.2 for parabolic equations, the proof of which is due to Friedman [57].

Theorem 1.6 *Let u satisfy* (1.12) *in D and where $c(x, t) \leq 0$ for (x, t) in D. Suppose that $u \leq M$ in D, $M \geq 0$, and $u(\xi, \tau) = M$ where $(\xi, \tau) \in \partial D$ and* *the normal to ∂D at (ξ, τ) is not parallel to the t-axis. Suppose, further, that at (ξ, τ) a circle tangent to ∂D can be constructed whose interior lies entirely in D and such that $u<M$ in this interior. Then if u is continuous in $D \cup \{(\xi, \tau)\}$ and the outward directional derivative $\partial u/\partial \mu$ exists at (ξ, τ), we have*

$$\partial u/\partial \mu > 0 \quad \text{at} \quad (\xi, \tau).$$

Proof Let Ω be an open disk of radius R tangent to ∂D at (ξ, τ). Without loss of generality assume the center of Ω is at the origin. Let Ω_1 be an open disk with center at (ξ, τ) and radius less than R. Let $\sigma_1 = \partial\Omega_1 \cap \bar{\Omega}$, $\sigma_2 = \partial\Omega \cap \Omega_1$ and G the region bounded by $\sigma_1 \cup \sigma_2$ (see Fig. 1.6).

By choosing Ω sufficiently small we can make $u<M$ on $\partial\Omega$ except at (ξ, τ). Then $u<M$ on σ_1 and

(1) $u<M$ on σ_2 except at (ξ, τ);
(2) $u=M$ at (ξ, τ);
(3) There exists an $\varepsilon>0$ such that

$$u \leqslant M - \varepsilon \quad \text{on} \quad \sigma_1. \tag{1.36}$$

Now define the comparison function

$$U = e^{-\alpha r^2} - e^{-\alpha R^2} \tag{1.37}$$

where $r = (x^2 + t^2)^{1/2}$ and note that for α sufficiently large

$$L[U] = \{4\alpha^2 a x^2 - 2\alpha a - 2\alpha b x + 2\alpha t + c\} e^{-\alpha r^2} - c e^{-\alpha R^2} > 0 \tag{1.38}$$

in $\bar{G}$. (1.37), (1.38) and the three conditions listed above now imply that

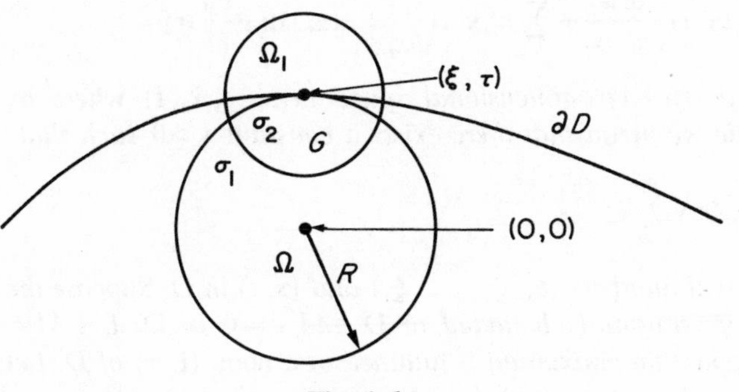

Fig. 1.6

for $\delta > 0$ sufficiently small

(1′) $L[u+\delta U] > 0$ in D;
(2′) $u+\delta U < M$ on σ_1;
(3′) $u+\delta U < M$ on σ_2, except at (ξ, τ).

(These conclusions are deduced in the same manner as in Theorem 1.1 and Lemma 1.1.) From Theorem 1.5 we can now conclude that the maximum of $u+\delta U$ in $\bar{D}$ is achieved at the single point (ξ, τ). Hence at (ξ, τ)

$$(\partial/\partial\mu)(u+\delta U) = \partial u/\partial\mu + \delta(\partial U/\partial\mu) \geq 0. \tag{1.39}$$

But at (ξ, τ) we have that the unit normal $\vec{\nu}$ points in the direction of the radius vector and hence

$$\partial U/\partial\mu = \vec{\mu}\cdot\vec{\nu}(\partial U/\partial r) = -2\alpha Re^{-\alpha R^2}\vec{\mu}\cdot\vec{\nu} < 0 \tag{1.40}$$

and hence from (1.39) we have $\partial u/\partial\mu > 0$ at (ξ, τ).

Note: The result of Theorem 1.6 may not hold if the normal to ∂D is parallel to the t-axis at a maximum point since u may be constant in a region part of whose boundary is perpendicular to the t direction and in this case $\partial u/\partial t = 0$ on this part of the boundary. Similarly, if there does not exist a disk tangent to (ξ, τ) such that $u < M$ in this disk, the result of Theorem 1.4 may fail (cf. Lemma 1.2). If $c \equiv 0$ then the condition $M \geq 0$ may be removed, and if u is replaced by $-u$ we obtain associated minimum principles for $L[u] \leq 0$.

As in the case of elliptic equations, the proofs of Theorems 1.5 and 1.6 can be easily carried over to prove analogous theorems in higher dimensions. We state these results below (cf. Protter and Weinberger [146]).

Theorem 1.7 *Let u satisfy the uniformly parabolic differential inequality*

$$L[u] \equiv \sum_{i,j=1}^{n} a_{ij}(\mathbf{x}, t)\frac{\partial^2 u}{\partial x_i\,\partial x_j} + \sum_{i=1}^{n} b_i(\mathbf{x}, t)\frac{\partial u}{\partial x_i} + c(\mathbf{x}, t)u - \frac{\partial u}{\partial t} \geq 0$$

in a domain D in $(n+1)-$ *dimensional space* $(x_1, \ldots, x_n, t)$ *where by uniformly parabolic we mean that there exists a constant* $\gamma > 0$ *such that*

$$\sum_{i,j=1}^{n} a_{ij}(\mathbf{x}, t)\xi_i\xi_j \geq \gamma \sum_{i=1}^{n} \xi_i^2$$

for all n-tuples of real numbers $(\xi_1, \xi_2, \ldots, \xi_n)$ *and* $(\mathbf{x}, t)$ *in D. Suppose the coefficients of L are uniformly bounded in D and* $c \leq 0$ *in D. Let* $M = \max u \geq 0$ *and suppose the maximum is attained at a point* $(\boldsymbol{\xi}, \tau)$ *of D. Let* $D(\tau)$ *be the connected component of the intersection of the hyperplane* $t = \tau$

with D which contains $(\boldsymbol{\xi}, \tau)$. *Then* $u \equiv M$ *in* $D(\tau)$. *Furthermore, if* $(\boldsymbol{\xi}_0, \tau_0)$ *is a point of D which can be connected to* $(\boldsymbol{\xi}, \tau)$ *by a path in D consisting only of horizontal segments and upward vertical segments, then* $u(\boldsymbol{\xi}_0, \tau_0) = M$.

Theorem 1.8 *Let L be as in Theorem* 1.7 *and suppose* $L[u] \geq 0$ *in D. Suppose further that the nonnegative maximum M of u is attained at a point* $(\boldsymbol{\xi}, \tau)$ *on* ∂D. *Assume that a sphere through* $(\boldsymbol{\xi}, \tau)$ *can be constructed whose interior lies entirely in D and in which* $u < M$. *Suppose also that the radial direction from the center of the sphere to* $(\boldsymbol{\xi}, \tau)$ *is not parallel to the t-axis. Then, if* $\partial/\partial\mu$ *denotes any directional derivative in an outward direction, we have*

$$\partial u/\partial \mu > 0 \quad \text{at} \quad (\boldsymbol{\xi}, \tau).$$

1.3 Phragmén–Lindelof principles

For solutions of elliptic equations defined in unbounded domains it is not necessarily true that the maximum is achieved on the boundary. For example, $u(x, y) = e^x \sin y$ satisfies Laplace's equation in the strip $|y| < \pi$, vanishes on the boundary $y = \pm\pi$, but achieves positive and negative values inside the strip $|y| < \pi$. However, the maximum principle can be rescued if we impose conditions on the solution at infinity, and in this section we shall address ourselves to this problem. The results we shall establish, based on what is called the Phragmén–Lindelof principle, will also enable us to obtain maximum principles for solutions in bounded domains when there are gaps in the prescription of the boundary data. Such situations arise in establishing uniqueness theorems for mixed boundary value problems (see Protter and Weinberger [146]).

To illustrate the basic ideas that are involved we first consider solutions of the inequality

$$\Delta_2 u \geq 0 \tag{1.41}$$

defined in a sector D of angle opening π/α (see Fig. 1.7). Note that if

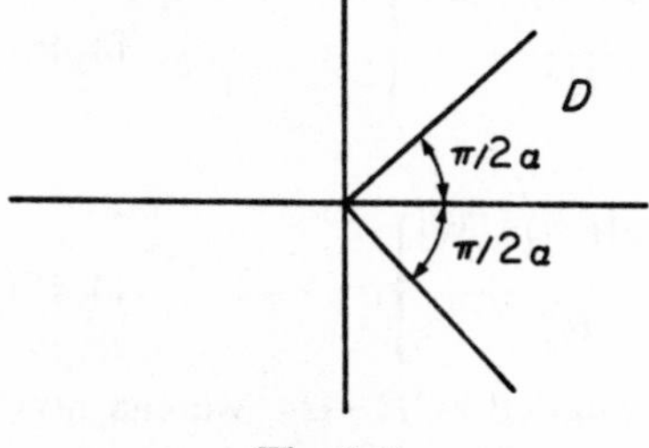

Fig. 1.7

(r, θ) are polar coordinates then the function $r^\alpha \cos \alpha\theta$ is harmonic in D and vanishes on ∂D. The following theorem will show that any solution of (1.41) which is bounded on ∂D and grows more slowly than r^α as $r \to \infty$ for $|\theta| < \pi/2\alpha$ is in fact bounded.

Theorem 1.9 (*Phragmén–Lindelof*) *Let u be continuous in $D \cup \partial D$ and satisfy* (1.41) *in a sector D of angle opening π/α. Then if $u \leq M$ on ∂D and*

$$\varliminf_{R\to\infty} \left\{ R^{-\alpha} \max_{\substack{r=R \\ |\theta| \leq \pi/2\alpha}} u(r, \theta) \right\} \leq 0,$$

we have $u \leq M$ in D.

Proof Let $D_R = D \cap \Omega_R$ where Ω_R is a disk of radius R and observe that the comparison function

$$U_R(r, \theta) = 1 + \frac{2}{\pi} R^\alpha \tan^{-1} \frac{2R^\alpha r^\alpha \cos \alpha\theta}{R^{2\alpha} - r^{2\alpha}} \tag{1.42}$$

is harmonic in D_R, equals one on $\partial D \cap \Omega_R$, and equals $1 + R^\alpha$ on $D \cap \partial\Omega_R$. It is easily verified that the function

$$v(r, \theta) = \frac{u(r, \theta) - M}{U_R(r, \theta)} \tag{1.43}$$

satisfies the differential inequality

$$\Delta_2 v + \frac{2}{U_R} \operatorname{grad} U_R \cdot \operatorname{grad} v \geq 0. \tag{1.44}$$

On $\partial D \cap \Omega_R$ we have $v \leq 0$ and on $D \cap \partial\Omega_R$ we have

$$v(R, \theta) \leq \frac{\max_{D \cap \partial\Omega_R} [u(R, \theta) - M]}{1 + R^\alpha} \tag{1.45}$$

and hence by the maximum principle for elliptic inequalities

$$v(r, \theta) = \frac{u(r, \theta) - M}{U_R(r, \theta)} \leq \max \left\{ 0, \frac{\max_{D \cap \partial\Omega_R} [u(r, \theta) - M]}{1 + R^\alpha} \right\} \tag{1.46}$$

that is,

$$u(r, \theta) \leq M + U_R(r, \theta) \max \left\{ 0, \frac{\max_{D \cap \partial\Omega_R} [u(r, \theta) - M]}{1 + R^\alpha} \right\}. \tag{1.47}$$

Now let $R \to \infty$. Since $U_R(r, \theta)$ remains bounded as $R \to \infty$, we can now conclude from the hypothesis of the theorem and (1.47) that $u(r, \theta) \leq M$.

Note: If D is any domain contained in a sector of angle opening π/α, and u satisfies (1.41) and is bounded by M on ∂D, then the above proof shows that if

$$\varliminf_{R\to\infty} \left\{ R^{-\alpha} \max_{D\cap\partial\Omega_R} u(r,\theta) \right\} \leqslant 0$$

then $u \leqslant M$ in D. This is due to the fact that $1 \leqslant U_R \leqslant 1 + R^\alpha$ in $D \cap \Omega$.

We now wish to generalize Theorem 1.9 to more general domains and more general elliptic inequalities. For the sake of simplicity we shall restrict our attention to

$$L[u] \equiv \Delta_2 u + a(x,y)u_x + b(x,y)u_y + c(x,y)u \geqslant 0 \tag{1.48}$$

defined in a domain D where the coefficients of (1.48) are assumed to be uniformly bounded. D may be bounded or unbounded. Let Γ be a subset of ∂D which may be all of D. If $u \leqslant 0$ on Γ we want to impose sufficient conditions such that we can conclude $u \leqslant 0$ in D (cf. Theorem 1.7). Note that if $\Gamma = \partial D$ and D is bounded then we can apply Theorem 1.1. Hence the interesting cases are when $\Gamma \neq \partial D$ or when D is unbounded.

Theorem 1.10 (*Phragmén–Lindelof*) *Let u satisfy* (1.48) *in a domain D and assume $u \leqslant 0$ on $\Gamma \subset \partial D$. Suppose there exists a function w defined and continuous in $D \cup \Gamma$ such that:*

(1) *$w > 0$ on $D \cup \partial D$;*
(2) *$L[w] \leqslant 0$ in D;*
(3) $\displaystyle\lim_{(x,y)\to\partial D\setminus\Gamma} w(x,y) = \infty$ *if $\Gamma \neq \partial D$;*
(4) $\displaystyle\lim_{x^2+y^2\to\infty} w(x,y) = \infty$ *if D is unbounded.*

Let $\Omega_R = \{(x,y) : w(x,y) = R\}$. Then if u is continuous in $D \cup \Gamma$ and

$$\varliminf_{R\to\infty} \left\{ \sup_{D\cap\Omega_R} \frac{u(x,y)}{w(x,y)} \right\} \leqslant 0$$

we have $u \leqslant 0$ in D.

Note: If $L[u] = \Delta_2 u$, $\Gamma = \partial D$, D is as in Theorem 1.9, and $\alpha > \delta > 0$, we can choose $w = r^{\alpha-\delta} \cos(\alpha-\delta)\theta$.

Proof Let $v = u/w$. Then in D

$$(1/w)L[u] = \Delta_2 v + (2/w) \operatorname{grad} w \cdot \operatorname{grad} v + a v_x + b v_y + (1/w)L[w]v \geqslant 0 \tag{1.49}$$

and hence, by condition (2), v satisfies a maximum principle in D. Let $D_R=\{(x, y): (x, y)\in D, w(x, y)<R\}$. Then D_R is an open subset of D and $\partial D_R=(D\cap\Omega_R)\cup(\partial D\cap\{(x, y): w(x, y)\leq R\})$. By conditions (1), (3) we have that $v\leq 0$ on $\partial D\cap\{(x, y): w(x, y)\leq R\}$ and hence, since by condition (4) we have that D_R is bounded,

$$v\leq \max\left\{0, \sup_{D\cap\Omega_R} u/w\right\}. \tag{1.50}$$

But the hypothesis of the theorem now implies that $v\leq 0$ and by condition (1) the theorem is proved.

In order to apply Theorem 1.10 it is of course necessary to construct an appropriate comparison function w. As an example we do this now for a bounded domain D with a single exceptional point on ∂D. We shall assume that this exceptional point is the origin and restrict ourselves to the case when $c(x, y)\leq 0$ in D. We shall show that a suitable w can be found in the form

$$w(x, y)=-\log r-r^{\varepsilon}-e^{\beta x}+\gamma \tag{1.51}$$

where $r=(x^2+y^2)^{1/2}$, $0<\varepsilon<1$, and β, γ are appropriate positive constants. Since

$$\begin{aligned} L[r^{\varepsilon}]&=\varepsilon(\varepsilon+ax+by)r^{\varepsilon-2}+cr^{\varepsilon}\\ L[\log r]&=(ax+by)r^{-2}+c\log r \end{aligned} \tag{1.52}$$

we have that

$$L[-r^{\varepsilon}-\log r]\to-\infty \quad \text{as} \quad r\to 0. \tag{1.53}$$

Since

$$L[-e^{\beta x}+\gamma]=-(\beta^2+\beta a+c)e^{\beta x}+\gamma c \tag{1.54}$$

we can choose β sufficiently large such that

$$L[w]\leq 0 \quad \text{in} \quad D. \tag{1.55}$$

Note that since $c\leq 0$ we can choose β independent of γ. Finally, by choosing γ sufficiently large we have that

$$w>0 \quad \text{in} \quad \bar{D}. \tag{1.56}$$

Thus, all the conditions of Theorem 1.10 are satisfied and we can conclude that if $L[u]\geq 0$ in D, $u\leq 0$ on ∂D except at the origin, and u grows more slowly than $\log 1/r$ as $r\to 0$, then $u\leq 0$ in D.

Example The growth condition as $r \to 0$ is essential. For example,

$$u(x, y) = \frac{1 - x^2 - y^2}{(x+1)^2 + y^2} \tag{1.57}$$

is harmonic in $D = \{(x, y): x^2 + y^2 < 1\}$, equals zero on ∂D except at $(-1, 0)$, but is positive in D.

Remark By adding several functions of the same type, the same result can be obtained as above if there is a finite number of exceptional points on the boundary.

We now consider an analogue of Theorems 1.9, 1.10 for the parabolic inequality

$$L[u] \equiv a(x, t)\frac{\partial^2 u}{\partial x^2} + b(x, t)\frac{\partial u}{\partial x} + c(x, t)u - \frac{\partial u}{\partial t} \geq 0 \tag{1.58}$$

defined in the half-strip $D = (0, \infty) \times (0, T)$. Similar results can also be obtained for (1.58) defined in domains of the form $\{(x, t): s(t) < x < \infty, 0 < t < T\}$ if we first make a preliminary change of variables of the form $\xi = x - s(t)$. We assume that L is uniformly parabolic in D and the coefficients in (1.58) are uniformly bounded in D.

Theorem 1.11 *Let u be continuous in* $[0, \infty) \times [0, T]$ *and satisfy* (1.58) *in D such that for some positive constant* α

$$\varliminf_{R \to \infty} e^{-\alpha R^2} \max_{0 \leq t \leq T} u(R, t) \leq 0.$$

Then if $u(x, 0) \leq 0$ *for* $0 \leq x < \infty$ *and* $u(0, t) \leq 0$ *for* $0 \leq t \leq T$, $u \leq 0$ *in D.*

Proof Let

$$v(x, t) = u(x, t) \exp\left(-\frac{\gamma\alpha x^2}{\gamma - \alpha t} - \beta t\right) \tag{1.59}$$

where γ and β are constants to be determined. We have

$$\exp\left(-\frac{\gamma\alpha x^2}{\gamma - \alpha t} - \beta t\right)L[u] = a\frac{\partial^2 v}{\partial x^2} + \left(b + \frac{4\alpha\gamma x a}{\gamma - \alpha t}\right)\frac{\partial v}{\partial x} + h(x, t)v - \frac{\partial v}{\partial t} \geq 0 \tag{1.60}$$

where

$$h(x, t) = \frac{4\alpha^2\gamma^2 x^2}{(\gamma - \alpha t)^2}a + \frac{2\alpha\gamma}{\gamma - \alpha t}(a + xb) + c - \beta - \frac{\alpha^2\gamma^2 x^2}{(\gamma - \alpha t)^2} \tag{1.61}$$

and hence for $\gamma > \alpha t$,

$$h(x,t) \leqslant -\frac{\alpha^2 x^2 \gamma}{(\gamma-\alpha t)^2}\left[1-4\gamma M-\frac{2(\gamma-\alpha t)}{\alpha}B\right]-\left[\beta-\frac{2\alpha\gamma}{\gamma-\alpha t}(A+M-c)\right] \tag{1.62}$$

where

$$M=\sup_D |a(x,t)|$$

$$A=\sup_{\substack{0\leqslant x\leqslant 1\\ 0\leqslant t\leqslant T}} |xb(x,t)| \tag{1.63}$$

$$B=\sup_{\substack{x\geqslant 1\\ 0\leqslant t\leqslant T}} \left|\frac{b(x,t)}{x}\right|.$$

Now let γ be small enough such that the expression in the first bracket of (1.62) is positive for $t\in[0,\gamma/2\alpha]$, and β large enough such that the expression in the second bracket is positive for $t\in[0,\gamma/2\alpha]$. Then $h(x,t)\leqslant 0$ for t in this interval and we can apply Theorem 1.5 to the inequality (1.60). Let $D_R=\{(x,t): 0<x<R, 0<t<\gamma/2\alpha\}$. By Theorem 1.5, v cannot have a positive maximum in D_R. Hence from the hypothesis of the theorem we can conclude by taking R sufficiently large that for any $\varepsilon>0$, $v\leqslant\varepsilon$ in D_R for $R=R(\varepsilon)$. Letting $\varepsilon\to 0$ now shows that $v\leqslant 0$ in $\{(x,t): 0\leqslant x<\infty, 0\leqslant t\leqslant\gamma/2\alpha\}$. Now the whole argument can be repeated with $t=\gamma/2\alpha$ as the initial line instead of $t=0$ and we obtain $v\leqslant 0$ in $\{(x,t): 0\leqslant x<\infty, \gamma/2\alpha\leqslant t\leqslant\gamma/\alpha\}$. After a finite number of steps we can conclude that $v\leqslant 0$ in D and hence $u\leqslant 0$ in D.

For applications of Theorem 1.11 to uniqueness theorems for initial-boundary value problems see the Exercises at the end of this chapter.

As with the case of maximum principles, the Phragmén–Lindelof principles of this section can be easily extended to higher dimensions. We state the results below (cf. Protter and Weinberger [146]).

Theorem 1.12 *Let u satisfy the uniformly elliptic inequality*

$$L[u]\equiv\sum_{i,j=1}^{n} a_{ij}(\mathbf{x})\frac{\partial^2 u}{\partial x_i\,\partial x_j}+\sum_{i=1}^{n} b_i(\mathbf{x})\frac{\partial u}{\partial x_i}+h(\mathbf{x})u\geqslant 0$$

in a domain D and assume that the coefficients of L are uniformly bounded in D. Assume $u\leqslant 0$ on $\Gamma\subset\partial D$ and suppose there exists a function w continuous in $D\cup\Gamma$ such that:

(1) *$w>0$ on $D\cup\partial D$;*
(2) *$L[w]\leqslant 0$ in D;*

(3) $\lim_{\mathbf{x}\to\partial D\setminus\Gamma} w(\mathbf{x})=\infty$ *if* $\Gamma\neq\partial D$;

(4) $\lim_{|\mathbf{x}|\to\infty} w(\mathbf{x})=\infty$ *if D is unbounded.*

Let $\Omega_R=\{\mathbf{x}: w(\mathbf{x})=R\}$. Then if u is continuous on $D\cup\Gamma$ and

$$\varliminf_{R\to\infty}\left\{\sup_{D\cap\Omega_R}\frac{u(\mathbf{x})}{w(\mathbf{x})}\right\}\leqslant 0,$$

we have $u\leqslant 0$ in D.

Theorem 1.13 *Let D be an unbounded domain in* R^n *(Euclidean n space) and suppose u satisfies the parabolic inequality*

$$L[u]\equiv\sum_{i,j=1}^{n} a_{ij}(\mathbf{x},t)\frac{\partial^2 u}{\partial x_i\,\partial x_j}+\sum_{i=1}^{n} b_i(\mathbf{x})\frac{\partial u}{\partial x_i}+c(\mathbf{x},t)u-\frac{\partial u}{\partial t}\geqslant 0$$

in $D\times(0,T)$ *where L is uniformly parabolic in D with bounded coefficients in* $D\times(0,T)$. *Assume that u satisfies*

$$\varliminf_{R\to\infty} \mathrm{e}^{-\alpha R^2}\left[\max_{\substack{|\mathbf{x}|=R\\ 0\leqslant t\leqslant T}} u(\mathbf{x},t)\ \leqslant 0\right]$$

for some positive constant α *and is continuous in* $(D\cup\partial D)\times[0,T]$. *Then if* $u\leqslant 0$ *for* $t=0$ *and* $u\leqslant 0$ *on* $\partial D\times(0,T)$, $u\leqslant 0$ *in* $D\times(0,T)$.

1.4 Applications to nuclear reactor theory

Maximum principles, particularly in the form of the Phragmén–Lindelof theorem, have had a wide success in illuminating many problems in applied mathematics, especially in the area of fluid mechanics. For examples and specific references to the literature we refer the reader to the already mentioned book of Protter and Weinberger [146]. In this section we shall show by way of illustration how the maximum principle can be applied to a particular problem in nuclear reactor theory. For details of the physical background of the problem we are about to consider, we suggest consulting the basic textbook in the area by Glasstone and Edlund [78]. The example below was first presented by Payne and Stakgold [144] and has since been extended by Payne, Stakgold, Schaefer and Sperb to nonlinear equations (see the References).

The basic problem can be described as follows: let u denote the neutron density of a homogeneous, monoenergetic reactor operating at criticality. Then u is the fundamental eigenfunction of the boundary value problem

$$\begin{aligned} \Delta_2 u + \lambda u = 0 \quad &\text{in} \quad D \\ u = 0 \quad &\text{on} \quad \partial D \end{aligned} \tag{1.64}$$

where the reactor is assumed to be cylindrical with cross-section D. We assume that D is an open, bounded region in R^2 with $C^{2+\varepsilon}$ boundary ∂D, and since u does not vanish in D, we can choose u to be positive in D. Note that D may not be contractible, e.g. D may be an annulus. We are interested in choosing the shape D of the nuclear reactor such that the average-to-peak neutron density, E, defined by

$$E = \frac{M_D(u)}{u_M} \tag{1.65}$$

where

$$\begin{aligned} u_M &= \max_D u \\ M_D(u) &= \frac{1}{A} \iint_D u \, dx \, dy \\ A &= \text{area of } D, \end{aligned} \tag{1.66}$$

is maximized, thus achieving the reasonably flat neutron profile which is desired in practice. To this end we shall use the maximum principle to obtain inequalities for E in terms of A and the length of ∂D. We also introduce the related nondimensional quantity

$$\tilde{E} = -\frac{M_{\partial D}(\partial u/\partial \nu)}{\sqrt{\lambda}\, u_M} \tag{1.67}$$

where ν is the exterior normal and

$$M_{\partial D}(u) = \frac{1}{S} \iint_{\partial D} u \, ds \qquad S = \text{length of } \partial D. \tag{1.68}$$

If one of the objectives is to maximize the radiative heat transfer at the boundary, then $\tilde{E}$ itself is of physical interest. From (1.64) and Green's identities we have

$$E = -\frac{\iint_{\partial D} (\partial u/\partial \nu) \, ds}{\lambda A u_M} = \frac{S M_{\partial D}(\partial u/\partial \nu)}{\lambda A u_M} \tag{1.69}$$

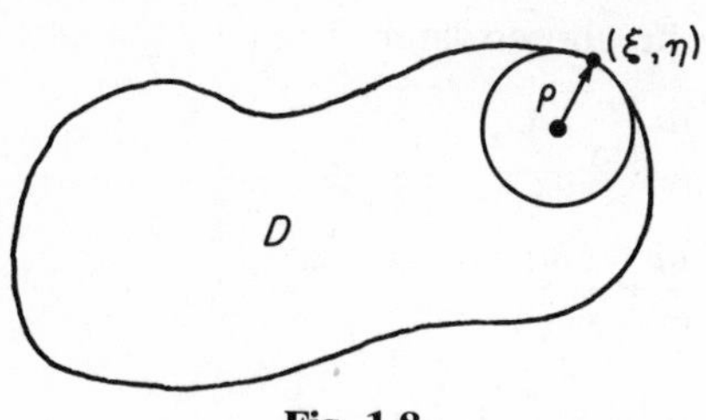

Fig. 1.8

and hence

$$\tilde{E} = (\sqrt{\lambda}\, A/S)E. \tag{1.70}$$

Due to the fact that ∂D is in class $C^{2+\varepsilon}$, the curvature K exists at each point on ∂D and is continuous, and hence there exists a positive constant K_0 such that

$$K_0 = \max_{\partial D} (-K). \tag{1.71}$$

Note that if D is convex we have $K_0 \leqslant 0$. Let $(\xi, \eta) \in \partial D$ and ρ the radius of the osculating circle at (ξ, η) (see Fig. 1.8). Then $\rho = 1/|K|$ and $\partial/\partial\rho = \pm\partial u/\partial\nu$, the sign depending on whether K is positive or negative. Hence, since $u = 0$ on ∂D we have at (ξ, η) that

$$\begin{aligned} \Delta_2 u + \lambda u = \Delta_2 u &= \frac{\partial^2 u}{\partial \rho^2} + \frac{1}{\rho}\frac{\partial u}{\partial \rho} \\ &= \frac{\partial^2 u}{\partial \nu^2} + K\frac{\partial u}{\partial \nu} = 0 \end{aligned} \tag{1.72}$$

where we have made use of the fact that the differential equation is satisfied on ∂D (cf. Vekua [172], p. 321).

We now proceed to derive the desired inequalities for E and $\tilde{E}$.

Lemma 1.4 *Let u be a solution of (1.64). Then*

$$|\text{grad } u|^2 \leqslant \lambda(u_M^2 - u^2) + \beta(u_M - u) \quad \textit{in} \quad D$$

$$|\text{grad } u|^2 \leqslant \lambda u_M^2 + \beta u_M = \lambda G^2 u_M^2 \quad \textit{on} \quad \partial D$$

where

$$G = \lambda^{-1/2}\{[\lambda + K_0^2]^{1/2} + K_0\}$$

$$\beta = 2K_0(\lambda)^{1/2} G u_M$$

if $K_0 > 0$, and $G = 1$, $\beta = 0$, otherwise.

Proof Let

$$\Phi = |\text{grad } u|^2 + \lambda u^2 + \alpha u \tag{1.73}$$

where α is a nonnegative constant to be chosen later. Then

$$\frac{\partial \Phi}{\partial x}=2\frac{\partial u}{\partial x}\frac{\partial^2 u}{\partial x^2}+2\frac{\partial u}{\partial y}\frac{\partial^2 u}{\partial x\,\partial y}+2\lambda u\frac{\partial u}{\partial x}+\alpha\frac{\partial u}{\partial x}$$

$$\frac{\partial \Phi}{\partial y}=2\frac{\partial u}{\partial y}\frac{\partial^2 u}{\partial y^2}+2\frac{\partial u}{\partial x}\frac{\partial^2 u}{\partial x\,\partial y}+2\lambda u\frac{\partial u}{\partial y}+\alpha\frac{\partial u}{\partial y} \tag{1.74}$$

$$\Delta_2\Phi=2\left[\left(\frac{\partial^2 u}{\partial x^2}\right)^2+\left(\frac{\partial^2 u}{\partial x\,\partial y}\right)^2+\left(\frac{\partial^2 u}{\partial y^2}\right)^2\right]-2\lambda^2u^2-\alpha\lambda u.$$

Then at points where grad $u\neq 0$, we have from (1.74) that

$$\Delta_2\Phi\geqslant\frac{\left(\frac{\partial\Phi}{\partial x}-2\lambda u\frac{\partial u}{\partial x}-\alpha\frac{\partial u}{\partial x}\right)^2+\left(\frac{\partial\Phi}{\partial y}-2\lambda u\frac{\partial u}{\partial y}-\alpha\frac{\partial u}{\partial y}\right)^2}{2\,|\text{grad}\,u|^2}-2\lambda^2u^2-\alpha\lambda u$$

$$=\frac{\left(\frac{\partial\Phi}{\partial x}\right)^2+\left(\frac{\partial\Phi}{\partial y}\right)^2}{2\,|\text{grad}\,u|^2}-\frac{(2\lambda u+\alpha)\left(\frac{\partial u}{\partial x}\frac{\partial\Phi}{\partial x}+\frac{\partial u}{\partial y}\frac{\partial\Phi}{\partial y}\right)}{|\text{grad}\,u|^2}+\alpha\lambda u+\frac{\alpha^2}{2}. \tag{1.75}$$

Since $\lambda>0$, $u\geqslant 0$ in D, we have

$$\Delta_2\Phi+\frac{(2\lambda u+\alpha)}{|\text{grad}\,u|^2}\,\text{grad}\,u\cdot\text{grad}\,\Phi\geqslant 0. \tag{1.76}$$

Hence, from the maximum principles for elliptic equations, we have that Φ assumes its maximum value either on ∂D or at a point in D where grad $u=0$. If the maximum occurs at a point (ξ,η) on ∂D, then either $\Phi\equiv$constant in D or $\partial\Phi/\partial\nu>0$ at (ξ,η).

We shall now show that for a suitable choice of α, $\partial\Phi/\partial\nu$ cannot be positive on ∂D. Since at an arbitrary point of ∂D we have $u=0$, it follows from (1.73) that

$$\frac{\partial\Phi}{\partial\nu}=2\frac{\partial u}{\partial\nu}\frac{\partial^2 u}{\partial\nu^2}+\alpha\frac{\partial u}{\partial\nu}, \tag{1.77}$$

where we have made use of the fact that grad u is normal to ∂D. Hence from (1.72) and (1.77) we have

$$\frac{\partial\Phi}{\partial\nu}=\frac{\partial u}{\partial\nu}\left(\alpha-2K\frac{\partial u}{\partial\nu}\right)=-q(\alpha+2Kq) \tag{1.78}$$

where $q=|\text{grad}\,u|$ on ∂D. Note that since $u>0$ in D we have $\partial u/\partial\nu<0$ on ∂D. From (1.71) we now have

$$\partial\Phi/\partial\nu\leqslant q(2K_0q-\alpha), \tag{1.79}$$

and hence if $K_0>0$ and we set

$$\alpha=2K_0q_0, \tag{1.80}$$

where q_0 is the maximum value of q on ∂D, we have $\partial\Phi/\partial\nu \leq 0$ on ∂D.

Therefore, either $\Phi \equiv$constant or the maximum of Φ occurs where grad $u = 0$. In the latter case we have from (1.73), (1.80) that

$$|\text{grad } u|^2 + \lambda u^2 + 2K_0 q_0 u \leq \max_D (\lambda u^2 + 2K_0 q_0 u) = \lambda u_M^2 + 2K_0 q_0 u_M, \tag{1.81}$$

the inequality being trivially true if $\Phi \equiv$constant. Hence on ∂D

$$q^2 = |\text{grad } u|^2 \leq \lambda u_M^2 + 2K_0 q_0 u_M, \tag{1.82}$$

and hence

$$q_0^2 \leq \lambda u_M^2 + 2K_0 q_0 u_M, \tag{1.83}$$

which implies

$$q_0 \leq \sqrt{\lambda}\, G u_M. \tag{1.84}$$

This proves the first inequality in the lemma (for $K_0 > 0$). Since (1.82) remains valid when an upper bound for q_0 is used, we may substitute from (1.84) to obtain the second inequality. Finally, if $K_0 \leq 0$ we can set $\alpha = 0$ in (1.80), yielding the inequalities in the lemma with $G = 1$, $\beta = 0$.

Theorem 1.14 *The following inequalities hold:*

$$E \leq G(S/\sqrt{\lambda}\, A)$$
$$\tilde{E} \leq G.$$

If ∂D has positive curvature everywhere (as is the case for a convex region) then

$$E \leq S/\sqrt{\lambda}\, A$$
$$\tilde{E} \leq 1.$$

Proof From (1.64), (1.84) and Green's identities, we have

$$\lambda \iint_D u \, dx \, dy = -\iint_D \Delta_2 u \, dx \, dy = -\int_{\partial D} (\partial u/\partial \nu) \, ds$$
$$= \int_{\partial D} |\text{grad } u| \, ds \leq \sqrt{\lambda}\, G S u_M, \tag{1.85}$$

which is the first inequality. The second inequality follows from (1.70). If ∂D has positive curvature, then $G = 1$, and this yields the final set of inequalities.

We now show how the estimates of Theorem 1.14 may be improved by using level curves of u.

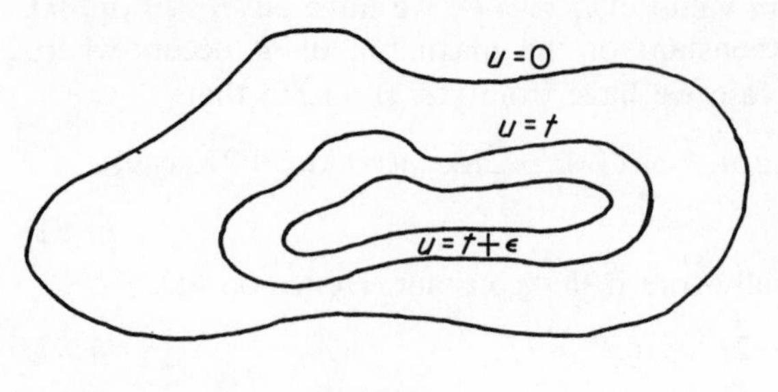

Fig. 1.9

Theorem 1.15 *$E \leqslant (\pi/4)G$ and if D is convex $E \leqslant \pi/4$.*

Proof Let $D_+(t) = \{(x, y)\colon\ (x, y) \in D,\ \ u(x, y) \geqslant t\}$ and $\partial D_+(t) = \{(x, y)\colon (x, y) \in D,\ u(x, y) = t\}$. Let $a(t)$ be the area of $D_+(t)$ and $s(t)$ the length of $\partial D_+(t)$. Then $D_+(0) = D$, $a(0) = A$, $a(u_M) = 0$. To each value of a in $0 \leqslant a \leqslant A$ there corresponds a value of $t = t(a)$ such that $D_+(t)$ has area a. Clearly, $t(a)$ is the inverse function of $a(t)$ and is strictly decreasing with $t(0) = u_M$ and $t(A) = 0$. Now divide the region of integration into thin annuli bounded by the level curves of u (see Fig. 1.9) and deduce that for any integrable function h

$$\iint\limits_{D_+(t)} h(u(x, y))\,\mathrm{d}x\,\mathrm{d}y = \int_t^{u_M} \iint\limits_{\partial D_+(\tau)} h(\tau)\frac{\mathrm{d}s\,\mathrm{d}\tau}{|\mathrm{grad}\, u|}, \tag{1.86}$$

where we have used the fact that an element of area of the annulus $D_+(t)\backslash D_+(t + \mathrm{d}t)$ is given by

$$\mathrm{d}x\,\mathrm{d}y = -\mathrm{d}s\,\mathrm{d}\nu = -\frac{\mathrm{d}s\,\mathrm{d}t}{\partial u/\partial \nu} = \frac{\mathrm{d}s\,\mathrm{d}t}{|\mathrm{grad}\, u|}. \tag{1.87}$$

If we set $h(t) = 1$ we have

$$a(t) = \int_t^{u_M} \int_{\partial D_+(\tau)} \frac{\mathrm{d}s\,\mathrm{d}\tau}{|\mathrm{grad}\, u|} \tag{1.88}$$

and $h(t) = t$ gives

$$F(a(t)) = \iint\limits_{D_+(t)} u(x, y)\,\mathrm{d}x\,\mathrm{d}y = \int_t^{u_M} \int_{\partial D_+(\tau)} \frac{\tau\,\mathrm{d}s\,\mathrm{d}\tau}{|\mathrm{grad}\, u|}. \tag{1.89}$$

We note that as a function of a,

$$F(a)=\int_0^a t(\mu)\,d\mu$$

$$F(0)=0$$

$$F(A)=\iint_D u(x, y)\,dx\,dy \tag{1.90}$$

$$F'(a)=t(a)$$

$$F'(0)=u_M$$

$$F'(A)=0$$

and

$$F''(a)=t'(a)=\frac{1}{a'(t)}=-\left[\int_{\partial D_+(t)}\frac{ds}{|\text{grad } u|}\right]^{-1}. \tag{1.91}$$

Hence F and F' are positive and F'' is negative. We now make use of the differential equation satisfied by u, integrate (1.64), and apply Green's formula to obtain (cf. (1.85)):

$$\lambda F(a)=\lambda\iint_{D_+(t)} u(x, y)\,dx\,dy=\int_{\partial D_+(t)}|\text{grad } u|\,ds \tag{1.92}$$

for $t=t(a)$. Hence from (1.91), (1.92) we have

$$-\lambda F(a)F''(a)=\left[\int_{\partial D_+(t)}|\text{grad } u|\,ds\right]\left[\int_{\partial D_+(t)}\frac{ds}{|\text{grad } u|}\right]^{-1}. \tag{1.93}$$

But from Schwarz's inequality

$$s^2(t)=\left(\int_{\partial D_+(t)}ds\right)^2\leqslant\left[\int_{\partial D_+(t)}|\text{grad } u|\,ds\right]\left[\int_{\partial D_+(t)}\frac{ds}{|\text{grad } u|}\right] \tag{1.94}$$

and hence

$$-F(a)F''(a)\leqslant\frac{1}{\lambda s^2(t)}\left[\int_{\partial D_+(t)}|\text{grad } u|\,ds\right]^2$$

$$\leqslant\frac{1}{\lambda}\max_{\partial D_+(t)}|\text{grad } u|^2\leq\left[(u_M^2-t^2)+\frac{\beta}{\lambda}(u_M-t)\right] \tag{1.95}$$

from Lemma 1.4, where $t=F'(a)$. Dropping the nonpositive term $-\beta t/\lambda$ on the right, we obtain

$$(F')^2-FF''\leqslant\left(u_M^2+\frac{\beta}{\lambda}u_M\right), \tag{1.96}$$

that is,

$$-\left[\left(\frac{F'}{F}\right)^2\right]' \leqslant -\left(u_M^2 + \frac{\beta}{\lambda} u_M\right)\left(\frac{1}{F^2}\right)'. \tag{1.97}$$

We now integrate (1.97) from a to A and use (1.90) to obtain

$$F'(a) \leqslant \left[\left(u_M^2 + \frac{\beta}{\lambda} u_M\right)\left(1 - \left(\frac{F(a)}{F(A)}\right)^2\right)\right]^{1/2}. \tag{1.98}$$

From (1.98) and Schwarz's inequality we can deduce that

$$\begin{aligned}
(F(A))^2 &= \left(\int_0^A F'(a)\,da\right)^2 \leqslant A \int_0^A (F'(a))^2\,da \\
&\leqslant A\left(u_M^2 + \frac{\beta}{\lambda} u_M\right)^{1/2} \int_0^A \left(1 - \left(\frac{F(a)}{F(A)}\right)^2\right)^{1/2} F'(a)\,da \\
&= AF(A)\left(u_M^2 + \frac{\beta}{\lambda} u_M\right)^{1/2} \int_0^1 (1-\mu^2)^{1/2}\,d\mu \\
&= F(A)\frac{\pi A}{4}\left(u_M^2 + \frac{\beta}{\lambda} u_M\right)^{1/2}
\end{aligned} \tag{1.99}$$

and hence

$$\frac{1}{A}F(A) = \frac{1}{A}\iint_D u(x, y)\,dx\,dy \leqslant \frac{\pi}{4}\left(u_M^2 + \frac{\beta}{\lambda} u_M\right)^{1/2} = \frac{\pi}{4} G u_M \tag{1.100}$$

which is the desired inequality.

Example When D is a square of side one we have $\lambda = 2\pi^2$, $S = 4$, $A = 1$, and hence $S/(\sqrt{\lambda}\, A) = (2/\pi)\sqrt{2} > \pi/4$ and hence at least in this case Theorem 1.15 is an improvement over Theorem 1.14.

The estimates used in Lemma 1.4 can also be used to obtain an estimate of the distance from the boundary to a point where u assumes its maximum, as well as an estimate for the behavior of u near ∂D. Let $P = (\xi, \eta) \in D$ and $P_1 = (\xi_1, \eta_1)$ be the point on ∂D nearest P. (If there is more than one such point on ∂D, any of them may be taken as P_1.) Let r be the distance measured from P along the ray from P to P_1. Then from Lemma 1.4 we have

$$-du/dr \leqslant |\text{grad } u| \leqslant [\lambda(u_M^2 - u^2) + \beta(u_M - u)]^{1/2}. \tag{1.101}$$

Integrating from P to P_1 we have (for $\overline{PP_1} \subset \bar{D}$)

$$-\int_{P_1}^{P} \frac{du}{[\lambda(u_M^2 - u^2) + \beta(u_M - u)]^{1/2}} \leqslant \int_{P_1}^{P} dr = d(P, P_1) \tag{1.102}$$

where $d(P, P_1)$ is the distance from P to P_1, and hence

$$d(P, P_1) \geq \frac{1}{\sqrt{\lambda}} \left\{ \sin^{-1} \left[\frac{\lambda u + \beta/2}{\lambda u_M + \beta/2} \right] - \sin^{-1} \left[\frac{\beta/2}{\lambda u_M + \beta/2} \right] \right\}. \tag{1.103}$$

Hence if D is convex (and hence $\beta = 0$) we have

$$d(P, P_1) \geq \frac{1}{\sqrt{\lambda}} \sin^{-1} \frac{u}{u_M} \geq \frac{1}{\sqrt{\lambda}} \frac{u}{u_M} \tag{1.104}$$

so that

$$u \leq \sqrt{\lambda}\, u_M d(P, P_1). \tag{1.105}$$

On the other hand if we choose P to be a point where u assumes its maximum, then for $K_0 > 0$ we have

$$d(P, P_1) \geq \frac{1}{\sqrt{\lambda}} \left\{ \frac{\pi}{2} - \sin^{-1} \left[\frac{K_0}{(\lambda + K_0^2)^{1/2}} \right] \right\} \tag{1.106}$$

and for $K_0 \leq 0$

$$d(P, P_1) \geq \pi/(2\sqrt{\lambda}). \tag{1.107}$$

These inequalities give the nearest distance that a point can be to the boundary such that u achieves its maximum at that point.

All of the above inequalities depend on a knowledge of the eigenvalue λ. Estimates for this quantity are well known (cf. Garabedian [61]). In particular, if k_{01} denotes the first zero of the Bessel function $J_0(k)$, D is star-shaped, and ∂D is described by $r = f(\theta)$ we have

$$\lambda_1 \leq \frac{k_{01}^2}{2A} \int_0^{2\pi} \left[1 + \frac{f'(\theta)}{f(\theta)^2} \right] d\theta \tag{1.108}$$

whereas the so-called Faber–Krahn inequality states that

$$\lambda_1 \geq \frac{\pi k_{01}^2}{A}. \tag{1.109}$$

The results of this section can again be easily extended to higher dimensions (Payne and Stakgold [144]).

Exercises

(1) Show that the problem

$$\Delta_2 u = u^3 \quad \text{for} \quad x^2 + y^2 < 1, \qquad u = 0 \quad \text{for} \quad x^2 + y^2 = 1$$

has no solution other than $u \equiv 0$.

(2) If (1.1) is quasilinear in the sense that the coefficients depend on x, y, u, u_x and u_y, show that the maximum principle is still valid for any solution such that (1.2) remains valid.

(3) Show that if the problem

$$\begin{aligned} \Delta_2 u = 0 \quad &\text{for} \quad |x|<1,\ |y|<1 \\ u = 0 \quad &\text{for} \quad |x| = 1 \\ |\text{grad}\ u| = 1 \quad &\text{for} \quad |y| = 1 \end{aligned}$$

has any solution at all, then it has at least two solutions.

(4) By considering the auxiliary function $v(\mathbf{x}, t) = u(\mathbf{x}, t) + \varepsilon r^2$ where $r = |\mathbf{x} - \mathbf{x}_0|$, ε is an appropriate scalar, and $\mathbf{x}_0$ a constant vector, show that if u is a solution of $\Delta_n u - u_t \geq 0$ in a cylinder $D \times (0, T)$ then the maximum of u must occur on the base or vertical sides of this cylinder.

(5) Show that if u is a solution of $\Delta_2 u = 0$ in $0 < x^2 + y^2 < R^2$ and $u/\log(x^2 + y^2) \to 0$ as $x^2 + y^2 \to 0$, then u can be extended to a harmonic function in the whole disk $0 \leq x^2 + y^2 < R^2$.

(6) Let u be a solution of

$$\begin{aligned} u_{xx} = u_t \quad &\text{in} \quad (0, \infty) \times (0, T) \\ u(x, 0) = 0 \quad &\text{in} \quad (0, \infty) \\ u(0, t) = 0 \quad &\text{in} \quad (0, T) \end{aligned}$$

where u is continuous in $[0, \infty) \times [0, T]$ and satisfies

$$\varliminf_{R \to \infty} e^{-\alpha R^2} \max_{0 \leq t \leq T} u(R, t) \leq 0$$

for some positive constant α. Show that $u \equiv 0$.

(7) Establish a maximum principle for the weakly coupled system

$$\begin{aligned} \partial^2 u_1/\partial x^2 - \partial u_1/\partial t + a_1(x, t)u_1 + a_2(x, t)u_2 = 0 \\ \partial^2 u_2/\partial x^2 - \partial u_2/\partial t + a_3(x, t)u_1 + a_4(x, t)u_2 = 0. \end{aligned}$$

(8) Derive the results of Section 1.4 when the problem (1.69) is replaced by

$$\begin{aligned} \Delta_2 u + \lambda u = 0 \quad &\text{in} \quad D \\ \partial u/\partial \nu + hu = 0 \quad &\text{on} \quad \partial D \end{aligned}$$

where $h > 0$ along ∂D and ν is the exterior normal to ∂D.

2

Entire solutions of elliptic equations

In this chapter we shall consider two problems in the theory of entire solutions of elliptic equations. The first of these is to derive an analogue of Liouville's theorem in analytic function theory for elliptic equations in the plane. This result is based on obtaining the so-called Harnack inequalities for elliptic equations, which in turn is a consequence of the maximum principle for elliptic equations. Our proof follows that of Protter and Weinberger [146] and Serrin [158]. The second topic we shall consider is entire solutions of the reduced wave equation in a nonhomogeneous medium satisfying a so called 'radiation condition' at infinity. This is a problem that has attracted considerable attention in recent years, particularly in the context of potential scattering in quantum mechanics (cf. De Alfaro and Regge [45]). We shall concentrate on the case of acoustic scattering in a spherically stratified medium, in which case the problem essentially becomes two-dimensional since the solution is now axially symmetric. The study of this problem gives us the opportunity to introduce in a simple way the concept of an integral operator for solutions of partial differential equations, a technique that has proved useful in much of the recent work in the analytic theory of partial differential equations (cf. Bergman [12]; Colton [27], [28]; Gilbert [70], [71]; Vekua [172]).

2.1 Harnack inequalities

We first consider the problem of deriving Harnack's inequality for the case of the two-dimensional Laplace equation

$$\Delta_2 u = 0. \tag{2.1}$$

For solutions of (2.1) defined in the open disk Ω_R of radius R and center at the origin we have *Poisson's formula* (cf. Garabedian [61])

$$u(r\cos\phi, r\sin\phi) = \frac{R^2 - r^2}{2\pi} \int_0^{2\pi} \frac{u(R\cos\theta, R\sin\theta)\,d\theta}{R^2 - 2Rr\cos(\theta - \phi) + r^2} \tag{2.2}$$

where (r, ϕ) are polar coordinates. (2.2) gives the solution $u \in C^2(\Omega_R) \cap C(\bar{\Omega}_R)$ of (2.1) in terms of the boundary value of u on $\partial\Omega_R$. From (2.2) we can deduce the mean value theorem

$$u(0, 0) = \frac{1}{2\pi} \int_0^{2\pi} u(R \cos \theta, R \sin \theta) \, d\theta. \tag{2.3}$$

Lemma 2.1 *Let $u \geq 0$ be a solution of (2.1) in Ω_R. Then for $(x, y) \in \Omega_R$, $r = (x^2 + y^2)^{1/2}$, we have*

$$\frac{R-r}{R+r} u(0, 0) \leq u(x, y) \leq \frac{R+r}{R-r} u(0, 0).$$

Proof We have

$$R - r \leq (R^2 - 2Rr \cos (\theta - \phi) + r^2)^{1/2} \leq R + r \tag{2.4}$$

and hence from (2.2), (2.3) we have

$$\begin{aligned} u(x, y) &\leq \frac{R^2 - r^2}{2\pi(R-r)^2} \int_0^{2\pi} u(R \cos \theta, R \sin \theta) \, d\theta \\ &= \frac{R+r}{R-r} u(0, 0). \end{aligned} \tag{2.5}$$

Similarly,

$$\begin{aligned} u(x, y) &\geq \frac{R^2 - r^2}{2\pi(R+r)^2} \int_0^{2\pi} u(R \cos \theta, R \sin \theta) \, d\theta \\ &= \frac{R+r}{R-r} u(0, 0). \end{aligned} \tag{2.6}$$

Theorem 2.1 *(Harnack's inequality) Let $u \geq 0$ be a solution of (2.1) in a domain D and $\bar{D}_0$ a compact subset of D. Then there exists a positive constant A depending on D and $\bar{D}_0$ but not on u such that for every pair of points (x, y) and (ξ, η) in $\bar{D}_0$ we have*

$$Au(\xi, \eta) \leq u(x, y) \leq A^{-1} u(\xi, \eta).$$

Proof Suppose P_1 and P_2 are two points such that $r = |P_1 - P_2| < R/2$ and that u is harmonic in the disk $\Omega_R(P_1)$ of radius R with center at P_1. Then from Lemma 2.1 we have

$$\frac{R-r}{R+r} u(P_1) \leq u(P_2) \leq \frac{R+r}{R-r} u(P_1) \tag{2.7}$$

and since $r < \frac{1}{2}R$ we have

$$\tfrac{1}{3} u(P_1) \leq u(P_2) \leq 3u(P_1). \tag{2.8}$$

Now if P_3 is a third point at distance less than $\frac{1}{2}R$ from P_2 we can repeat the above analysis to obtain

$$\tfrac{1}{3}u(P_2) \leq u(P_3) \leq 3u(P_2), \tag{2.9}$$

from which we have

$$(\tfrac{1}{3})^2 u(P_1) \leq u(P_2) \leq (3)^2 u(P_1). \tag{2.10}$$

This process may be continued for any finite number of points. Now choose R such that R is less than the distance of $\bar{D}_0$ to ∂D. By the Heine–Borel theorem we can cover $\bar{D}_0$ by a finite number k of disks of radius $\frac{1}{2}R$ with centers in $\bar{D}_0$. Now let $P=(\xi, \eta)$ and $Q=(x, y)$ be points in $\bar{D}_0$. Then P and Q may be connected by a chain of k disks of radius R such that each disk is contained in D, the first has center at P, the last has center at Q, and the centers of two successive disks are always at a distance less than $\frac{1}{2}R$ from each other. To see this simply replace the chain of circles from P to Q arising from the Heine–Borel theorem by larger concentric circles (with shifted centers). By applying (2.8) repeatedly we have

$$(\tfrac{1}{3})^{k-1} u(P) \leq u(Q) \leq 3^{k-1} u(P), \tag{2.11}$$

where $k = k(D, \bar{D}_0)$.

Corollary 2.1 *Let $\{u_k\}_{k=1}^{\infty}$ be a nonincreasing sequence of harmonic functions in a domain D. Then if $\{u_k\}_{k=1}^{\infty}$ converges at a single point $(\xi, \eta) \in D$ it converges uniformly on every compact subset of D.*

Proof If $i<j$ the function $v_{ij} = u_i - u_j$ is harmonic and nonnegative. If $(x, y) \in \bar{D}_0 \subset D$ then from Harnack's inequality

$$0 \leq v_{ij}(x, y) \leq A^{-1} v_{ij}(\xi, \eta) = A^{-1}[u_i(\xi, \eta) - u_j(\xi, \eta)]. \tag{2.12}$$

Hence, since $\{u_k(\xi, \eta)\}_{k=1}^{\infty}$ converges, $v_{ij}(x, y) \to 0$ as $i, j \to \infty$ and the result follows.

Corollary 2.2 (*Liouville's Theorem*) *Any harmonic function bounded either above or below in the entire plane is a constant.*

Proof Suppose $u \geq 0$. Letting $R \to \infty$ in Lemma 2.1 shows that $u(x, y) = u(0, 0)$ for any $(x, y) \in R^2$. If $u \geq m$ or $u \leq M$ the same argument can be applied to $u - m$ and $M - u$, respectively.

We now want to obtain an analogue of Theorem 2.1 for the uniformly elliptic equation

$$L[u]\equiv a(x, y)\frac{\partial^2 u}{\partial x^2}+b(x, y)\frac{\partial^2 u}{\partial x\,\partial y}+c(x, y)\frac{\partial^2 u}{\partial y^2}+d(x, y)\frac{\partial u}{\partial x}+e(x, y)\frac{\partial u}{\partial y}+h(x, y)u=0 \qquad (2.13)$$

in an open disk Ω_R of radius R centered at the origin where the coefficients are assumed to be uniformly bounded in Ω_R and $h\leqslant 0$ in Ω_R.

Theorem 2.2 *Let u be a solution of (2.13) in Ω_R and assume $u\geqslant 0$ in Ω_R. Then there exists a constant $A>0$ depending only on the ellipticity constant γ and the bound M for the coefficients of L such that*

$$u(x, y)\geqslant Au(0, 0) \quad \text{whenever } r=(x^2+y^2)^{1/2}<\tfrac{1}{2}R.$$

Proof Assume that $u\not\equiv 0$. Then by the maximum principle $u(0, 0)>0$. Now consider the set of all points P such that $u(P)>\frac{1}{2}u(0)$. The component of this set which contains the origin must extend to $\partial\Omega_R$ since otherwise its boundary would be a set where $u(P)=\frac{1}{2}u(0)<u(0)$, contrary to the maximum principle. Therefore, there exists a curve Γ connecting the origin to a point Q on the boundary such that $u(P)\geqslant\frac{1}{2}u(0)$ along Γ. Without loss of generality we can assume that Q is on the y-axis since a rotation of coordinates preserves the ellipticity and ellipticity constant γ of the elliptic operator L.

Now define the two comparison functions $z_\pm$ by

$$z_\pm=e^{-\alpha T_\pm}-1 \qquad (2.14)$$

where

$$T_\pm=[(y-\tfrac{1}{2}R)^2\pm\tfrac{1}{3}Rx-\tfrac{1}{4}R^2]R^{-2} \qquad (2.15)$$

and α is a positive constant to be chosen later. Then

$$L[z_\pm]=\alpha^2R^{-2}e^{-\alpha T_\pm}[\tfrac{1}{9}a\pm\tfrac{4}{3}bR^{-1}(y-\tfrac{1}{2}R)+4cR^{-2}(y-\tfrac{1}{2}R)^2]-\alpha e^{-\alpha T_\pm}[2cR^{-2}\pm\tfrac{1}{3}\,dR^{-1}+2R^{-2}e(y-\tfrac{1}{2}R)]+hz_\pm. \qquad (2.16)$$

By the uniform ellipticity of L we have for all real numbers ξ, η,

$$a\xi^2+2b\xi\eta+c\eta^2\geqslant\gamma(\xi^2+\eta^2) \qquad (2.17)$$

and choosing $\xi=\pm\frac{1}{3}$, $\eta=2R^{-1}(y-\frac{1}{2}R)$, gives

$$[\tfrac{1}{9}a\pm\tfrac{4}{3}bR^{-1}(y-\tfrac{1}{2}R)+4cR^{-2}(y-\tfrac{1}{2}R)^2]\geqslant\gamma[\tfrac{1}{9}+4R^{-2}(y-\tfrac{1}{2}R)^2] \qquad (2.18)$$

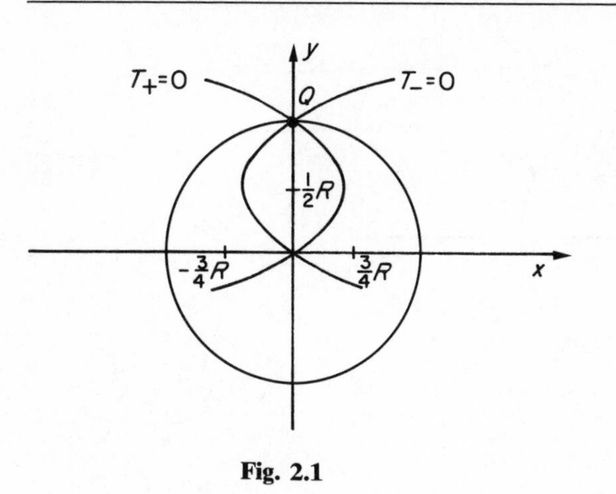

Fig. 2.1

and hence from (2.16) we have

$$L[z_\pm] \geq R^{-2} e^{-\alpha T_\pm} \{\alpha^2 \gamma [\tfrac{1}{9} + 4R^{-2}(y - \tfrac{1}{2}R)^2] - \alpha[2c \pm \tfrac{1}{3}dR + 2e(y - \tfrac{1}{2}R)] + hR^2\} - h. \quad (2.19)$$

Since the coefficients of L are bounded and $h \leq 0$ we have now that for α sufficiently large

$$L[z_\pm] > 0 \quad (2.20)$$

in Ω_R. Note that α depends only on γ and M and on the parabola $T_+ = 0$ we have $z_+ = 0$ whereas on the parabola $T_- = 0$, $z_- = 0$ (see Fig. 2.1). To the left of the parabola $T_+ = 0$ we have $R^2 T_+ < 0$ and hence $z_+ > 0$ and similarly to the right of the parabola $T_- = 0$ we have $z_- > 0$. Since in Ω_R we have

$$-R^2 T_\pm = \tfrac{1}{4}R^2 \pm \tfrac{1}{3}Rx - (y - \tfrac{1}{2}R)^2 \leq \tfrac{1}{4}R^2 \pm \tfrac{1}{3}Rx \leq \tfrac{7}{12}R^2 \quad (2.21)$$

we can now conclude from the above and (2.14) that to the left of the parabola $T_+ = 0$ in Ω_R

$$0 < z_+ < e^{7\alpha/12} \quad (2.22)$$

and to the right of the parabola $T_- = 0$ in Ω_R

$$0 < z_- < e^{7\alpha/12}. \quad (2.23)$$

Now consider the curve Γ again. Since Γ starts at the origin and ends up at Q, there must exist a subarc Γ^* which goes directly from one of the portions of $T_+ = 0$ or $T_- = 0$ in $y \leq 0$ to Q. Let G be the region bounded by T_+ and T_- and let $V \in G$. Assume for the moment that V lies to the right of Γ^* and let G_V be the smallest domain containing V and bounded

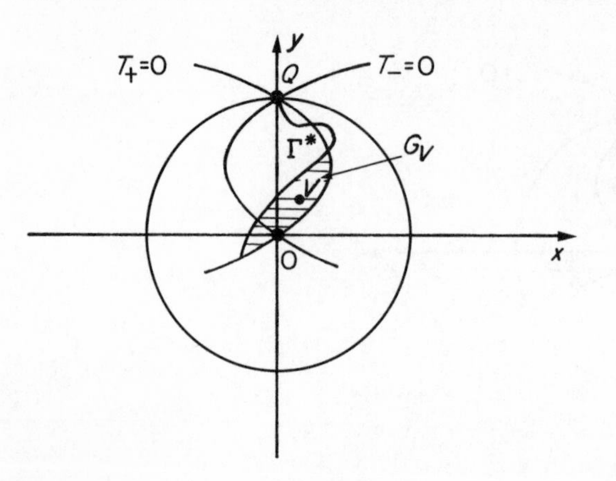

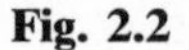
Fig. 2.2

by Γ^* and $T_+=0$ (see Fig. 2.2). Since $u(P)\geqslant\frac{1}{2}u(0)$ on Γ^* we have from (2.22) that

$$u(P)\geqslant\tfrac{1}{2}e^{-7\alpha/12}u(0)z_+(P) \tag{2.24}$$

on $\Gamma^*\cap\partial G_V$, and since $z_+=0$ on $T_+=0$ and $u\geqslant 0$ by assumption, we can conclude that (2.24) is valid on all of ∂G_V. Finally, since

$$L[\tfrac{1}{2}e^{-7\alpha/12}u(0)z_+-u]=\tfrac{1}{2}e^{-7\alpha/12}u(0)L[z_+]>0 \tag{2.25}$$

we have by the maximum principle that (2.24) is valid for all P in G_V, and in particular for $P=V$. Hence

$$u(V)\geqslant\tfrac{1}{2}e^{-7\alpha/12}u(0)z_+(V). \tag{2.26}$$

Similarly, if $V\in G$ is to the left of Γ^* we find that

$$u(V)\geqslant\tfrac{1}{2}e^{-7\alpha/12}u(0)z_-(V) \tag{2.27}$$

and hence for all $V\in G$

$$u(V)\geqslant\tfrac{1}{2}e^{-7\alpha/12}u(0)\min\,(z_+(V),z_-(V)). \tag{2.28}$$

Now consider the line segment $I=\{(x,y)\colon -\frac{3}{4}R<x<\frac{3}{4}R,\ y=\frac{1}{2}R\}$ which is contained in G. On I we have from (2.28) that

$$u(x,\tfrac{1}{2}R)\geqslant\tfrac{1}{2}e^{-7\alpha/12}u(0)\left[\exp\left(\alpha\left[\frac{1}{4}-\frac{1}{3}\frac{|x|}{R}\right]\right)-1\right]. \tag{2.29}$$

Let G_I be the region bounded by I and the parabola $x^2-\frac{3}{8}R(y+R)=0$ (see Fig. 2.3). Then the function

$$z(x,y)=\exp\left[-\tfrac{2}{9}\alpha R^{-2}(x^2-\tfrac{3}{8}R(y+R)\right]-1 \tag{2.30}$$

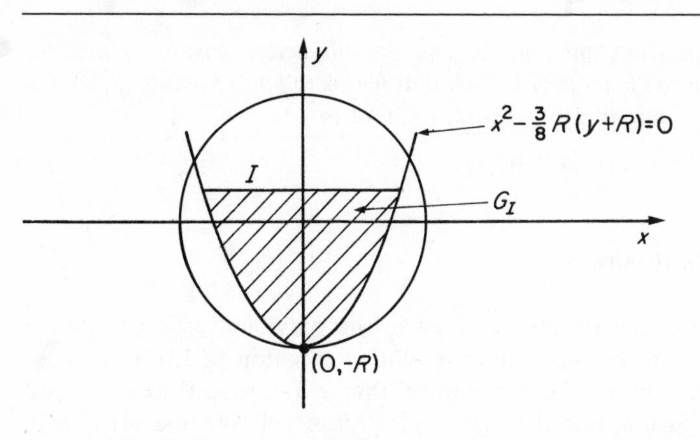

Fig. 2.3

vanishes on this parabola and for $\alpha = \alpha(\gamma, M, R)$ sufficiently large $L[z] > 0$ in that part of Ω_R in which $z \geqslant 0$. Let α be large enough such that both $L[z] > 0$ and $L[z_\pm] > 0$. On I we have

$$\begin{aligned} z(x, \tfrac{1}{2}R) &= \exp[-\tfrac{2}{9}\alpha R^{-2}(x^2 - \tfrac{9}{16}R^2)] - 1 \\ &= \exp\left[\frac{\alpha}{8} - \frac{2}{9}\alpha x^2 R^{-2}\right] - 1 \\ &\leqslant \exp[\alpha(\tfrac{1}{4} - \tfrac{1}{3}|x| R^{-1})] - 1 \end{aligned} \tag{2.31}$$

and hence since $z = 0$ on $\partial G_I \backslash I$, $u \geqslant 0$, and (2.29), we have

$$u(P) \geqslant \tfrac{1}{2}e^{-7\alpha/12}u(0)z(P) \tag{2.32}$$

for P on ∂G_I. By the maximum principle the inequality (2.32) is valid in G_I, and in particular for the disk $\Omega_{R/2} = \{(x, y): x^2 + y^2 \leqslant \frac{1}{4}R^2\} \subset G_I$. In $\Omega_{R/2}$ it can be shown that

$$z \geqslant e^{\alpha/51} - 1 \tag{2.33}$$

and hence from (2.32), (2.33)

$$u(P) \geqslant \tfrac{1}{2}e^{-7\alpha/12}(e^{\alpha/51} - 1)u(0) \tag{2.34}$$

for $P \in \Omega_{R/2}$. This proves Theorem 2.2.

Noting that the ellipticity constant γ is unchanged by translations of the coordinate axis, we can use Theorem 2.2 to argue exactly as in Theorem 2.1 to obtain the following theorem.

Theorem 2.3 (*Harnack's inequality*) *Let $u \geqslant 0$ be a solution of $L[u] = 0$ in a domain D and $\bar{D}_0$ a compact subset of D. Then there exists a positive* *constant A depending only on D, $\bar{D}_0$, the ellipticity constant γ and the bound M for the coefficients of L such that for every pair of points (x, y) and (ξ, η) in $\bar{D}_0$ we have $u(x, y) \geq Au(\xi, \eta)$, that is*

$$Au(\xi, \eta) \leq u(x, y) \leq A^{-1}u(\xi, \eta).$$

2.2 Liouville's theorem

We shall now use the results obtained in the previous section to derive Liouville's theorem for the uniformly elliptic equation (2.13) in the case when $h \equiv 0$. By making the assumption that $r^2h(x, y) \to 0$ as $r \to \infty$ we could drop the assumption that $h \equiv 0$ (cf. Protter and Weinberger [146]). However, to avoid going into technical details we shall make the stronger assumption that h vanishes identically.

Theorem 2.4 (*Liouville's theorem*) *Let u be a solution of* (2.13) *in the entire plane where $h \equiv 0$ and the quantity $a^2 + 2b^2 + c^2 + (x^2 + y^2)(d^2 + e^2)$ is bounded. Then if u is bounded either above or below in the entire plane, u is a constant.*

Proof Let $C_{2R} = \{(x, y): x^2 + y^2 = (2R)^2\}$. Then any two points on C_{2R} can be connected by a chain of at most 13 points on the circle at most $\frac{1}{2}R$ from each other. Applying Theorem 2.2 repeatedly (shifting the origin to each of these 13 points) we have that for any two points P and Q on C_{2R},

$$u(Q) \geq A^{13}u(P), \tag{2.35}$$

where A depends only on the ellipticity constant γ and the bounds for the coefficients of L. In particular from (2.34) we have

$$A = \tfrac{1}{2}e^{-7\alpha/12}(e^{\alpha/51} - 1), \tag{2.36}$$

and for disks centered on C_{2R} we can show that a suitable choice for α is given by (see the Exercises)

$$\alpha = \sup_{\substack{-\infty<x<\infty \\ -\infty<y<\infty}} \{2(288)^2\gamma^{-2}[a^2 + 2b^2 + c^2 + (x^2 + y^2)(d^2 + e^2)]\}. \tag{2.37}$$

Hence from the hypothesis of the theorem, A is independent of R.

We now note that from the maximum principle

$$\varliminf_{x^2+y^2\to\infty} u(x, y) \leq u(\xi, \eta) \leq \varlimsup_{x^2+y^2\to\infty} u(x, y) \tag{2.38}$$

where (ξ, η) is an arbitrary point in the plane, and hence the theorem will be proved if we can show $\lim_{x^2+y^2\to\infty} u(x, y)$ exists. To show this first assume that u is bounded above and hence

$$\overline{\lim_{x^2+y^2\to\infty}}\ u(x, y)=M<\infty. \tag{2.39}$$

Then for any $\varepsilon>0$ we have a number R_ε such that $M-u+\varepsilon>0$ for $x^2+y^2>R_\varepsilon^2$ and there is a sequence of points P_k such that $R_k=|P_k|\to\infty$ and $u(P_k)>M-\varepsilon$. Without loss of generality we can assume that the sequence of moduli R_k is monotone increasing. Since $M-u+\varepsilon$ is a solution of (2.13) (for $h\equiv 0$) we have from (2.35) that for $P\in C_{R_k/2}$,

$$M-u(P)+\varepsilon\leqslant A^{-13}[M-u(P_k)+\varepsilon]\leqslant 2\varepsilon A^{-13}. \tag{2.40}$$

Hence by the maximum principle

$$u>M-\varepsilon(1+2A^{-13}) \tag{2.41}$$

in the annular regions between these circles. Since A is independent of R_k and ε is arbitrary, we can conclude from (2.39) and (2.41) that $u\to M$ as $x^2+y^2\to\infty$.

If u is bounded below we arrive at the same conclusion by considering $-u$. This completes the proof of the theorem.

Example The assumption of Liouville's theorem implies that d and e must tend to zero as $x^2+y^2\to\infty$. Such a restriction is necessary as can be seen by the fact that $u=e^x$ is a solution of

$$u_{xx}+u_{yy}-u_x=0$$

but $e^x\geqslant 0$ for all (real) values of x.

2.3 Wave propagation in a nonhomogeneous medium

In this section we shall consider entire solutions of the elliptic equation

$$\Delta_3 u+k^2(1-m(x, y, z))u=0, \tag{2.42}$$

where m is a real-valued continuously differentiable function of x, y and z with compact support. The assumptions on the smoothness of m as well as the assumption of compact support can be weakened, but for the sake of simplicity we shall not present such generalizations here. Our purpose, rather, is to examine some of the interesting mathematical problems which arise in the study of (2.42), particularly in the case when m is spherically symmetric and the problem becomes essentially two-dimensional. Entire solutions of (2.42) have been the subject of numerous investigations over a period of many years, motivated in part by the fact that (2.42) models acoustic wave propagation in a nonhomogeneous medium. In particular, if one considers a simple harmonic plane wave passing through a pocket of rarefied or condensed air then the velocity potential satisfies an equation of the form (2.42) where $k>0$ is the wave number and m is related to the speed of sound in the nonhomogeneous air pocket. If we assume the velocity potential of the incoming wave is given by $u_i = e^{ikz}$ and let u_s denote the scattered wave then the mathematical problem to be solved is to find an entire solution u of (2.42) such that

$$u(\mathbf{x}) = e^{ikz} + u_s(\mathbf{x}) \tag{2.43}$$

and

$$\lim_{r\to\infty} r\left(\frac{\partial u_s}{\partial r} - iku_s\right) = 0 \tag{2.44}$$

where (2.44) is the so-called Sommerfeld radiation condition and is understood to hold uniformly as $r = |\mathbf{x}|$ tends to infinity. This radiation condition guarantees that u_s is in fact an 'outgoing' wave. For modern treatments of these problems the reader is referred to Lax and Phillips [113] and Wilcox [180]. Similar problems also occur in potential scattering in quantum mechanics (cf. De Alfaro and Regge [45]).

We first note that if $m \equiv 0$ the only solution of (2.42)–(2.44) is $u = e^{ikz}$.

Theorem 2.5 *Let $m \equiv 0$. Then if $u = e^{ikz} + u_s(\mathbf{x})$ is a solution of* (2.42)–(2.44), $u_s \equiv 0$.

Proof Since e^{ikz} is a solution of $\Delta_3 u + k^2 u = 0$ we have that u_s is also. Let (r, θ, ϕ) denote spherical coordinates and for fixed r expand u_s in a series of spherical harmonics

$$u_s(r, \theta, \phi) = \sum_{n=0}^{\infty} \sum_{m=-n}^{n} a_{nm}(r) S_{nm}(\theta, \phi) \tag{2.45}$$

$$a_{nm}(r) = \int_0^{\pi} \int_0^{2\pi} u_s(r, \theta, \phi) S_{nm}(\theta, \phi)\, d\theta\, d\phi. \tag{2.46}$$

Since u_s satisfies the Helmholtz equation $\Delta_3 u + k^2 u = 0$ we can differentiate (2.46) underneath the integral sign and integrate by parts to conclude that a_{nm} is a solution of Bessel's equation

$$\frac{d^2 a_{nm}(r)}{dr^2} + \frac{2}{r}\frac{da_{nm}(r)}{dr} + \left(k^2 - \frac{n(n+1)}{r^2}\right) a_{nm}(r) = 0, \tag{2.47}$$

that is,

$$a_{nm}(r) = b_{nm} j_n(kr) + c_{nm} h_n^{(1)}(kr) \tag{2.48}$$

where b_{nm} and c_{nm} are constants, j_n denotes a spherical Bessel function (see (2.78)) and $h_n^{(1)}$ a spherical Hankel function (see Exercise 7). Since u_s is regular at the origin so is a_{nm}, and hence $c_{nm} \equiv 0$ for all n and m. From (2.44) and (2.46) we have that a_{nm} satisfies the Sommerfeld radiation condition and since asymptotically

$$j_n(kr) \sim \frac{1}{kr} \sin\left(kr - \frac{n\pi}{2}\right) \tag{2.49}$$

we have that $b_{nm} \equiv 0$ for all n and m. Hence $a_{nm} \equiv 0$ and from (2.45) we have that $u_s \equiv 0$.

We now assume that m is not constant, but $m \equiv 0$ for $r \geq a$, and let

$$\Omega = \{\mathbf{x} \colon |\mathbf{x}| \leq a\}. \tag{2.50}$$

Since e^{ikz} is a solution of the Helmholtz equation, we see from the elementary properties of volume potentials in potential theory (cf. Courant and Hilbert [43]) that (2.42)–(2.44) can be written in the equivalent form

$$u(\mathbf{x}) = e^{ikz} - \frac{k^2}{4\pi} \iiint_{\Omega} \frac{e^{ikR}}{R} m(\mathbf{y}) u(\mathbf{y}) \, dV_y \tag{2.51}$$

where $R = |\mathbf{x} - \mathbf{y}|$. Following Ahner [5] (see also Rorres [148]) we shall now show that for k sufficiently small the solution of (2.51) can be represented as a Neumann series for $\mathbf{x} \in \Omega$. Once u has been determined for $r \leq a$ we can easily determine u for $r \geq a$ by noting that the right-hand side of (2.51) provides the unique continuation of u into all of R^3. In particular for $r \geq a$ it is seen that u_s can be represented in the form

$$u_s(\mathbf{x}) = \sum_{n=0}^{\infty} \sum_{m=-n}^{n} a_{nm} h_n^{(1)}(kr) S_{nm}(\theta, \phi). \tag{2.52}$$

To produce our promised Neumann series we first define the linear operator $\mathbf{T}$ by

$$(\mathbf{T}u)(\mathbf{x}) = -\frac{1}{4\pi} \iiint_{\Omega} \frac{e^{ikR}}{R} m(y) u(y) \, dV_y \tag{2.53}$$

where the function space $\mathbf{T}$ operates on the space of continuous complex functions defined on Ω with the norm

$$\|f\| = \max_{\mathbf{x} \in \Omega} |f(\mathbf{x})|. \tag{2.54}$$

We denote this space by $C(\Omega)$ and note that it is a Banach space. We shall make use of the fact (cf. Taylor [164]) that if $\mathbf{T}$ is a bounded linear operator on $C(\Omega)$, then for values of k such that $k^2 \|\mathbf{T}\| < 1$, the series

$$u(\mathbf{x}) = \sum_{j=0}^{\infty} (k^2\mathbf{T})^j e^{ikz} \tag{2.55}$$

converges uniformly in Ω to the unique solution of (2.51). We note that it follows from potential theory that $\mathbf{T}$ as defined by (2.53) is a bounded linear operator mapping $C(\Omega)$ into itself. Hence our problem is to determine how small k must be such that $k^2 \|\mathbf{T}\| < 1$.

Theorem 2.6 *Let* $\mu = \max_{\mathbf{x}\in\Omega} |m(\mathbf{x})|$. *Then* $k^2 \|\mathbf{T}\| < 1$ *whenever* $k^2 < 2/\mu a^2$.

Proof From (2.53) we have

$$|(\mathbf{T}u)(\mathbf{x})| = \frac{1}{4\pi} \left| \iiint_\Omega \frac{e^{ikR}}{R} m(y)u(y)\, dV_y \right|$$

$$\leqslant \frac{\mu}{4\pi} \|u\| \iiint_\Omega \frac{1}{R}\, dV_y. \tag{2.56}$$

We now evaluate the integral in (2.56). We first note that

$$h(\mathbf{x}) = \iiint_\Omega \frac{1}{R}\, dV_y \tag{2.57}$$

is a solution of

$$\Delta_3 h = -4\pi \tag{2.58}$$

in Ω and that, by symmetry, h depends only on $r = |\mathbf{x}|$. Hence $h(\mathbf{x}) = h(r)$ is a solution of

$$\frac{\partial^2 h}{\partial r^2} + \frac{2}{r}\frac{\partial h}{\partial r} = -4\pi, \tag{2.59}$$

that is, $h(r) = -(2\pi/3)r^2 + \text{constant}$ since h is regular at $r = 0$. To determine the constant we simply note that

$$h(0) = \iiint_\Omega \frac{1}{|y|}\, dV_y = \int_0^a \int_0^{2\pi} \int_0^{\pi} \rho \sin\phi\, d\phi\, d\theta = 2\pi a^2. \tag{2.60}$$

Hence $h(r) = 2\pi a^2 - (2\pi/3)r^2$, and from (2.56) it follows that

$$|(\mathbf{T}u)(\mathbf{x})| \leqslant (\mu/2)a^2 \|u\| \tag{2.61}$$

that is $\|\mathbf{T}\| \leqslant (\mu/2)a^2$, and hence $k^2 \|\mathbf{T}\| < 1$ whenever $k^2 < 2/\mu a^2$.

We now assume that $m(\mathbf{x})=m(r)$ is spherically symmetric. From Erdélyi *et al.* [53] we have the expansions

$$e^{ikz}=e^{ikr\cos\theta}=\sum_{n=0}^{\infty} i^n(2n+1)j_n(kr)P_n(\cos\theta)$$
$$\frac{e^{ikR}}{R}=ik\sum_{n=0}^{\infty}(2n+1)j_n(kr_<)h_n^{(1)}(kr_>)P_n(\cos\gamma) \tag{2.62}$$

where P_n denotes Legendre's polynomial, $r_<=\min\{|\mathbf{x}|,|\mathbf{y}|\}$, $r_>=\max\{|\mathbf{x}|,|\mathbf{y}|\}$, and γ is the angle between $\mathbf{x}$ and $\mathbf{y}$. Since e^{ikz} is independent of ϕ we can expand $u(\mathbf{x})$ in a Legendre series

$$u(\mathbf{x})=\sum_{n=0}^{\infty}u_n(r)P_n(\cos\theta). \tag{2.63}$$

Substituting into the integral equation (2.51) we have for $\mathbf{x}\in\Omega$

$$\sum_{n=0}^{\infty}u_n(r)P_n(\cos\theta)=\sum_{n=0}^{\infty}i^n(2n+1)j_n(kr)P_n(\cos\theta)$$
$$-\frac{k^2}{4\pi}\int_0^a\int_0^{2\pi}\int_0^{\pi}\left[ik\sum_{n=0}^{\infty}(2n+1)\zeta_n(r,r_y)P_n(\cos\gamma)\right]$$
$$\cdot\left[m(r_y)\sum_{n=0}^{\infty}u_n(r_y)P_n(\cos\theta_y)\right]r_y^2\,d\phi_y\,d\theta_y\,dr_y \tag{2.64}$$

where $r_y=|\mathbf{y}|$ and

$$\zeta_n(r,r_y)=\begin{cases}j_n(kr)h_n^{(1)}(kr_y) & \text{if } r_y\geq r\\ j_n(kr_y)h_n^{(1)}(kr) & \text{if } r_y<r\end{cases}. \tag{2.65}$$

If we now use the addition formula for Legendre polynomials (cf. Erdélyi *et al.* [53])

$$P_n(\cos\gamma)=\sum_{l=0}^{n}\varepsilon_l\frac{(n-l)!}{(n+l)!}P_n^l(\cos\theta)P_n^l(\cos\theta_y)\cos l(\phi-\phi_y) \tag{2.66}$$

where $\varepsilon_0=1$, $\varepsilon_n=2$ for $n\geq 1$, P_n^l are the associated Legendre polynomials and (r_y,θ_y,ϕ_y) are the spherical coordinates of $\mathbf{y}$, we can integrate (2.64) termwise with respect to θ and ϕ and use the orthogonality properties of the Legendre polynomials to obtain

$$\sum_{n=0}^{\infty}u_n(r)P_n(\cos\theta)=\sum_{n=0}^{\infty}i^n(2n+1)j_n(kr)P_n(\cos\theta)$$
$$-ik^3\sum_{n=0}^{\infty}P_n(\cos\theta)\int_0^a\zeta_n(r,r_y)m(r_y)u_n(r_y)r_y^2\,dr_y. \tag{2.67}$$

Hence we can conclude that u_n is a solution of the integral equation

$$u_n(r) = \mathrm{i}^n(2n+1)j_n(kr) - \mathrm{i}k^3 \int_0^a \zeta_n(r, r_y)m(r_y)u_n(r_y)r_y^2 \, \mathrm{d}r_y. \tag{2.68}$$

Note that in contrast to the integral equation (2.51) which is three-dimensional, the integral equation (2.68) is one-dimensional. From a computational point of view this is a decided advantage, although the disadvantage is that (2.68) must be solved anew for each value of n, $n = 0, 1, 2, \ldots$, and the kernel is not independent of n. We shall now establish that for k sufficiently small (2.68) can be solved by iteration.

Let $k^2\mathbf{T}_n$ denote the integral operator in (2.68), that is

$$(\mathbf{T}_n f)(r) = -\mathrm{i}k \int_0^a \zeta_n(r, r_y)m(r_y)f(r_y)r_y^2 \, \mathrm{d}r_y, \tag{2.69}$$

where $\mathbf{T}_n$ is defined on the Banach space $C[0, a]$ of continuous complex valued functions defined in $[0, a]$ with norm

$$\|f\| = \max_{0 \leq r \leq a} |f(r)|. \tag{2.70}$$

It can be easily verified that $\mathbf{T}_n$ is a bounded linear operator mapping $C[0, a]$ into itself (see the Exercises). As was the case for the integral equation (2.51), the existence of a unique solution to (2.68) will follow if we can show that $k^2 \|\mathbf{T}_n\| < 1$ for k less than a given real number.

Theorem 2.7 *Let* $\mu = \max_{0 \leq r \leq a} |m(r)|$. *Then for each integer n we have* $k^2 \|\mathbf{T}_n\| < 1$ *whenever* $k^2 < 2/\mu a^2$.

Proof For $f \in C[0, a]$ we have

$$\mathbf{T}[f(r)P_n(\cos\theta)] = P_n(\cos\theta)(\mathbf{T}_n f)(r). \tag{2.71}$$

Hence, since $\max_{0 \leq \theta \leq 2\pi} |P_n(\cos\theta)| = 1$ (Erdélyi *et al.* [53]) we have from Theorem 2.6

$$\begin{aligned} \|\mathbf{T}_n f\| &= \|\mathbf{T}[f(r)P_n(\cos\theta)]\| \\ &\leq \|\mathbf{T}\| \, \|f(r)P_n(\cos\theta)\| \\ &= \|\mathbf{T}\| \, \|f(r)\| \\ &\leq (\mu a^2/2) \, \|f\| \end{aligned} \tag{2.72}$$

and hence $\|\mathbf{T}_n\| \leq \mu a^2/2$. The theorem now follows.

From our previous discussion we have that the solution of the integral equation (2.68) can be expressed in the form

$$u_n(r) = \sum_{j=0}^{\infty} (k^2\mathbf{T}_n)^j u_n^i(r) \tag{2.73}$$

where

$$u_n^i(r) = \mathrm{i}^n(2n+1)j_n(kr). \tag{2.74}$$

Hence if $k^2 < 2/\mu a^2$ the integral equation can be solved by the iteration scheme

$$\begin{aligned} u_n^{(q)} &= u_n^i + k^2\mathbf{T}_n u_n^{(q-1)}, \qquad q = 1, 2, 3, \ldots \\ u_n^{(0)} &= u_n^i. \end{aligned} \tag{2.75}$$

Since

$$\begin{aligned} u_n(r) - u_n^{(q)}(r) &= (k^2\mathbf{T}_n)^q[u_n(r) - u_n^i(r)] \\ &= (k^2\mathbf{T}_n)^q \sum_{j=1}^{\infty} (k^2\mathbf{T}_n)^j u_n^i(r) \end{aligned} \tag{2.76}$$

we have that

$$\|u_n - u_n^{(q)}\| \leq k^{2q+2} \|\mathbf{T}_n\|^{q+1} \frac{\|u_n^i\|}{1 - k^2 \|\mathbf{T}_n\|}. \tag{2.77}$$

From (2.74) and the fact that (cf. Erdélyi *et al.* [53])

$$j_n(kr) = \frac{\sqrt{\pi}}{2}\left(\frac{kr}{2}\right)^n \sum_{p=0}^{\infty} \frac{(-1)^p (kr/2)^{2p}}{p!\,\Gamma(n+p+\frac{3}{2})} \tag{2.78}$$

we have $\|u_n^i\| = 0(k^n)$ as $k \to 0$ and hence from (2.77)

$$\|u_n - u_n^{(q)}\| = 0(k^{2q+n+2}) \quad \text{as} \quad k \to 0. \tag{2.79}$$

We shall make use of the estimate (2.79) in our discussion of the far field or scattering amplitude which we take up next.

The far field pattern gives the asymptotic behavior of the scattered wave and is defined (in the case of spherical symmetry) by

$$\begin{aligned} F(\theta, k) &= \lim_{r\to\infty} r\mathrm{e}^{-\mathrm{i}kr} u_s(\mathbf{x}) \\ &= \lim_{r\to\infty} r\mathrm{e}^{-\mathrm{i}kr}\left[-\frac{k^2}{4\pi}\iiint_{\Omega} \frac{\mathrm{e}^{\mathrm{i}kR}}{R} m(|\mathbf{y}|)u(\mathbf{y})\,\mathrm{d}V_y\right]. \end{aligned} \tag{2.80}$$

We shall show that the limit in (2.80) exists uniformly for $0 \leq \theta \leq 2\pi$ and k on compact subsets of the disk $|k| < (1/a)(2/u)^{1/2}$. We first note that for $\mathbf{x} = r\vec{\eta}$, $\mathbf{y} = r_y\vec{\sigma}$, $|\vec{\eta}| = |\vec{\sigma}| = 1$, we have

$$R = |\mathbf{x} - \mathbf{y}| = (r^2 - 2rr_y\vec{\eta}\cdot\vec{\sigma} + r_y^2)^{1/2} = r - r_y\vec{\eta}\cdot\vec{\sigma} + 0(1/r) \tag{2.81}$$

and hence

$$\frac{1}{R}\exp(ikR) = \frac{1}{r}\exp(ik(r - r_y\vec{\eta}\cdot\vec{\sigma})) + 0(1/r^2). \tag{2.82}$$

Therefore from (2.80), (2.82) we can conclude that

$$F(\theta, k) = \lim_{r\to\infty} re^{-ikr}\left[-\frac{k^2}{4\pi}\iiint_\Omega \frac{e^{ikR}}{R} m(|\mathbf{y}|)u(\mathbf{y})\,dV_y\right]$$

$$= -\frac{k^2}{4\pi}\iiint_\Omega \exp(-ikr_y\vec{\eta}\cdot\vec{\sigma})]m(r_y)u(\mathbf{y})\,dV_y, \tag{2.83}$$

where, since u is defined and analytic for $|k| < (1/a)(2/\mu)^{1/2}$ (see (2.55)), we can conclude that the limit in (2.83) holds uniformly for $0 \leq \theta \leq 2\pi$, $|k| < (1/a)(2/\mu)^{1/2}$. In particular, we can conclude that F is an analytic function of k for $|k| < (1/a)(2/\mu)^{1/2}$.

We now use (2.62), (2.63) and observe that $\vec{\eta}\cdot\vec{\sigma} = \cos\gamma$ to arrive at

$$F(\theta, k) = -\frac{k^2}{4\pi}\int_0^a\int_0^{2\pi}\int_0^{\pi}\left[\sum_{n=0}^{\infty}(-i)^n(2n+1)j_n(r_y)P_n(\cos\gamma)\right]$$

$$\cdot\left[m(r_y)\sum_{n=0}^{\infty}u_n(r)P_n(\cos\theta)\right]d\phi\,d\theta\,dr_y \tag{2.84}$$

and from (2.66) and the orthogonality of Legendre polynomials we have

$$F(\theta, k) = \sum_{n=0}^{\infty} A_n(k)P_n(\cos\theta) \tag{2.85}$$

where (replacing r_y by r)

$$A_n(k) = -k^2(-i)^n\int_0^a j_n(kr)u_n(r)m(r)r^2\,dr. \tag{2.86}$$

Note that $A_n(k)$ is analytic for $|k| < (1/a)(2/\mu)^{1/2}$. If we define

$$A_n^{(q)}(k) = -k^2(-i)^n\int_0^a j_n(kr)u_n^{(q)}(r)m(r)r^2\,dr \tag{2.87}$$

we have from (2.78) and (2.79) that

$$|A_n(k) - A_n^{(q)}(k)| = 0(k^{2n+2q+4}) \quad \text{as} \quad k \to 0. \tag{2.88}$$

We are now in a position to prove the following theorem which relates the Legendre coefficients of the far field pattern to the function m.

Theorem **2.8** *Let $A_n(k)$ be defined as in* (2.86). *Then*

$$A_n(k)=\left\{\frac{-1}{2n+1}\left(\frac{2^n n!}{(2n)!}\right)^2\int_0^a m(r)r^{2n+2}\,dr\right\}k^{2n+2}+0(k^{2n+4}).$$

Proof From (2.74), (2.75) and (2.86) we have

$$A_n^0(k)=-k^2(2n+1)\int_0^a [j_n(kr)]^2 m(r)r^2\,dr \tag{2.89}$$

and hence from (2.78) we see, after some simplification, that

$$A_n^{(0)}(k)=\left\{\frac{-1}{2n+1}\left(\frac{2^n n!}{(2n)!}\right)^2\int_0^a m(r)r^{2n+2}\,dr\right\}k^{2n+2}+0(k^{2n+4}). \tag{2.90}$$

The theorem now follows from (2.88).

The expression for $A_n(k)$ given in Theorem 2.8 is known as the Born approximation. The coefficients of the higher-order terms in the Taylor series for $A_n(k)$ can be found by using (2.75), (2.87) and (2.88) for $q>0$ as was done in Theorem 2.8 for $q=0$. Finally, an estimate on the rate of convergence of these higher Born approximations can be determined from the fact that $A_n(k)$ is analytic for $|k|<(1/a)(2/\mu)^{1/2}$. The Taylor series for $A_n(k)$ is often known as the Born series for $A_n(k)$. A knowledge of the location of the singularities of $A_n(k)$, or equivalently of the far field pattern F as an analytic function of k, is of considerable interest in a variety of situations arising in scattering theory, particularly in the case when an obstacle is present, or the nonhomogeneous medium is no longer of compact support. For a readable survey of this intriguing aspect in the theory of wave propagation we refer the reader to Dolph [47] and Dolph and Scott [48] and the references contained therein.

The above-described method of integral equations for solving the scattering problem (2.42)–(2.44) in the case when m is spherically symmetric has several disadvantages. One of these is the fact that the method is limited to values of the wave number k such that $|k|<(1/a)(2/\mu)^{1/2}$. Secondly, the actual computation of the functions u_n from the Fredholm integral equation (2.68) can be quite tedious, especially since the kernel depends on $n, n=0,1,2,\ldots$. We shall now present an alternate approach to this problem that is based on the simple idea of separation of variables and the construction of the radial wave function u_n by means of a so-called transformation operator. Our discussion is based on the work of Colton and Kress [40] and, in addition to being valid for unrestricted (positive) values of k, reduces the problem of the construction of the functions u_n, $n=0,1,2,\ldots$, to that of solving a characteristic initial-value problem for a hyperbolic equation in two independent variables.

We consider the scattering problem

$$\Delta_3 u + k^2(1-m(r))u = 0 \quad \text{in} \quad R^3 \tag{2.91a}$$

$$u(\mathbf{x}) = e^{ikz} + u_s(\mathbf{x}) \tag{2.91b}$$

$$\lim_{r\to\infty} r\left(\frac{\partial u_s}{\partial r} - iku_s\right) = 0 \tag{2.91c}$$

where $m(r) \equiv 0$ for $r \geq a$ and look for a solution of (2.91) in the form

$$u(\mathbf{x}) = \sum_{n=0}^{\infty} a_n y_n(r) P_n(\cos\theta) \tag{2.92}$$

for $r \leq a$ and

$$u(\mathbf{x}) = e^{ikz} + \sum_{n=0}^{\infty} b_n h_n^{(1)}(kr) P_n(\cos\theta) \tag{2.93}$$

for $r \geq a$. In (2.92) y_n is the (unique) solution of

$$\frac{d^2y}{dr^2} + \frac{2}{r}\frac{dy}{dr} + \left[k^2(1-m(r)) - \frac{n(n+1)}{r^2}\right] y = 0 \tag{2.94}$$

which behaves like $j_n(kr)$ as $r \to 0$, that is

$$\lim_{r\to\infty} r^{-n} y_n(r) = \frac{\sqrt{\pi}\, k^n}{2^{n+1}\Gamma(n+\frac{3}{2})}. \tag{2.95}$$

The coefficients a_n and b_n in (2.92), (2.93) are to be determined from the condition that u as defined by (2.92), (2.93) must be continuously differentiable across the circle $r = a$. From the regularity properties of solutions to elliptic equations (cf. Friedman [58]) we can conclude then that u is twice continuously differentiable and a 'classical' solution of (2.91a). The problems we face are thus three-fold:

(1) To give a constructive method for determining the functions y_n;
(2) To determine the constants a_n and b_n;
(3) To show that the series (2.92), (2.93) converge and satisfy (2.91a)–(2.91c).

Having answered the above three questions we have only to convince ourselves that the solution to (2.91a)–(2.91c) is unique. This will be assured if we assume that k is real, since in this case y_n is real and if two solutions existed we deduce by the orthogonality of the Legendre polynomials that y_n and $h_n^{(1)}$ are linearly dependent solutions of (2.94) for $r \geq a$. But this is impossible since y_n is real-valued and not identically zero.

We first consider question (1). One method of constructing the functions y_n is to renormalize the functions u_n as determined by the integral equation (2.68). However, for reasons already discussed we must discard this approach. A related approach is to construct each $y_n(r)$ as a solution of the Volterra integral equation (cf. De Alfaro and Regge [45])

$$y_n(r)=\frac{\sqrt{\pi}\,k^n r^n}{2^{n+1}\Gamma(n+\frac{3}{2})}+\frac{k^2}{2n+1}\int_0^r \sqrt{r\rho}\left(\left(\frac{\rho}{r}\right)^{n+1/2}-\left(\frac{r}{\rho}\right)^{n+1/2}\right)(1-m(\rho))y_n(\rho)\,\mathrm{d}\rho. \tag{2.96}$$

It can be shown (see the Exercises) that (2.96) can be solved by successive approximations for any value of k, thus removing one of our main objections to the use of the Fredholm integral equations (2.68). However, in addition to the numerical problems involved in computing y_n as a solution of (2.96), we have the unfortunate fact that the kernel of (2.96) depends on n, and hence all computations must essentially begin anew for each value of n, $n=0, 1, 2, \ldots$. This last disadvantage can be overcome if we make use of transformation operators to construct y_n and we now proceed to describe this method (Colton and Kress [40]).

We look for a solution of (2.91a) in the form

$$u(r,\theta,\phi)=h(r,\theta,\phi)+\int_0^r G(r,s;k)h(s,\theta,\phi)\,\mathrm{d}s \tag{2.97}$$

where (r,θ,ϕ) are spherical coordinates, h is a solution of

$$\Delta_3 h+k^2h=0 \tag{2.98}$$

in $\{\mathbf{x}: |\mathbf{x}|<a\}$, and G is a function to be determined. We assume that m is continuously differentiable on $[0, a]$. If we now substitute (2.97) into (2.91a) and integrate by parts we find that (2.97) will be a solution of (2.91a) provided G satisfies

$$r^2\left[\frac{\partial^2 G}{\partial r^2}+\frac{2}{r}\frac{\partial G}{\partial r}+k^2(1-m(r))G\right]=s^2\left[\frac{\partial^2 G}{\partial s^2}+\frac{2}{s}\frac{\partial G}{\partial s}+k^2 G\right] \tag{2.99}$$

for $0\leqslant s\leqslant r\leqslant a$ and the boundary conditions

$$G(r,r;k)=\frac{k^2}{2r}\int_0^r sm(s)\,\mathrm{d}s \tag{2.100a}$$

$$G(r,s;k)=0((rs)^{1/2}) \tag{2.100b}$$

where the order relation (2.100b) is understood to be twice differentiable with respect to r and s, e.g. $(\partial G/\partial r)(r,s;k)=0((s/r)^{1/2})$, $(\partial G/\partial s)(r,s;k)=0((r/s)^{1/2})$, etc. for $0\leqslant s\leqslant r\leqslant a$. Now let

$$\xi=(rs)^{1/2},\qquad \eta=(r/s)^{1/2} \tag{2.101}$$

and define M by

$$M(\xi, \eta; k) = \xi G(\xi\eta, \xi/\eta; k). \tag{2.102}$$

Then for $0<\xi<a$, $1<\eta<\infty$, $\xi\eta<a$, M satisfies the hyperbolic equation

$$\partial^2 M/\partial\xi\partial\eta + k^2\xi\eta(1-1/\eta^4 - m(\xi\eta))M = 0 \tag{2.103}$$

and the initial conditions

$$M(\xi, 1; k) = \frac{k^2}{2}\int_0^\xi sm(s)\,ds, \qquad M(\xi, \eta; k) = 0(\xi^2). \tag{2.104}$$

We now need to establish the existence of the function M satisfying (2.103), (2.104). To this end we reformulate (2.103), (2.104) as the integral equation

$$\begin{aligned} M(\xi, \eta; k) &= \frac{k^2}{2}\int_0^\xi sm(s)\,ds \\ &\quad - k^2\int_1^\eta\int_0^\xi st\left(1-\frac{1}{t^4}-m(st)\right)M(s, t; k)\,ds\,dt \end{aligned} \tag{2.105}$$

and look for a solution of (2.105) in the form

$$\begin{aligned} M(\xi, \eta; k) &= M_0(\xi, \eta; k) + \sum_{n=1}^\infty [M_n(\xi, \eta; k) - M_{n-1}(\xi, \eta; k)] \\ &= \lim_{n\to\infty} M_n(\xi, \eta; k) \end{aligned} \tag{2.106}$$

where

$$\begin{aligned} M_0(\xi, \eta; k) &= \frac{k^2}{2}\int_0^\xi sm(s)\,ds \\ \tilde{M}_n(\xi, \eta; k) &= M_n(\xi, \eta; k) - M_{n-1}(\xi, \eta; k) \\ &= -k^2\int_1^\eta\int_0^\xi st\left(1-\frac{1}{t^4}-m(st)\right)\tilde{M}_{n-1}(s, t; k)\,ds\,dt. \end{aligned} \tag{2.107}$$

If we now let $\max_{0\le s\le a} |sm(s)| = A$ then one can easily verify that

$$\begin{aligned} |M_0(\xi, \eta; k)| &\le \frac{k^2\xi}{2}(A+a) \\ |\tilde{M}_n(\xi, \eta; k)| &\le k^2(A+a)\int_0^\eta\int_0^\xi |\tilde{M}_{n-1}(s, t; k)|\,ds\,dt \end{aligned} \tag{2.108}$$

and hence, by induction,

$$|\tilde{M}_n(\xi, \eta; k)| \le \frac{k^2\xi}{2}(A+a)\frac{[k^2\xi\eta(A+a)]^n}{(n+1)!\,n!} \tag{2.109}$$

for $0 \leqslant \xi \leqslant a$, $1 \leqslant \eta < \infty$, $n \geqslant 1$. This implies that the series (2.106) is uniformly convergent and defines a solution of the integral equation (2.105). It is easily verified that since m is continuously differentiable, G is twice continuously differentiable and that the boundary condition (2.100b) is satisfied. Hence we have established the existence of the operator (2.97).

Example Let $m(r)=1$ for $0 \leqslant r \leqslant a$. Then from (2.107) we can deduce that

$$G(r, s; k) = \frac{kr^{1/2}}{2(r-s)^{1/2}} I_1(k[s(r-s)]^{1/2}) \tag{2.110}$$

where I_1 denotes a modified Bessel function of order one, and hence in this case (2.97) can be rewritten in the form

$$u(r, \theta, \phi) = h(r, \theta, \phi) + \frac{kr}{2} \int_0^1 I_1(kr[t(1-t)]^{1/2}) \frac{h(rt, \theta, \phi)}{(1-t)^{1/2}} \, dt. \tag{2.111}$$

Except in special cases (such as the above example) the construction of M from the recursion formula (2.107) can be computationally quite difficult. Hence it is desirable to have other methods available for constructing M. Since M is a solution of a characteristic initial-value problem for a hyperbolic equation (with data given on $\eta = 1$ and $\xi = 0$), a variety of such methods present themselves. One such method is the Cauchy–Euler polygon method as described by Diaz [46] (see also Gilbert and Colton [75]) and we now briefly outline this approach. For details the reader is referred to Diaz [46]. To implement this method we first choose a sequence of subdivisions of the rectangle $R = [0, a] \times [1, b]$ where b is a large fixed number, i.e. for each ordered pair (m, n) we form a subdivision

$$\begin{aligned} &0 = \xi_{0m} < \xi_{1m} < \cdots < \xi_{mm} = a \\ &1 = \eta_{0n} < \eta_{1n} < \cdots < \eta_{nn} = b, \end{aligned} \tag{2.112}$$

and on each of the subrectangles $R_{kl} = [\xi_{km}, \xi_{k+1,m}] \times [\eta_{ln}, \eta_{l+1,n}]$ we consider the 'miniature problem'

$$\begin{aligned} &\partial^2 M/\partial\xi\, \partial\eta = A_{kl} M; \qquad A_{kl} = A(\xi_{km}, \eta_{ln}) \\ &\quad A(\xi, \eta) = k^2 \xi\eta(m(\xi\eta) + 1/\eta^4 - 1) \end{aligned} \tag{2.113}$$

where

$$\begin{aligned} &M(\xi, \eta_l) = D_{kl} + B_{kl}(\xi - \xi_k); \qquad \xi_k \leqslant \xi \leqslant \xi_{k+1} \\ &M(\xi_k, \eta) = D_{kl} + C_{kl}(\eta - \eta_l); \qquad \eta_l \leqslant \eta \leqslant \eta_{l+1} \end{aligned} \tag{2.114}$$

and we have used the abbreviated notation $\xi_k = \xi_{km}$, $\eta_l = \eta_{ln}$. Hence, in

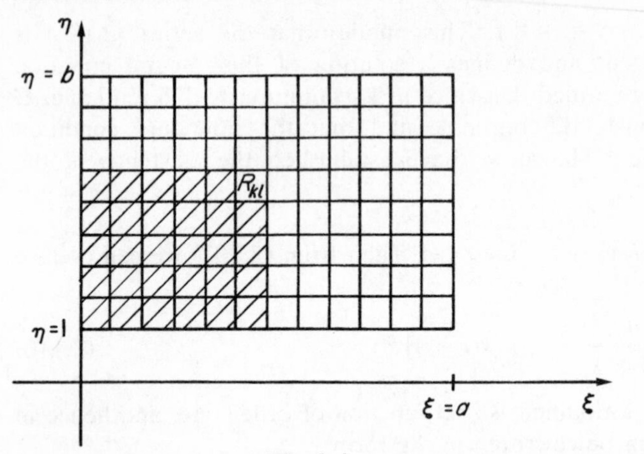

Fig. 2.4

R_{kl}, we have

$$M(\xi, \eta) = A_{kl}(\xi - \xi_k)(\eta - \eta_l) + B_{kl}(\xi - \xi_k) + C_{kl}(\eta - \eta_l) + D_{kl}. \quad (2.115)$$

Note that the constants B_{kl}, C_{kl} and D_{kl} are determined from the solution in the rectangles beneath and to the left of the rectangle R_{kl} (see Fig. 2.4). On the rectangles having sides on $\eta = 1$ the initial data $M(\xi, 1) = (k^2/2)\int_0^\xi sm(s)\,ds$ is linearly approximated and on rectangles having sides on $\xi = 0$ the data is chosen to be zero. The general form of the solution in R_{kl} can now be computed to be (see Diaz [46])

$$\begin{aligned} M(\xi, \eta) = M_{k0} &+ \frac{M_{k+1,0} - M_{k0}}{\xi_{k+1} - \xi_k}(\xi - \xi_k) \\ &+ \sum_{i=1}^{k} \sum_{j=1}^{l} A_{i-1,j-1}(\xi_i - \xi_{i-1})(\eta_j - \eta_{j-1}) \\ &+ \sum_{j=1}^{l} A_{k,j-1}(\xi - \xi_k)(\eta_j - \eta_{j-1}) \\ &+ \sum_{i=1}^{k} A_{i-1,l}(\xi_i - \xi_{i-1})(\eta - \eta_l) + A_{kl}(\xi - \xi_k)(\eta - \eta_l), \end{aligned} \quad (2.116)$$

where $M_{ij} = M(\xi_i, \eta_j)$. From this approximate solution we immediately obtain an approximation to the kernel G from (2.102).

Given the fact that we know G we can now use (2.97) to compute the functions y_n appearing in (2.92). To do this we simply observe that $j_n(kr)P_n(\cos\theta)$ is a solution of the Helmholtz equation (2.98) and hence from (2.97) (factoring out the Legendre polynomial)

$$y_n(r) = j_n(kr) + \int_0^r G(r, s; k) j_n(ks)\,ds. \quad (2.117)$$

Note that the kernel G is independent of n and if m is 'small' the integral in (2.117) represents a 'small' perturbation of the spherical Bessel function j_n.

Although we shall not make use of it in this chapter, we can also construct a transformation operator for (2.91a) which is valid in exterior domains and is in a certain sense complimentary to the operator defined in (2.97). For applications of this operator see Colton [28] and the exercises of Chapter 8. Although the assumption that $m(r) \equiv 0$ for $r \geq a$ can be relaxed (cf. Colton and Kress [39]) we shall not discuss this more general case here.

We look for a solution of (2.19a) in the form

$$u(r, \theta, \phi) = h(r, \theta, \phi) + \int_r^\infty E(r, s; k) h(s, \theta, \phi)\, ds \tag{2.118}$$

where $h(\mathbf{x}) = h(r, \theta, \phi)$ is again a solution of the Helmholtz equation (2.98), but this time in the exterior of a star-like domain Ω and satisfying the radiation condition (2.91c). We assume that E vanishes identically for $rs \geq a^2$ and note that this implies that u as defined by (2.118) also satisfies the radiation condition (2.91c). Substituting (2.118) into (2.91a) and integrating by parts shows that (2.118) will be a solution of (2.91a) provided E satisfies the hyperbolic equation (2.99) for $s > r > R$ (where R is the radius of the smallest sphere centered at the origin and contained in Ω) and the initial condition

$$E(r, r; k) = \frac{k^2}{2r} \int_r^\infty s m(s)\, ds. \tag{2.119}$$

The complimentary nature of the operators (2.118) and (2.97) is indicated in Fig. 2.5, where L denotes the differential operator defined by (2.99).

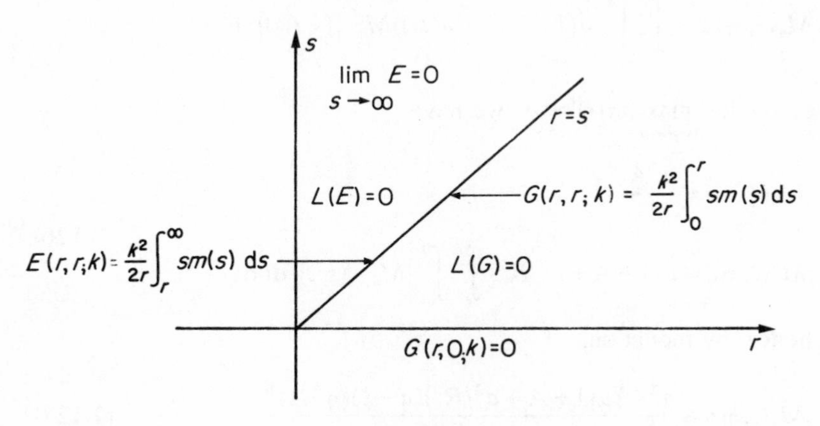

Fig. 2.5

Making the change of variables (2.101) and defining M by

$$M(\xi, \eta; k) = \xi E(\xi\eta, \xi/\eta; k) \tag{2.120}$$

shows that for $R < \xi < \infty$, $(R/a)^{1/2} < \eta < 1$, M satisfies the hyperbolic equation

$$\partial^2 M/\partial\xi\partial\eta + k^2\xi\eta(1 - 1/\eta^4 - m(\xi\eta))M = 0, \tag{2.121}$$

and the 'initial' conditions

$$\begin{aligned} M(\xi, 1; k) &= \frac{k^2}{2}\int_\xi^a sm(s)\,\mathrm{d}s \\ M(\xi, \eta; k) &= 0 \quad \text{for} \quad \xi \geqslant a, \end{aligned} \tag{2.122}$$

where we have made use of the fact that $m(r) \equiv 0$ for $r > a$. We can reformulate (2.121), (2.122) as the integral equation

$$\begin{aligned} M(\xi, \eta; k) = {} & \frac{k^2}{2}\int_\xi^a sm(s)\,\mathrm{d}s \\ & - k^2\int_\xi^a\int_\eta^1 st\left(1 - \frac{1}{t^4} - m(st)\right)M(s, t; k)\,\mathrm{d}t\,\mathrm{d}s \end{aligned} \tag{2.123}$$

and look for a solution of (2.123) in the form

$$M(\xi, \eta; k) = \sum_{n=0}^{\infty} k^{2n+2}M_n(\xi, \eta) \tag{2.124}$$

where

$$\begin{aligned} M_0(\xi, \eta) &= \frac{1}{2}\int_\xi^a sm(s)\,\mathrm{d}s \\ M_n(\xi, \eta) &= -\int_\xi^a\int_\eta^1 st(1 - 1/t^4 - m(st))M_{n-1}(s, t)\,\mathrm{d}t\,\mathrm{d}s. \end{aligned} \tag{2.125}$$

If we now let $\max\limits_{0 \leqslant r \leqslant a} |m(r)| \leqslant A$, we have

$$\begin{aligned} |M_0(\xi, \eta)| &\leqslant \frac{a^2 A}{4} \\ |M_n(\xi, \eta)| &\leqslant a(1 + A + a^2/R^2)\int_\xi^a\int_\eta^1 |M_{n-1}(s, t)|\,\mathrm{d}t\,\mathrm{d}s \end{aligned} \tag{2.126}$$

and hence, by induction,

$$|M_n(\xi, \eta)| \leqslant \frac{a^2 A}{4}\frac{[a(1 + A + a^2/R^2)(a - \xi)(\eta - 1)]^n}{(n!)^2} \tag{2.127}$$

for $R \leq \xi \leq a$, $(R/a)^{1/2} \leq \eta \leq 1$. This implies that the series (2.124) is uniformly convergent, thus establishing the existence of the kernel E and the operator (2.118). Since M satisfies a characteristic initial-value problem for a hyperbolic equation (with initial data given on $\eta = 1$ and $\xi = a$) we can again construct numerical approximations to this function (and hence E) by using the Cauchy–Euler polygon method. We finally note that there is a relationship between the operators (2.97) and (2.118) and Gilbert's 'method of ascent' (cf. Gilbert [71], [74]).

Now that we have constructed the functions y_n appearing in (2.92) we turn to the problem of computing the constants a_n and b_n appearing in (2.92), (2.93). This is in fact quite simple. Using the first expansion in (2.62) and making the requirement that u is continuously differentiable across $r = a$, we see from (2.92), (2.93) and the orthogonality of Legendre polynomials that

$$\begin{aligned} &a_n y_n(a) - b_n h_n^{(1)}(ka) = (2n+1)\mathrm{i}^n j_n(ka) \\ &a_n \frac{\mathrm{d}}{\mathrm{d}r} y_n(a) - b_n \frac{\mathrm{d}}{\mathrm{d}r} h_n^{(1)}(ka) = (2n+1)\mathrm{i}^n \frac{\mathrm{d}}{\mathrm{d}r} j_n(ka). \end{aligned} \tag{2.128}$$

Since for $r \geq a$, $h_n^{(1)}$ and y_n are linearly independent solutions of (2.94), we conclude that the algebraic system (2.128) has a unique solution, and thus we can determine the constants a_n and b_n, $n = 0, 1, 2, \ldots$.

In order to show now that the series (2.92), (2.93) converge and satisfy (2.91a)–(2.91c) we must obtain estimates on the coefficients a_n and b_n. But from the estimates (cf. Erdélyi *et al.* [53] and (2.117))

$$\begin{aligned} h_n^{(1)}(ka) &= \frac{\mathrm{i}\Gamma(n+\frac{1}{2})2^n}{\sqrt{\pi}\,(ka)^{n+1}}[1+0(1/n)] \\ \frac{\mathrm{d}}{\mathrm{d}r} h_n^{(1)}(ka) &= \frac{\mathrm{i}\Gamma(n+\frac{3}{2})2^n}{\sqrt{\pi}\,(ka)^{n+1}}[1+0(1/n)] \\ (2n+1)\mathrm{i}^n j_n(ka) &= \frac{\sqrt{\pi}\,(ka)^n \mathrm{i}^n}{2^n\Gamma(n+\frac{1}{2})}[1+0(1/n)] \\ (2n+1)\mathrm{i}^n \frac{\mathrm{d}}{\mathrm{d}r} j_n(ka) &= \frac{\sqrt{\pi}\,(ka)^n \mathrm{i}^n}{2^n\Gamma(n-\frac{1}{2})}[1+0(1/n)] \\ y_n(a) &= \frac{\sqrt{\pi}\,(ka)^n}{2^{n+1}\Gamma(n+\frac{3}{2})}[1+0(1/n)] \\ \frac{\mathrm{d}}{\mathrm{d}r} y_n(a) &= \frac{\sqrt{\pi}\,(ka)^n}{2^{n+1}\Gamma(n+\frac{1}{2})}[1+0(1/n)] \end{aligned} \tag{2.129}$$

we can deduce from (2.128) and Cramer's rule that

$$a_n = 0\left[\frac{(ka)^n}{2^n\Gamma(n+\frac{1}{2})}\right] \quad \text{and} \quad b_n = 0\left[\frac{(ka)^n}{2^n\Gamma(n+\frac{1}{2})}\right]. \tag{2.130}$$

Since the estimates in (2.129) are uniformly valid on compact subsets we can conclude from (2.129) and (2.130) that the series (2.92) and (2.93) are uniformly convergent for r on compact subsets of $[0, a)$ and (a, ∞), respectively. A similar analysis shows that the series may be differentiated termwise and define a solution to (2.91a) for $r<a$ and $r>a$, respectively, and since u is by construction continuously differentiable across $r=a$ we can now conclude that (2.92), (2.93) define a solution to (2.91a) for $0 \leqslant r < \infty$.

All that remains now is to show that

$$u_s(\mathbf{x}) = \sum_{n=0}^{\infty} b_n h_n^{(1)}(kr) P_n(\cos\theta); \qquad r>a \tag{2.131}$$

satisfies the radiation condition (2.91c), where we are given the estimate (2.130) on the b_n, $n=0, 1, 2, \ldots$. Showing this is somewhat more delicate than our previous arguments showing that (2.92), (2.93) defines a solution to (2.91a), since now both n and r tend to infinity. However, the following theorem yields the desired conclusion (Colton [27]; Karp [100]).

Theorem 2.9 *Let u_s be given by* (2.131) *where the coefficients b_n satisfy the estimate in* (2.130). *Then for $r \geqslant a' > a$ we have that u_s has the representation*

$$u_s(\mathbf{x}) = h_0^{(1)}(kr) \sum_{n=0}^{\infty} \frac{F_n(\cos\theta)}{r^n} + h_1^{(1)}(kr) \sum_{n=0}^{\infty} \frac{G_n(\cos\theta)}{r^n}$$

where the series converge absolutely and uniformly for $r \geqslant a'$, $0 \leqslant \theta \leqslant \pi$, and may be differentiated termwise as often as desired. The functions F_n and G_n are analytic functions of $\cos\theta$ uniquely determined by the function u_s.

Proof If the recursion formula

$$h_{n-1}^{(1)}(kr) + h_{n+1}^{(1)}(kr) = (2n/kr) h_n^{(1)}(kr) \tag{2.132}$$

is used repeatedly, one obtains

$$h_n^{(1)}(kr) = h_0^{(1)}(kr) R_{n,1/2}(kr) + h_1^{(1)}(kr) R_{n-1,3/2}(kr) \tag{2.133}$$

where $R_{n,\nu}$ are Lommel polynomials (cf. Erdélyi *et al.* [53]) defined by

$$R_{n,\nu}(z) = \sum_{m=0}^{[n/2]} \frac{(-1)^n \Gamma(n-m+1)\Gamma(\nu+n-m)}{m!\,\Gamma(n-2m+1)\,\Gamma(\nu+m)} \left(\frac{z}{2}\right)^{-n+2m}. \tag{2.134}$$

Now consider the series ($z = kr$, $\xi = \cos\theta$)

$$E(z,\xi) = \sum_{n=0}^{\infty} b_n R_{n,1/2}(z) P_n(\xi)$$
$$Q(z,\xi) = \sum_{n=0}^{\infty} b_n R_{n-1,3/2}(z) P_n(\xi) \tag{2.135}$$

which result when (2.133) is substituted into the series representation (2.131) of u_s. We can rewrite E as

$$E(z,\xi) = \sum_{n=0}^{\infty} \frac{b_n P_n(\xi) \dfrac{1}{\Gamma(n+\frac{1}{2})} \left(\dfrac{z}{2}\right)^{n+1/2} R_{n,1/2}(z)}{\dfrac{1}{\Gamma(n+\frac{1}{2})} \left(\dfrac{z}{2}\right)^{n+1/2}} \tag{2.136}$$

From Erdélyi *et al.* [53] we have

$$\lim_{n\to\infty} \frac{\left(\dfrac{z}{2}\right)^{n+1/2} R_{n,1/2}(z)}{\Gamma(n+\frac{1}{2})} = \frac{z^{3/2}}{(2\pi)^{1/2}} j_0(z), \tag{2.137}$$

and hence from (2.130), (2.137) and (2.136) we have that E is uniformly and absolutely convergent for $|z| = ka'$, $\xi \in [-1, +1]$. But the series defining E is a series of polynomials in $1/z$, and since it converges uniformly and absolutely on the circle $|1/z| = 1/ka'$, it is analytic for $|1/z| < 1/ka'$, $\xi \in [-1, +1]$. A similar result holds for Q. Hence

$$u_s(\mathbf{x}) = h_0^{(1)}(kr) E(1/r, \cos\theta) + h_1^{(1)}(kr) Q(1/r, \cos\theta) \tag{2.138}$$

where E and Q are analytic functions of $1/r$ and are regular in the interior of the circle $|1/r| = 1/a'$ in the complex $1/r$ plane. The theorem now follows from this statement.

Since the spherical Hankel functions satisfy the Sommerfeld radiation condition (2.91c), we can now conclude that u_s as defined by (2.131) also satisfies this condition and we have completed our construction of the solution to (2.91a)–(2.91c) valid for any positive value of the wave number k.

In this section we have only briefly touched on some of the mathematical problems arising in scattering theory, e.g. analyticity of the far field pattern in the complex k plane, construction of the 'wave functions' $y_n(r)$, and the asymptotic behavior of 'outgoing' solutions of the reduced wave equation (Theorem 2.9). Of the many areas we have not mentioned are scattering by obstacles (Kleinman and Wendland [103]; Leis [114]; Ursell [170]), high-frequency approximations (Bloom and Kazarinoff [18]; Fock

[56]; Keller and Lewis [101]; Ursell [169]), and inverse problems (Gelfand and Levitan [68]; Majda [122]; Prosser [145]). (The references just cited are by no means intended to be definitive and are merely intended to give the reader a starting point in searching through the literature!) In later chapters we shall return to problems in scattering theory, and in fact discuss some aspects of the areas listed above (cf. Chapters 5 and 8). However, it is by no means our intention to give any kind of systematic survey of mathematical problems in scattering theory, but rather simply to point out some of the interesting analytical problems that arise in this area in the hope that the reader's interest will be aroused to search further on his own. For a more comprehensive treatment of the mathematical problems in scattering theory the reader is referred to one or more of the many excellent monographs in this field, for example: De Alfaro and Regge [45]; Jones [98]; Lax and Phillips [113]; Müller [131]; Newton [136]; Wilcox [180].

Exercises

(1) Prove the analogue of Corollary 2.1 for the elliptic equation (2.13).

(2) Verify formula (2.37).

(3) Suppose the function h in (2.13) is not nonpositive, but for R sufficiently small there is a positive function u such that $L[u] \leq 0$ in Ω_R. Under this assumption prove Theorem 2.3 without making any assumptions on the nonpositivity of h.

(4) Let λ be a positive constant. Show that any entire solution of $\Delta_2 u - \lambda^2 u = 0$ which grows more slowly than $(1/\sqrt{r})e^{\lambda r}$ as $r \to \infty$ (uniformly in θ) must be identically zero.

(5) Consider the integral equation (2.51) where $m(\mathbf{y}) = m(r)$ is spherically symmetric and satisfies $k^2 \int_0^\infty rm(r)\,dr < 1$, but where it is no longer assumed that $m(r)$ has compact support. Show that the solution of (2.51) can be represented as a Neumann series and in the case where $m(r)$ is of compact support compare this result with Theorem 2.6.

(6) Verify that the operator $\mathbf{T}_n$ defined by equation (2.69) is a bounded linear operator mapping $C[0, a]$ into itself.

(7) Deduce Theorem 2.8 by using (2.117), (2.78), (2.128), and the series representation

$$h_n^{(1)}(kr) = \frac{1}{k^{n+1}r} \exp\left\{-i\frac{\pi}{2}(n+1) + ikr\right\} \sum_{p=0}^{n} \frac{i^p (n+p)!\, k^{n-p}}{(n-p)!\, p!\, (2r)^p}$$

to obtain the leading term of the Taylor series representation for $b_n = b_n(k)$.

(8) Show that the integral equation (2.96) can be solved by the method of successive approximations for any value of the wave number k.

(9) Construct an integral operator of the form (2.97) where h is a solution of $\Delta_3 h + k_0^2 h = 0$ with k_0 not necessarily equal to k.

(10) In Theorem 2.9 show that F_0 and G_0 uniquely determine the functions F_n and G_n for $n \geqslant 1$ and hence uniquely determine $u_s(\mathbf{x})$.

3

Analytic solutions of partial differential equations

This chapter is devoted to the study of two problems, namely the construction of solutions to analytic partial differential equations satisfying prescribed analytic Cauchy data and the problem of showing that solutions of elliptic equations with analytic coefficients are analytic functions of their independent variables. These problems are two of the classical problems in the theory of partial differential equations, and as such have a long history with important contributions having been made by many mathematicians. Here we shall only touch the surface of this topic and refer the reader to more advanced textbooks for further developments and references (cf. Friedman [57]; Hörmander [92]; John [93]; Nirenberg [137]). The classic result in this area is, of course, the Cauchy–Kowalewski theorem. Here we shall only state this theorem, since its proof is straightforward and can be found in almost any book on partial differential equations. However, in the special case of semilinear hyperbolic and elliptic equations in two independent variables we shall show how the solution to Cauchy's problem can be constructed by a simple iteration procedure. Our reason for considering these special cases is that the results for hyperbolic equations are needed to prove the analyticity of solutions to elliptic equations, and in the case of elliptic equations it provides us with a method for constructing the complex Riemann function which we shall define and use in Chapter 4. Our discussion of these topics related to the Cauchy–Kowalewski theorem will be followed by a proof of Holmgren's uniqueness theorem, which has been the forerunner of a flood of papers dealing with the uniqueness of solutions to Cauchy's problem (for references see Bers, John and Schechter [17] and Nirenberg [137]). We shall show that the uniqueness of the solution to Cauchy's problem for elliptic equations is in fact equivalent to the Runge approximation property. These results on unique continuation are then followed by Lewy's proof (Lewy [116]; see also Garabedian [61]) that classical solutions of semilinear elliptic equations in two independent variables are analytic functions of their independent variables.

We also state without proof the analogous results for linear elliptic and parabolic equations in several independent variables. Finally, we conclude this chapter by exhibiting a linear partial differential equation whose solutions, far from being analytic, do not even exist. The proof of this nonexistence result is based on the identity theorem for analytic functions.

3.1 The Cauchy–Kowalewski theorem

The Cauchy–Kowalewski theorem states that there is exactly one analytic solution to Cauchy's problem for an analytic partial differential equation. Since there is no basic dependence on dimension or order, we only state the result for second order equations in two independent variables. Analogous results hold for higher order equations and systems in several independent variables.

Theorem 3.1 (*Cauchy–Kowalewski theorem*) *Let ϕ and ψ be analytic functions in a neighborhood of the origin and set*

$$\phi(0) = u_0, \qquad \psi(0) = p_0, \quad \phi'(0) = q_0,$$
$$\phi''(0) = t_0, \qquad \psi'(0) = s_0.$$

Let $F(x, y, u, p, q, s, t)$ be analytic in a neighborhood of $(0, 0, u_0, p_0, q_0, s_0, t_0)$. Then the equation

$$u_{xx} = F(x, y, u, u_x, u_y, u_{xy}, u_{yy})$$

has exactly one solution u which is analytic in a neighborhood of the origin and satisfies the initial conditions

$$u(0, y) = \phi(y), \qquad u_x(0, y) = \psi(y).$$

The Cauchy–Kowalewski theorem is proved by representing u as a Taylor series and then determining the coefficients of the Taylor series by substituting into the differential equation and making use of the initial conditions (cf. Courant and Hilbert [43]). Needless to say, such a procedure is far too tedious for practical application. Furthermore, the series for u may not converge in the region needed in a specific example. For this reason more constructive methods are needed to solve Cauchy's problem for partial differential equations, and we now present one such approach for semilinear hyperbolic and elliptic equations in two independent variables.

We first consider the hyperbolic Cauchy problem

$$u_{xx} - u_{tt} = f(t, x, u, u_t, u_x) \tag{3.1a}$$

$$u(0, x) = \phi(x), \qquad u_t(0, x) = \psi(x) \tag{3.1b}$$

where we assume that ϕ is twice continuously differentiable, ψ is continuously differentiable, and f is a continuous function satisfying the Lipschitz condition

$$|f(t, x, u, p, q)-f(t, x, u_0, p_0, q_0)| \leqslant M[|u-u_0|+|p-p_0|+|q-q_0|], \quad (3.2)$$

where M is a positive constant. The continuity conditions on ϕ, ψ and f are assumed to hold in a region D containing the t-axis and for u, p and q sufficiently small. Due to the fact that (3.1a) is of hyperbolic type, no assumptions on analyticity are needed. If we make the change of variables

$$\xi = x+t, \qquad \eta = x-t \quad (3.3)$$

(3.1a) is transformed into an equation of the form

$$u_{\xi\eta} = F(\xi, \eta, u, u_\xi, u_\eta) \quad (3.4a)$$

where F is a continuous function satisfying a Lipschitz condition and the Cauchy data become

$$\begin{aligned} &u(\xi, \eta) = \phi(\xi) \quad \text{on} \quad \xi = \eta \\ &(\partial u/\partial \xi)(\xi, \eta) - (\partial u/\partial \eta)(\xi, \eta) = \psi(\xi) \quad \text{on} \quad \xi = \eta. \end{aligned} \quad (3.4b)$$

Without loss of generality we assume that the domain $G = \{(\xi, \eta): (x, t) \in D,\ x = \frac{1}{2}(\xi+\eta),\ t = \frac{1}{2}(\xi-\eta)\}$ contains the origin. Now define a new function s by

$$s(\xi, \eta) = (\partial^2 u/\partial \xi \partial \eta)(\xi, \eta). \quad (3.5)$$

Then

$$\begin{aligned} u(\xi, \eta) &= \int_0^\xi \int_0^\eta s(\xi, \eta)\, d\eta\, d\xi + \int_0^\xi \rho(\xi)\, d\xi + \int_0^\eta \omega(\eta)\, d\eta + u(0, 0) \\ u_\xi(\xi, \eta) &= \int_0^\eta s(\xi, \eta)\, d\eta + \rho(\xi) \\ u_\eta(\xi, \eta) &= \int_0^\xi s(\xi, \eta)\, d\xi + \omega(\eta) \end{aligned} \quad (3.6)$$

where $\rho(\xi) = u_\xi(\xi, 0)$, $\omega(\eta) = u_\eta(0, \eta)$. Note that s must satisfy the equation

$$\begin{aligned} s(\xi, \eta) = F\Big(\xi, \eta, &\int_0^\xi \int_0^\eta s(\xi, \eta)\, d\eta\, d\xi + \int_0^\xi \rho(\xi)\, d\xi + \int_0^\eta \omega(\eta)\, d\eta \\ &+ u(0, 0), \int_0^\eta s(\xi, \eta)\, d\eta + \rho(\xi), \int_0^\xi s(\xi, \eta)\, d\xi + \omega(\eta)\Big) \end{aligned} \quad (3.7)$$

and, conversely, if s satisfies (3.7) then a solution of (3.4) is given by

(3.6). The initial conditions (3.4b) become

$$\int_0^\xi \int_0^\xi s(\xi,\eta)\,d\eta\,d\xi + \int_0^\xi \rho(\xi)\,d\xi + \int_0^\xi \omega(\eta)\,d\eta + u(0,0) = \phi(\xi)$$

or, differentiating, (3.8)

$$\int_0^\xi s(\xi,\eta)\,d\eta + \int_0^\xi s(\tau,\xi)\,d\tau + \rho(\xi) + \omega(\xi) = \phi'(\xi) \tag{3.9}$$

and

$$\int_0^\xi s(\xi,\eta)\,d\eta + \rho(\xi) - \int_0^\xi s(\tau,\xi)\,d\tau - \omega(\xi) = \psi(\xi). \tag{3.10}$$

Equations (3.9) and (3.10) now yield the following expressions for ρ and ω in terms of the function s:

$$\rho(\xi) = \tfrac{1}{2}[\phi'(\xi) + \psi(\xi)] - \int_0^\xi s(\xi,\eta)\,d\eta \tag{3.11}$$

$$\omega(\eta) = \tfrac{1}{2}[\phi'(\eta) - \psi(\eta)] - \int_0^\eta s(\xi,\eta)\,d\xi. \tag{3.12}$$

Hence, we can express the functions ρ and ω as operators on the function s. In particular, if we define the operators $\mathbf{B}_i$, $i = 1, 2, 3$, by the right-hand sides of the three equations in (3.6) where ρ and ω are given by (3.11), (3.12) and $u(0,0) = \phi(0)$, then $s(\xi,\eta)$ satisfies the equation

$$s(\xi,\eta) = F(\xi,\eta,\mathbf{B}_1[s], \mathbf{B}_2[s], \mathbf{B}_3[s]). \tag{3.13}$$

We now want to show the existence of a unique continuous function s that satisfies (3.13). Let Ω be a disk contained in G with center at the origin and let $B(\Omega)$ be the Banach space of all continuous functions defined on Ω and continuous in $\bar{\Omega}$ with the norm

$$\|s\|_\lambda = \sup_\Omega \{e^{-\lambda(|\xi|+|\eta|)}\,|s(\xi,\eta)|\} \tag{3.14}$$

where $\lambda > 0$ is fixed. We shall show that the operator $\mathbf{T}$ defined by

$$\mathbf{T}[s] = F(\xi,\eta,\mathbf{B}_1[s], \mathbf{B}_2[s], \mathbf{B}_3[s]) \tag{3.15}$$

maps the Banach space $B(\Omega)$ into itself and is a contraction mapping. Hence, by the Banach contraction mapping theorem there exists a unique $s \in B(\Omega)$ such that $\mathbf{T}[s] = s$ and this will establish the existence of a unique function s satisfying (3.13). From the Lipschitz condition satisfied by F we have that for $s_1, s_2 \in B(\Omega)$ and Ω sufficiently small,

$$\|\mathbf{T}[s_1] - \mathbf{T}[s_2]\|_\lambda \leq M\{\|\mathbf{B}_1[s_1] - \mathbf{B}_1[s_2]\|_\lambda + \|\mathbf{B}_2[s_1] - \mathbf{B}_2[s_2]\|_\lambda + \|\mathbf{B}_3[s_1] - \mathbf{B}_3[s_2]\|_\lambda\}, \tag{3.16}$$

for some positive constant M. From estimates of the form

$$\left|\int_0^\xi s(\xi, \eta)\, d\xi\right| \leq \int_0^{|\xi|} \|s\|_\lambda e^{\lambda|\xi|+\lambda|\eta|}\, d\xi \leq \frac{1}{\lambda} e^{\lambda|\xi|+\lambda|\eta|} \|s\|_\lambda \tag{3.17}$$

that is

$$\left\|\int_0^\xi s(\xi, \eta)\, d\xi\right\|_\lambda \leq \frac{\|s\|_\lambda}{\lambda} \tag{3.18}$$

it can be seen that

$$\|\mathbf{B}_i[s_1] - \mathbf{B}_i[s_2]\|_\lambda \leq (N_i/\lambda)\, \|s_1 - s_2\|_\lambda, \qquad i = 1, 2, 3 \tag{3.19}$$

where the N_i are positive constants independent of λ. Hence

$$\|\mathbf{T}[s_1] - \mathbf{T}[s_2]\|_\lambda \leq (C/\lambda)\, \|s_1 - s_2\|_\lambda \tag{3.20}$$

where C is a positive constant independent of λ. From (3.20) we have that for λ sufficiently large $\mathbf{T}$ is a contraction mapping, and the existence of a unique solution to the equation $Ts = s$ (and hence a solution to (3.1)) now follows from the Banach contraction mapping theorem.

By introducing complex characteristic or conjugate coordinates we can treat the elliptic Cauchy problem

$$u_{xx} + u_{yy} = f(x, y, u, u_x, u_y) \tag{3.21a}$$

$$u(x, 0) = \phi(x), \qquad u_y(x, 0) = \psi(x) \tag{3.21b}$$

in the same manner as we solved the hyperbolic problem (3.1). In order to do this it is necessary to make the assumption that f is an analytic function of its independent variables in some neighborhood of the origin and hence satisfies a Lipschitz condition with respect to its last three arguments (cf. Gilbert [70], pp. 38, 139). We further assume that ϕ and ψ are analytic functions in a neighborhood of the origin. We now introduce the *conjugate coordinates*

$$z = x + iy, \qquad z^* = x - iy \tag{3.22}$$

which map the space C^2 of two complex variables into itself. Note that $z^* = \bar{z}$ if and only if x and y are real. Under this change of variables (3.21) becomes

$$U_{zz^*} = F(z, z^*, U, U_z, U_{z^*}) \tag{3.23a}$$

$$\left.\begin{aligned} U(z, z^*) &= \phi(z) \quad \text{on} \quad z = z^* \\ \partial U(z, z^*)/\partial z - \partial U(z, z^*)/\partial z^* &= -i\psi(z) \quad \text{on} \quad z = z^* \end{aligned}\right\} \tag{3.23b}$$

where

$$u((z + z^*)/2, (z - z^*)/2i) = U(z, z^*). \tag{3.24}$$

Note that $z = z^*$ describes a plane in the four-dimensional space C^2. By considering the Banach space of functions of two complex variables which are holomorphic and bounded in $\Delta\rho \times \Delta\rho^*$ where $\Delta\rho = \{z : |z| < \rho\}$, $\Delta\rho^* = \{z : z^* \in \Delta\rho\}$, with the norm defined by

$$\|s\|_\lambda = \sup_{\Delta\rho \times \Delta\rho^*} \{e^{-\lambda(|z|+|z^*|)} |s(z, z^*)|\}, \tag{3.25}$$

the existence of a solution to (3.23) can be established in exactly the same way as in the hyperbolic case (cf. Colton [27]). For further methods of solving the elliptic Cauchy problem and the appearance of such problems in certain areas of fluid mechanics see Garabedian [61], [62], [65].

In closing this section we should note that the elliptic Cauchy problem (3.21) is improperly posed in the sense that the solution does not depend continuously on the initial data. To see this consider the Laplace equation

$$u_{xx} + u_{yy} = 0 \tag{3.26a}$$

subject to the initial data

$$u(x, 0) = 0, \qquad u_y(x, 0) = (1/n) \sin nx \tag{3.26b}$$

where $n > 0$. Then the solution of (3.26) is

$$u(x, y) = (1/n^2) \sinh ny \sin nx \tag{3.27}$$

and as n tends to infinity the initial data tend uniformly to zero whereas the solution (3.27) oscillates between arbitrarily large values. However, in the *complex* domain the elliptic Cauchy problem is well posed. (In the above example the initial data do not tend to zero as n tends to infinity for x complex.) In our construction of the solution to (3.21) the unstable dependence of the solution to the elliptic equation (3.21a) on the (real) Cauchy data (3.21b) occurs exclusively in the step where the data are extended to complex values of the independent variable x. When this can be done in an elementary way, for example by direct substitution via the transformation (3.22), no instabilities will occur when one uses the contraction mapping operator **T** corresponding to (3.23) to obtain approximations to the desired solution.

3.2 Unique continuation

The Cauchy–Kowalewski theorem asserts the existence of a unique analytic solution to the noncharacteristic Cauchy problem for an analytic partial differential equation. It does not exclude the possibility of other nonanalytic solutions to this problem. It is to this question that Holmgren's uniqueness theorem is directed. We consider the second order equation

$$L(u) = \sum_{i,j=1}^{n} a_{ij}(\mathbf{x}) \frac{\partial^2 u}{\partial x_i \, \partial x_j} + \sum_{i=1}^{n} b_i(\mathbf{x}) \frac{\partial u}{\partial x_i} + c(\mathbf{x})u = 0 \tag{3.28}$$

with analytic coefficients and assume that the solution u of (3.28) has vanishing Cauchy data on a smooth noncharacteristic surface S, which may not even be analytic. Then Holmgren's theorem says the following.

***Theorem* 3.2** (*Holmgren's uniqueness theorem*) *Let u be a solution of $L[u]=0$ such that u has zero Cauchy data on S. Then u is identically zero at all points in its domain of definition that cannot be separated from S by a characteristic surface.*

Proof Without loss of generality, we can assume that a preliminary analytic change of variables has been made such that S is tangent to the hyperplane $x_1 = 0$ at the origin and S is convex there, i.e. S is described by

$$x_1 = \sum_{i,j=2}^{n} \mathrm{d}_{ij} x_i x_j + \text{higher order terms} \tag{3.29}$$

with the finite sum being positive definite. Since S is noncharacteristic we also have that the coefficient $a_{11}(\mathbf{x})$ of the differential operator L is nonzero at the origin. Under these circumstances there exists a positive constant α such that S together with the hyperplane $x_1 = \alpha$ bounds a domain D (see Fig. 3.1) such that one can apply the Green's identity

$$\iint_D (vL[u] - uM[v]) \, \mathrm{d}x_1 \cdots \mathrm{d}x_n + \int_{\partial D} B[u, v] = 0 \tag{3.30}$$

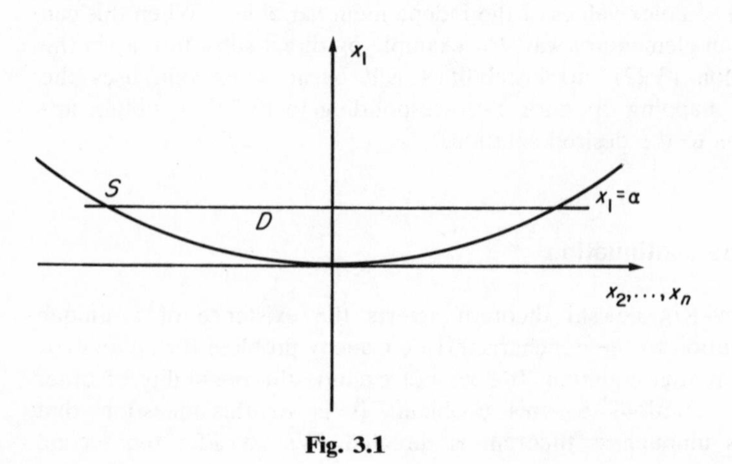

Fig. 3.1

where the adjoint operator M is defined by

$$M[v]=\sum_{i,j=1}^{n}\frac{\partial^2(a_{ij}(\mathbf{x})v)}{\partial x_i\,\partial x_j}-\sum_{i=1}^{n}\frac{\partial(b_i(\mathbf{x})v)}{\partial x_i}+c(\mathbf{x})v \tag{3.31}$$

and

$$B[u,v]=\sum_{i=1}^{n}(-1)^i\left\{\left[b_i-\sum_{j=1}^{n}\frac{\partial a_{ij}}{\partial x_j}\right]uv\right.$$
$$\left.+\sum_{j=1}^{n}a_{ij}\left[v\frac{\partial u}{\partial x_j}-u\frac{\partial v}{\partial x_j}\right]\right\}\mathrm{d}x_1\cdots\mathrm{d}x_{i-1}\,\mathrm{d}x_{i+1}\cdots\mathrm{d}x_n. \tag{3.32}$$

Note that u is defined in D but may have no meaning on the opposite side of S from D. Now let v be a solution of

$$M[v]=p \tag{3.33}$$

in D where p is a polynomial and v has zero Cauchy data on $x_1=\alpha$. The existence of v is assured by the Cauchy–Kowalewski theorem and for α small enough v will be defined throughout D. Since M is linear and the construction of v is obtained by representing v as a Taylor series and substituting in (3.33) to determine the coefficients, it can be seen that α can be chosen independent of p. This follows from the fact that the Taylor series expansion of p converges everwhere. From (3.30) we now have that

$$\int\!\!\int_D uM[v]\,\mathrm{d}x_1\cdots\mathrm{d}x_n=\int\!\!\int_D up\,\mathrm{d}x_1\cdots\mathrm{d}x_n=0 \tag{3.34}$$

for every polynomial p. But by the Weierstrass approximation theorem the set of polynomials is dense in $L_2(D)$ and hence from (3.34) we can conclude that $u\equiv 0$ in D. The above arguments can now be repeated for any smooth noncharacteristic initial surface lying in D, and in this manner we can conclude that $u\equiv 0$ for all points in its domain of definition that cannot be separated from S by a characteristic surface.

For more information on Holmgren's theorem see Bers, John and Schechter [17], and for recent developments on the problem of the uniqueness of Cauchy's problem see Nirenberg [137]. Note that Holmgren's theorem immediately implies unique continuation of solutions to second order partial differential equations with analytic coefficients, provided that in the process of continuation no characteristics are crossed.

We now want to establish the remarkable result that for elliptic equations the uniqueness of the solution to Cauchy's problem is equivalent to the Runge approximation property defined below. In the definition L is the operator (3.28) which is assumed to be uniformly elliptic with smooth coefficients.

Definition 3.1 *Solutions of $L[u]=0$ are said to have the Runge approximation property if, whenever D_1 and D_2 are two bounded simply connected domains, and D_1 is a subset of D_2, any solution in D_1 can be approximated uniformly on compact subsets of D_1 by a sequence of solutions which can be extended as solutions to D_2.*

In order to show the equivalence of the uniqueness of solutions to Cauchy's problem and the Runge approximation property we need the following results on the existence and uniqueness of the solution to Dirichlet's problem for the elliptic operator L defined in a bounded simply connected domain D with smooth boundary ∂D.

Definition 3.2 *Let $(f, g)=\iint_D fg$ for $f, g \in L_2(D)$. u is said to be a weak solution of $Lu=f$ if for every $\phi \in C_0^\infty(D)$, $(M\phi, u)=(\phi, f)$ where M is the adjoint operator defined by* (3.31).

Theorem 3.3 *Let $\phi \in C(\partial D)$. Then the boundary value problem*

$$L[u]=f \quad \text{in} \quad D, \qquad u=\phi \quad \text{on} \quad \partial D \tag{a}$$

where $f \in L_2(D)$ has a weak solution if and only if

$$\iint_D f\omega = \int_{\partial D} B[\phi, \omega] \tag{b}$$

for all solutions ω of

$$M[\omega]=0 \quad \text{in} \quad D \qquad \omega=0 \quad \text{on} \quad \partial D \tag{c}$$

where the boundary operator B is defined by (3.32). *The set of solutions to* (c) *is finite-dimensional.*

Corollary 3.1 *The orthogonal complement on ∂D of the space of all boundary values of all solutions of $L[u]=0$ is the finite-dimensional space spanned by $\partial\omega/\partial\nu$ where ω is a solution of equation* (c) *of Theorem* 3.3 *and $\partial/\partial\nu$ is the conormal derivative defined by*

$$\partial\omega/\partial\nu = \sum_{i,j=1}^{n} a_{ij}(\mathbf{x})\eta_i(\partial\omega/\partial x_j) \tag{3.35}$$

where $(\eta_1, \ldots, \eta_n)$ is the unit normal in an outward direction at a point on ∂D.

A proof of the above results can be found in any book on Hilbert space methods in partial differential equations, e.g. Agmon [1]; Friedman [58]; Showalter [159].

We are now in a position to show the equivalence of the uniqueness of solutions to Cauchy's problem for elliptic equations and the Runge approximation property (Browder [19]; Lax [112]; Malgrange [123]).

Theorem 3.4 *Let L denote the operator defined by* (3.28) *and assume that L is uniformly elliptic with smooth coefficients. Then solutions of* $L[u]=0$ *have the Runge approximation property if and only if solutions of the adjoint equation* $M[u]=0$ *are uniquely determined throughout their domain of existence by their Cauchy data along any smooth hypersurface.*

Remarks From a result of Friedrichs [59] the L_2 norm of a solution u over a domain bounds the maximum norm of u over any compact subset of this domain. Hence, in order to show that a solution over D_1 can be approximated uniformly over any compact subset by solutions in D_2, it is sufficient to show that it can be approximated in the L_2 sense over any subdomain whose closure lies in D_1. Let D_0 be such a subdomain and denote by S_1 the restriction of solutions in D_1 to D_0 and by S_2 the restriction of solutions in D_2 to D_0. S_2 is a subspace of S_1 and our aim in the first part of the theorem is to show that it is a dense subspace in the L_2 topology. By a classical criterion, S_2 is dense in S_1 if and only if every function v_0 in $L_2(D_0)$ that is orthogonal to S_2 is also orthogonal to S_1.

Proof We first show that the uniqueness of the solution to Cauchy's problem implies the Runge approximation property. Let $v_0 \in L_2(D_0)$, $v_0 \perp S_2$. We shall show that $v_0 \perp S_1$. Let ω_0 be a solution of

$$M[\omega_0]=\begin{cases} v_0 & \text{in} \quad D_0 \\ 0 & \text{in} \quad D_2 \backslash D_0 \end{cases}$$
$$\omega_0 = 0 \quad \text{on} \quad C_2 = \partial D_2. \tag{3.36}$$

By Theorem 3.3, ω_0 exists since $v_0 \perp S_2$. Let $u \in S_2$ and apply the Green's identity (3.30) to u and ω_0. Since $v_0 \perp S_2$ this shows that $\partial\omega_0/\partial\nu$ (as defined in (3.35)) is orthogonal to the boundary values of functions in S_2. By Corollary 3.1, $\partial\omega_0/\partial\nu = \partial\omega/\partial\nu$ where ω satisfies the homogeneous equation $M[\omega]=0$ in D_2, $\omega=0$ on C_2, which implies by Theorem 3.3 that (3.36) has a solution ω_0 such that $\partial\omega_0/\partial\nu=0$ on C_2 (e.g. $\omega-\omega_0$ is one such solution). By the uniqueness of the solution to Cauchy's problem this implies that $\omega_0=0$ in $D_2 \backslash D_0$. If we now apply (3.30) to $v=\omega_0$ and $u \in S_1$ over $D=D_1$ and use (3.36) we can conclude that $v_0 \perp S_1$.

We now show that the Runge approximation property implies the uniqueness of the solution to Cauchy's problem for elliptic equations. Let ω_0 be a solution of $M[\omega_0]=0$ with zero Cauchy data on a piece of a surface C. We shall show $\omega_0=0$ wherever it is defined. First, we assume that C is a closed surface and ω_0 is defined in a boundary strip of the

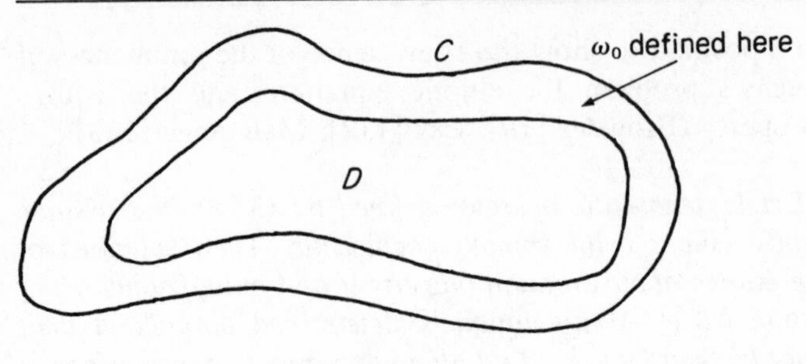

Fig. 3.2

domain D bounded by C (see Fig. 3.2). Since uniqueness is a local problem, without loss of generality we choose D so small that L and M are positive definite over D, i.e. $L[u]=0$ or $M[u]=0$ in D, $u=0$ on C, has only the trivial solution $u=0$ in D. We first extend ω_0 to the whole interior of C, not necessarily as a solution of $M[\omega_0]=0$ but such that $M[\omega_0]\in L_2(D)$. From the Green's identity (3.30) with $v=\omega_0$ and u a solution of $L[u]=0$ we have that $M[\omega_0]\perp u$ for every such u. By the Runge approximation property this implies that $M[\omega_0]$ is orthogonal to any solution of $L[u]=0$ in a domain $\tilde{D}\subset D$ which contains the support of $v_0=M[\omega_0]$. Let $\tilde{C}$ be the (smooth) boundary of $\tilde{D}$. Now define $\tilde{\omega}_0$ as the solution of

$$M[\tilde{\omega}_0]=0 \quad \text{in} \quad \tilde{D} \qquad \tilde{\omega}_0=\omega_0 \quad \text{on} \quad \tilde{C}. \tag{3.37}$$

$\tilde{\omega}_0$ exists by Theorem 3.3 and the fact that L is positive definite. In the Green's identity (3.30) we set $v=\omega_0-\tilde{\omega}_0$ and let u be any solution of $L[u]=0$ over $\tilde{D}$ to obtain

$$\iint_{\tilde{D}} uM[\omega_0]=\int_{\tilde{C}} B[u,v]. \tag{3.38}$$

Since $M[\omega_0]\perp u$, the left-hand side of (3.38) equals zero. Since the boundary values of u on $\tilde{C}$ are arbitrary, we can conclude that

$$\frac{\partial}{\partial \nu}(\omega-\tilde{\omega}_0)=0 \quad \text{on} \quad \tilde{C} \tag{3.39}$$

where ν is the conormal derivative. Now define ω_1 by

$$\omega_1=\begin{cases}\omega_0 & \text{in} \quad D\backslash\tilde{D}\\ \tilde{\omega}_0 & \text{in} \quad \tilde{D}\end{cases} \tag{3.40}$$

and note that ω_1 satisfies $M[\omega_1]=0$ in both domains and has continuous first derivatives across $\tilde{C}$. Hence ω_1 is a weak solution of $M[\omega_1]=0$ in D and by the regularity properties of solutions to elliptic equations (cf.

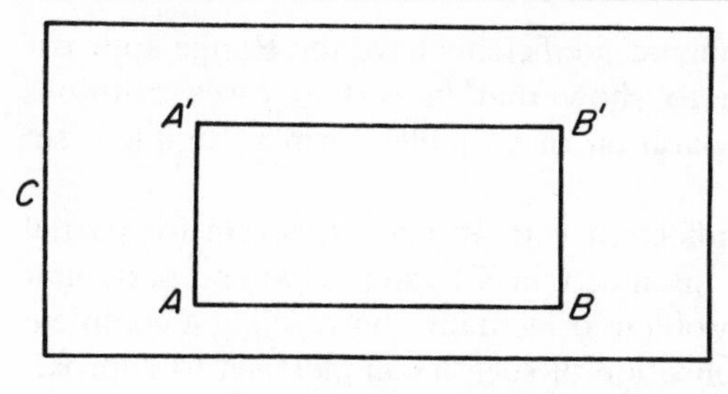

Fig. 3.3

Friedman [58]) we can conclude that ω_1 is a strong or classical solution, i.e. $\omega_1 \in C^2(D)$. Thus ω_1 is an extension of ω_0 to the whole interior of C as a solution of $M[\omega_1]=0$. Since M is positive definite over D and $\omega_1=0$ on C, $\omega_1 \equiv 0$ in D which implies $\omega_0=0$ in $D\backslash\tilde{D}$. Since the only restriction on $\tilde{D}$ was that $\tilde{C}$ should be contained in the boundary strip in which ω_0 was originally defined, we can conclude that $\omega_0=0$ in the whole boundary strip.

We now remove the restriction that C be a closed surface. Let F be a sufficiently small piece of a surface ($F=AB$ in Fig. 3.3). Let ω_0 be a solution of $M[\omega_0]=0$ in $ABB'A'$ of Fig. 3.3 which has zero Cauchy data on F. Let C be the boundary of the whole rectangle in Fig. 3.3. Define $\omega_0=0$ outside $ABB'A'$and consider ω_0 in the boundary strip consisting of those points inside C which lie outside the octagon $AA'B'BGHKL$ of Fig. 3.4. Since we have already shown that solutions with zero Cauchy data on a closed surface are identically zero, $\omega_0=0$ in $ABB'A'$ and the proof is complete. Note that ω_0 has continuous first derivatives across AB which implies that ω_0 is a weak solution of $M[\omega_0]=0$ and hence ω_0 is a strong solution of $M[\omega_0]=0$ in the region under consideration.

From Holmgren's uniqueness theorem and Theorem 3.4, we can now conclude that solutions of second order elliptic equations with analytic coefficients have the Runge approximation property. From the results of the following section it is also seen that such solutions are in fact analytic functions of their independent variables. The fact that solutions of second

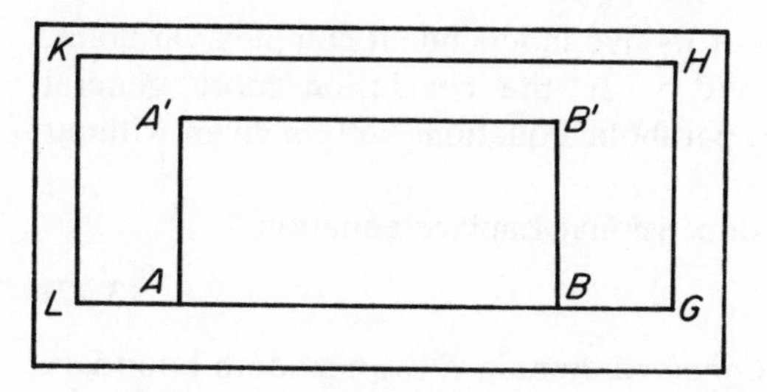

Fig. 3.4

order elliptic equations with analytic coefficients have the Runge approximation property can be used to show that in certain cases solutions obtained by the method of separation of variables form a complete set (see the Exercises).

Further developments in connection with Runge's theorem for partial differential equations will be discussed in Chapter 5 where particular attention will be given to the problem of actually constructing a complete family of solutions and the application of such a complete set to approximate solutions of boundary value problems. As is the case of Theorem 3.4, all such results are intimately connected with the continuation properties of solutions to partial differential equations.

3.3 Analyticity of solutions to partial differential equations

Solutions of elliptic and parabolic equations with analytic coefficients enjoy very strong regularity properties. In particular, solutions of elliptic equations with analytic coefficients are analytic functions of their independent variables, whereas in the case of parabolic equations they are analytic with respect to the space variables and infinitely differentiable with respect to the time variable. In the case of linear equations such results can be proved by using the fundamental solution in conjunction with Green's formula (cf. Garabedian [61]). The case of nonlinear equations is more difficult, but analogous results are seen to hold (where in the nonlinear case analytic 'coefficients' is understood to mean that the nonlinear terms are analytic functions of their independent variables). In this section we give the flavor of such results by using the fundamental solution for Laplace's equation to show that harmonic functions are analytic functions of their independent variables, and use the method of Lewy (Garabedian [61]; Lewy [116]) to show that solutions of the semilinear elliptic equation

$$u_{xx} + u_{yy} = f(x, y, u, u_x, u_y), \tag{3.41}$$

where f is an analytic function of its five independent complex variables, are analytic functions of x and y. In the conclusion more general theorems for linear elliptic and parabolic equations will be given without proof.

We first consider the three-dimensional Laplace equation

$$\Delta_3 u = 0 \tag{3.42}$$

defined in a bounded simply connected domain D with smooth boundary ∂D. Let ν be the unit outward normal to ∂D and assume that $u \in C^2(\bar{D})$

We shall make use of the Green's identity

$$\iint_D (v\Delta_3 u - u\Delta_3 v)\, \mathrm{d}V = \int_{\partial D} (v(\partial u/\partial \nu) - u(\partial v/\partial \nu))\, \mathrm{d}S \tag{3.43}$$

and the fundamental solution

$$\frac{1}{4\pi R} = \frac{1}{4\pi((x-\xi)^2+(y-\eta)^2+(z-\zeta)^2)^{1/2}} \tag{3.44}$$

to derive an elementary proof of the fact that solutions of (3.42) are analytic functions of x, y and z.

Theorem 3.5 *Solutions of* (3.42) *in D are analytic functions of their independent variables in D.*

Proof Let $P=(x, y, z)$ and $Q=(\xi, \eta, \zeta)$. Let D_ρ be a small sphere of radius ρ about Q such that $D_\rho \subset D$. Now apply (3.43) in the domain $D\backslash D_\rho$ where u is a solution of (3.42) and $v=1/R$ to obtain

$$0 = \int_{\partial D} \left(\frac{1}{4\pi R}\frac{\partial u}{\partial \nu} - u\frac{\partial}{\partial \nu}\left(\frac{1}{4\pi R}\right)\right) \mathrm{d}S - \int_{\partial D_\rho} \left(\frac{1}{4\pi R}\frac{\partial u}{\partial \nu} - u\frac{\partial}{\partial \nu}\left(\frac{1}{4\pi R}\right)\right) \mathrm{d}S. \tag{3.45}$$

Here we have made use of the fact that $\Delta_3(1/R)=0$ for $P \neq Q$. But

$$\frac{1}{4\pi}\int_{\partial D_\rho} \left(\frac{1}{R}\frac{\partial u}{\partial \nu} - u\frac{\partial}{\partial \nu}\left(\frac{1}{R}\right)\right) \mathrm{d}S = \frac{1}{4\pi}\int_0^{2\pi}\int_0^{\pi} \left(\frac{1}{\rho}\frac{\partial u}{\partial \rho} - u\frac{\partial}{\partial \rho}\left(\frac{1}{\rho}\right)\right) \rho^2 \sin\theta \,\mathrm{d}\theta\, \mathrm{d}\phi \tag{3.46}$$

and since $\partial u/\partial \rho$ is bounded we have

$$\lim_{\rho\to 0}\frac{1}{4\pi}\int_{\partial D_\rho} \left(\frac{1}{R}\frac{\partial u}{\partial \nu} - u\frac{\partial}{\partial \nu}\left(\frac{1}{R}\right)\right) \mathrm{d}S = \lim_{\rho\to 0}\frac{1}{4\pi}\int_0^{2\pi}\int_0^{\pi} u \sin\theta\, \mathrm{d}\theta\, \mathrm{d}\phi = u(Q). \tag{3.47}$$

Hence from (3.45) we have

$$u(\xi, \eta, \zeta) = \frac{1}{4\pi}\int_{\partial D} \left(\frac{1}{R}\frac{\partial u}{\partial \nu} - u\frac{\partial}{\partial \nu}\left(\frac{1}{R}\right)\right) \mathrm{d}S \tag{3.48}$$

and since $1/R$ depends analytically on ξ, η and ζ for (x, y, z) on ∂D we can conclude that $u(\xi, \eta, \zeta)$ is an analytic function of ξ, η and ζ in D.

Analogous results of course hold for solutions of the n-dimensional Laplace equation, and the same proof can be extended to linear elliptic equations with analytic coefficients once we have a fundamental solution at our disposal which is known to depend analytically on its independent variables (cf. Garabedian [61]).

We now consider the more difficult case of equation (3.41) and for the sake of simplicity suppose f is entire.

Theorem 3.6 *Let u be a solution of* (3.41) *in a domain D where f is assumed to be an entire function of its independent variables. Suppose $u \in C^3(D)$. Then u is an analytic function of x and y in D.*

Proof We first assume that u is already known to be analytic in D and derive the hyperbolic equations which will turn out to be basic to the proof of the theorem. Let

$$x = x_1 + \mathrm{i}x_2, \qquad y = y_1 + \mathrm{i}y_2 \tag{3.49}$$

and define the operators $\partial/\partial x$, $\partial/\partial \bar{x}$, $\partial/\partial y$, $\partial/\partial \bar{y}$ by

$$\begin{aligned} \frac{\partial u}{\partial x} &= \frac{1}{2}\left(\frac{\partial u}{\partial x_1} - \mathrm{i}\frac{\partial u}{\partial x_2}\right), & \frac{\partial u}{\partial \bar{x}} &= \frac{1}{2}\left(\frac{\partial u}{\partial x_1} + \mathrm{i}\frac{\partial u}{\partial x_2}\right) \\ \frac{\partial u}{\partial y} &= \frac{1}{2}\left(\frac{\partial u}{\partial y_1} - \mathrm{i}\frac{\partial u}{\partial y_2}\right), & \frac{\partial u}{\partial \bar{y}} &= \frac{1}{2}\left(\frac{\partial u}{\partial y_1} + \mathrm{i}\frac{\partial u}{\partial y_2}\right). \end{aligned} \tag{3.50}$$

Since u is analytic, the Cauchy–Riemann equations imply that

$$\partial u/\partial \bar{x} = 0, \qquad \partial u/\partial \bar{y} = 0 \tag{3.51}$$

and u is a solution of both

$$\partial^2 u/\partial x_1^2 + \partial^2 u/\partial x_2^2 = 0 \tag{3.52}$$

and

$$\partial^2 u/\partial y_1^2 + \partial^2 u/\partial y_2^2 = 0. \tag{3.53}$$

From (3.41) we have that

$$u_{x_1x_1} + u_{y_1y_1} = f(x, y, u, u_{x_1}, u_{y_1}) \tag{3.54}$$

holds in a four-dimensional domain with coordinates (x_1, x_2, y_1, y_2) and hence, from (3.51) and (3.53), we have

$$u_{x_1x_1} - u_{y_2y_2} = f(x, y, u, u_{x_1}, -\mathrm{i}u_{y_2}). \tag{3.55}$$

Similarly, from (3.51) and (3.52) we have

$$-u_{x_2x_2} + u_{y_1y_1} = f(x, y, u, -\mathrm{i}u_{x_2}, u_{y_1}). \tag{3.56}$$

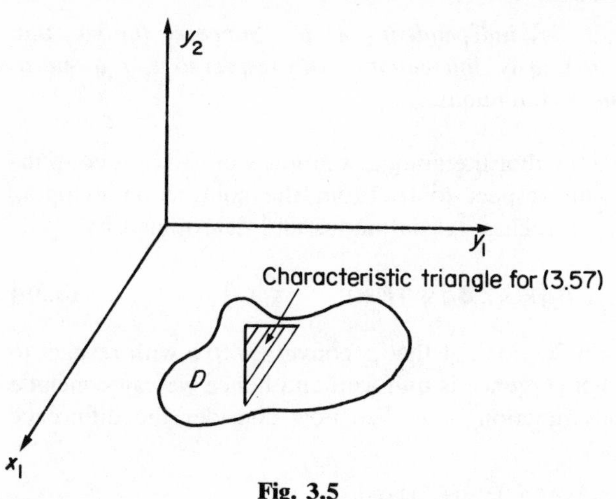

Fig. 3.5

Equations (3.55) and (3.56) are the basic hyperbolic equations which we shall use in order to show the analyticity of u.

Now assume only that $u \in C^3(D)$. Let U be the solution of the hyperbolic equation

$$U_{x_1x_1} - U_{y_2y_2} = f(x_1, y_1 + iy_2, U, U_{x_1}, -iU_{y_2}) \tag{3.57}$$

depending on the parameter y_1 such that on $y_2 = 0$, U satisfies the initial condition

$$\begin{aligned} U(x_1, y_1, 0) &= u(x_1, y_1) \\ U_{y_2}(x_1, y_1, 0) &= iu_{y_1}(x_1, y_1) \end{aligned} \tag{3.58}$$

for $(x_1, y_1) \in D$. (The initial conditions are motivated by the fact that if u were already known to be analytic then the Cauchy–Riemann equation (3.51) would imply that $u_{y_2} = iu_{y_1}$. Hence (3.57), (3.58) would provide a solution U in the (x_1, y_1, y_2) space that agrees with u and its first derivative in the real domain $y_2 = 0$; see Fig. 3.5.)

From the results of Section 3.1 we see that (3.57), (3.58) can be solved by the method of successive approximations (i.e. the contraction mapping theorem). By varying the parameter y_1 we can thus determine U throughout some neighborhood of D in the three-dimensional space spanned by x_1, y_1, y_2. We now want to show that U is an analytic function of the variable $y = y_1 + iy_2$ in this neighborhood. In order to show this we need the following lemma.

Lemma 3.1 *Let u be a solution of the hyperbolic Cauchy problem (3.1) where f depends continuously on a parameter α such that the Lipschitz* *constant M in (3.2) is independent of α. Suppose further that $f(t, x, u, p, q; \alpha)$ is continuously differentiable with respect to u, p, q and α. Then $\partial u/\partial\alpha$ exists and is continuous.*

Proof It suffices to show that the unique solution s of (3.12) is continuously differentiable with respect to α. From the contraction mapping theorem $s = \lim_{j\to\infty} s_j$ where each s_j is continuous and determined by

$$s_j = F(\xi, \eta, \mathbf{B}_1[s_{j-1}], \mathbf{B}_2[s_{j-1}], \mathbf{B}_3[s_{j-1}]; \alpha), \qquad s_0 = 0. \tag{3.59}$$

Since we have shown in Section 3.1 that s_j converges to s with respect to the norm (3.13), the convergence is uniform, and hence we can conclude that s is a continuous function of α. We now consider the difference quotient

$$v(\xi, \eta; \alpha_1, \alpha_2) = \frac{s(\xi, \eta; \alpha_2) - s(\xi, \eta; \alpha_1)}{\alpha_2 - \alpha_1}. \tag{3.60}$$

From (3.12) and the mean value theorem we have

$$v(\xi, \eta; \alpha_1, \alpha_2) = \frac{\partial F}{\partial\alpha} + \frac{\partial F}{\partial u}\mathbf{B}_1[v] + \frac{\partial F}{\partial p}\mathbf{B}_2[v] + \frac{\partial F}{\partial q}\mathbf{B}_2[v] \tag{3.61}$$

where the derivatives of $F(\xi, \eta, u, p, q; \alpha)$ are evaluated at values of u, p, q and α intermediate between the values corresponding to $\alpha = \alpha_1$ and $\alpha = \alpha_2$ with $u = \mathbf{B}_1[s]$, $p = \mathbf{B}_2[s]$, $q = \mathbf{B}_3[s]$. We now fix α_2 and view $\partial F/\partial\alpha$, $\partial F/\partial u$, $\partial F/\partial p$ and $\partial F/\partial q$ as known functions of the variables ξ, η and α_1. Note that these functions are continuous as α_1 tends to α_2 since s is a continuous function of α. From (3.18) we can conclude that the right-hand side of (3.61) defines a contraction mapping with respect to v and hence from the first part of the proof of the lemma v is a continuous function of α_1 as α_1 tends to α_2. From (3.60) we can conclude that $\partial s/\partial\alpha$ exists at $\alpha = \alpha_2$. Passing to the limit in (3.61) shows that $\partial s/\partial\alpha$ is the fixed point of a contraction mapping which depends continuously on the parameter α and hence we can conclude that $\partial s/\partial\alpha$ is continuous.

We now return to the proof of Theorem 3.6. In order to show that U is an analytic function of $y = y_1 + iy_2$ we must show that the Cauchy–Riemann equation

$$\partial U/\partial\bar{y} = 0 \tag{3.62}$$

is valid. Note that due to Lemma 3.1 and the differentiability assumptions on the initial data (3.58) it is permissible to differentiate U with respect to y_1 as well as x_1 and y_2. We shall show that (3.62) is valid by showing that $w = \partial U/\partial\bar{y}$ satisfies a homogeneous linear hyperbolic equation with homogeneous initial data and hence is identically zero. From (3.57) we have that w satisfies the hyperbolic equation

$$w_{x_1x_1} - w_{y_2y_2} = f_U w + f_p w_{x_1} - i f_Q w_{y_2} \tag{3.63}$$

where $P = U_x$, $Q = U_y$, and we have made use of the fact that f is an analytic function of y. Note that U is a known function and hence (3.63) is a linear hyperbolic equation in w. From (3.58) we have that

$$w(x_1, y_1, 0) = \tfrac{1}{2}[U_{y_1}(x_1, y_1, 0) + iU_{y_2}(x_1, y_1, 0)] = 0. \tag{3.64}$$

To compute w_{y_2} we note that from (3.57) we have

$$\begin{aligned} w_{y_2}(x_1, y_1, 0) &= \tfrac{1}{2}[U_{y_2y_2}(x_1, y_1, 0) + iU_{y_2y_2}(x_1, y_1, 0)] \\ &= \tfrac{1}{2}U_{y_1y_2} + \frac{i}{2}[U_{x_1x_1} - f(x_1, y_1, U, U_{x_1}, -iU_{y_2})] \end{aligned} \tag{3.65}$$

whereas from (3.54) we have that at $y_2 = 0$

$$U_{x_1x_1} - f(x_1, y_1, U, U_{x_1}, U_{y_1}) = -U_{y_1y_1}, \tag{3.66}$$

noting that from (3.58) $U(x_1, y_1, 0) = u(x_1, y_1)$. Hence from (3.64), (3.65), (3.66) we have

$$w_{y_2}(x_1, y_1, 0) = -\frac{i}{2}(U_{y_1y_1} + iU_{y_1y_2}) = -iw_{y_1}(x_1, y_1, 0) = 0. \tag{3.67}$$

Hence from (3.63), (3.64) and (3.67) we can conclude that $w = 0$ in the region where U has been defined (this follows from the uniqueness of the solution to Cauchy's problem for linear second order equations). Hence U is an analytic function of $y = y_1 + iy_2$. We have now shown that for each fixed x_1 there exists an analytic function of $y = y_1 + iy_2$ that agrees with u on the real axis $y_2 = 0$. We now want to show that in a similar fashion u can be continued off the real x_1-axis to an analytic function of $x = x_1 + ix_2$ such that u is an analytic function of the two complex variables x and y in a four-dimensional complex domain containing the real domain D where u was originally defined.

Let $\tilde{U}$ be the unique solution of the hyperbolic Cauchy problem

$$-\tilde{U}_{x_2x_2} + \tilde{U}_{y_1y_1} = f(x_1 + ix_2, y_1 + iy_2, \tilde{U}, -i\tilde{U}_{x_2}, \tilde{U}_{y_1}) \tag{3.68}$$

$$\begin{aligned} \tilde{U}(x_1, 0, y_1, y_2) &= U(x_1, y_1, y_2) \\ \tilde{U}_{x_2}(x_1, 0, y_1, y_2) &= iU_{x_1}(x_1, y_1, y_2) \end{aligned} \tag{3.69}$$

depending on the parameters x_1 and y_2. Note that if we already knew that U is an analytic function of x we would have $U_{x_2} = iU_{x_1}$ where $x = x_1 + ix_2$. From (3.68) and (3.69) we can conclude from Section 3.1 that $\tilde{U}$ is determined in a four-dimensional region containing the domain of definition of U. We now want to show that $\tilde{U}$ is an analytic function of

$x = x_1 + ix_2$ and $y = y_1 + iy_2$, i.e. the Cauchy–Riemann equations

$$\tilde{U}_{\bar{y}} = 0, \qquad \tilde{U}_{\bar{x}} = 0 \tag{3.70}$$

are satisfied. To this end we shall show that $\tilde{w} = \tilde{U}_{\bar{y}}$ and $\tilde{v} = \tilde{U}_{\bar{x}}$ satisfy homogeneous Cauchy problems for hyperbolic equations and hence are identically zero. From (3.68) we have that

$$-\tilde{w}_{x_2x_2} + \tilde{w}_{y_1y_1} = f_{\tilde{U}}\tilde{w} - if_P\tilde{w}_{x_2} + f_Q\tilde{w}_{y_1} \tag{3.71}$$

and

$$-\tilde{v}_{x_2x_2} + \tilde{v}_{y_1y_1} = f_{\tilde{U}}\tilde{v} - if_P\tilde{v}_{x_2} + f_Q\tilde{v}_{y_1} \tag{3.72}$$

where $P = \tilde{U}_x$, $Q = \tilde{U}_y$ and use has been made of the fact that f is an analytic function of x and y. From (3.69) we have that

$$\tilde{w}(x_1, 0, y_1, y_2) = w(x_1, y_1, y_2) = 0 \tag{3.73a}$$

and

$$\tilde{w}_{x_2}(x_1, 0, y_1, y_2) = iw_{x_1}(x_1, y_1, y_2) = 0. \tag{3.73b}$$

Hence w is a solution of a homogeneous hyperbolic equation satisfying zero Cauchy data and thus $w \equiv 0$. Therefore we can conclude that $\tilde{U}$ is an analytic function of y.

We now consider $\tilde{v}$. From (3.69) we have that

$$\tilde{v}(x_1, 0, y_1, y_2) = \tfrac{1}{2}(\tilde{U}_{x_1}(x_1, 0, y_1, y_2) + i\tilde{U}_{x_2}(x_1, 0, y_1, y_2)) = 0 \tag{3.74}$$

and from (3.68), (3.69), (3.57) we have that

$$\tilde{U}_{x_1x_1} + \tilde{U}_{x_2x_2} - \tilde{U}_{y_1y_1} - \tilde{U}_{y_2y_2} = 0 \tag{3.75}$$

at $x_2 = 0$. Since we have already shown that $\tilde{U}_{\bar{y}} = 0$, (3.75) implies that at $x_2 = 0$,

$$\tilde{U}_{x_1x_1} + \tilde{U}_{x_2x_2} = 0 \tag{3.76}$$

and hence

$$\begin{aligned} \tilde{v}_{x_2}(x_1, 0, y_1, y_2) &= \tfrac{1}{2}(\tilde{U}_{x_1x_2} + i\tilde{U}_{x_2x_2}) \\ &= \tfrac{1}{2}(\tilde{U}_{x_1x_2} - i\tilde{U}_{x_1x_1}) \\ &= -i\tilde{v}_{x_1}(x_1, 0, y_1, y_2) \\ &= 0. \end{aligned} \tag{3.77}$$

Thus, $\tilde{v}$ is a solution of a homogeneous hyperbolic equation with vanishing Cauchy data on $x_2 = 0$ and hence $\tilde{v} \equiv 0$. This implies that $\tilde{U}$ is an analytic function of $x = x_1 + ix_2$. We have therefore now shown that there exists a function $\tilde{U}$ defined in a complex neighborhood of D such that $\tilde{U}$ is an analytic function of x and y and in the real domain D, $\tilde{U}$ agrees with u. This completes the proof of the theorem.

The results of Theorem 3.6 can also be obtained for elliptic equations in more than two independent variables (cf. Hopf [90]) and if the boundary is analytic it can even be shown that the solution is analytic 'up to the boundary.' We content ourselves with stating several results in this direction for the case of linear elliptic and parabolic equations and refer the reader to Friedman [58] for proofs. Let

$$A(\mathbf{x}, D) = \sum_{|\alpha| \leq 2m} a_\alpha(\mathbf{x}) D^\alpha \tag{3.78}$$

where

$$\begin{aligned} &D_j = \partial/\partial x_j, \qquad D^\alpha = D_1^{\alpha_1} \cdots D_n^{\alpha_n} \\ &\alpha = (\alpha_1, \ldots, \alpha_n), \qquad \mathbf{x} = (x_1, \ldots, x_n) \end{aligned} \tag{3.79}$$

and

$$\sum_{|\alpha| \leq 2m} a_\alpha(\mathbf{x}) D^\alpha \equiv \sum_{k=0}^{2m} \sum_{\alpha_1 + \cdots + \alpha_n = k} a_{\alpha_1 \cdots \alpha_n}(\mathbf{x}) D_1^{\alpha_1} \cdots D_n^{\alpha_n}. \tag{3.80}$$

We suppose that the operator $A(\mathbf{x}, D)$ is uniformly elliptic with coefficients defined in the closure $\bar{D}$ of a bounded domain D. (By uniformly elliptic we mean that $\operatorname{Re}\left\{\sum_{|\alpha|=2m} a_\alpha(\mathbf{x}) \boldsymbol{\xi}^\alpha\right\} \geq \gamma\, |\boldsymbol{\xi}|^{2m}$ for all real $\boldsymbol{\xi}$ and $\mathbf{x} \in D$ where γ is a positive constant.)

Theorem 3.7 *Let u be a solution of*

$$A(\mathbf{x}, D)u = f(\mathbf{x})$$

in D where f and the coefficients of $A(\mathbf{x}, D)$ are analytic in D. Then u is analytic in D.

Theorem 3.8 *Let u be a solution of*

$$A(\mathbf{x}, D)u = f(\mathbf{x})$$

in D such that

$$\partial^j u/\partial \nu^j = 0 \quad \text{on} \quad \partial D, \qquad 0 \leq j \leq m-1$$

where ν is the outward normal to ∂D. Assume that ∂D is analytic and that f and the coefficients of $A(\mathbf{x}, D)$ are analytic in $\bar{D}$. Then u is analytic in $\bar{D}$.

Results analogous to Theorems 3.7 and 3.8 can also be obtained for the uniformly parabolic equation

$$(-1)^m (\partial u/\partial t) + A(\mathbf{x}, t, D)u = f(\mathbf{x}, t) \tag{3.81}$$

where by uniformly parabolic we mean that the elliptic operator $A(\mathbf{x}, t, D) = \sum_{|\alpha| \leq 2m} a_\alpha(\mathbf{x}, t) D^\alpha$ is uniformly elliptic in a cylinder $Q_T = D \times (0, T)$ for some positive constant T.

Theorem 3.9 *Let u be a solution of*

$$(-1)^m (\partial u/\partial t) + A(\mathbf{x}, t, D)u = f(\mathbf{x}, t)$$

in the cylinder $Q_T = D \times (0, T)$ where f and the coefficients of $A(\mathbf{x}, t, D)$ are analytic in Q_T. Then u is an analytic function of $\mathbf{x}$ in Q_T for each fixed t and is infinitely differentiable with respect to t for each fixed $\mathbf{x}$.

Theorem 3.10 *Let u be a solution of*

$$(-1)^m (\partial u/\partial t) + A(\mathbf{x}, t, D)u = f(\mathbf{x}, t)$$

in $Q_T = D \times (0, T)$ such that

$$\partial^j u/\partial \nu^j = 0 \quad on \quad S_T = \partial D \times (0, T), \qquad 0 \leq j \leq m-1,$$

where ν is the outward normal to ∂D. Assume that ∂D is analytic and that f and the coefficients of $A(\mathbf{x}, t, D)$ are analytic in $\bar{Q}_T$. Then u is analytic in $\bar{D} \times (0, T)$.

A further problem concerned with the analyticity of solutions to elliptic and parabolic equations is the precise domain of regularity of the solution in the complex domain. Some results in this direction will be given in Chapters 4 and 5 where they are needed in the derivation of reflection principles and Runge approximation theorems.

3.4 Equations with no solution

If the coefficients of a linear partial differential equation are analytic, then by the Cauchy–Kowalewski theorem we can always conclude that there exist plenty of solutions to the given differential equation. However, as pointed out by Lewy in 1957, if the coefficients are not analytic there may not exist any solutions at all (Lewy [117])! Since Lewy's paper, there has been an extensive investigation into the conditions under which a given linear differential equation admits solutions, and the reader is referred to Egorov [49], Hörmander [92], Nirenberg [137] and Nirenberg and Trèves [139], [140] for past and current developments in this area. Here our aims are more modest, and we simply wish to present an example of a linear partial differential equation having no solution (Garabedian [66] Grushin [79]). The connection of the present example with the topic of this chapter is in one sense negative—that is, instead of a solution being as nice as possible, i.e. analytic, it is as bad as possible, i.e. it does not even exist—and in another sense positive, due to the fact that the key result used in the proof of nonexistence is the identity theorem for analytic functions, i.e. the unique continuation property for the Cauchy–Riemann equations.

The equation we shall consider is

$$\partial w/\partial x + \mathrm{i}x(\partial w/\partial y) = f(x, y) \tag{3.82}$$

in a neighborhood of the origin and we shall show that for certain functions $f \in C^\infty$ no solution exists to (3.82). In particular, let D_n, $n = 1, 2, \ldots$, be a sequence of closed nonintersecting disks in the half plane $x > 0$ with centers at $(x_n, 0)$ such that $x_n \to 0$. Then let f be any infinitely differentiable function of compact support (i.e. $f \in C_0^\infty$) such that:

(1) f is an even function of x;

(2) $f = 0$ in $D = \{(x, y): x \geq 0, (x, y) \notin \bigcup_1^\infty D_n\}$;

(3) $\iint_{D_n} f(x, y)\,\mathrm{d}x\,\mathrm{d}y \neq 0$ for $n = 1, 2, \ldots$.

Theorem 3.11 *For f satisfying conditions (1)–(3) above there does not exist a solution $w \in C^1$ of (3.82) in any neighborhood of the origin.*

Proof Suppose there exists a solution $w \in C^1$ of (3.82) defined in a neighborhood Ω of the origin. Let u and v be the odd and even parts, respectively, of w with respect to x. Then, taking the even part of (3.82) we have

$$\partial u/\partial x + \mathrm{i}x(\partial u/\partial y) = f(x, y). \tag{3.83}$$

Now consider (3.83) in the half plane $x \geq 0$ and note that

$$u(0, y) = 0. \tag{3.84}$$

Making the change of variables $s = x^2/2$ we have from (3.83), (3.84) that $U(x, y) = u((2s)^{1/2}, y)$ satisfies

$$\frac{\partial U}{\partial s} + \mathrm{i}\frac{\partial U}{\partial y} = \frac{1}{(2s)^{1/2}} f((2s)^{1/2}, y) \tag{3.85a}$$

$$U(0, y) = 0 \tag{3.85b}$$

and hence, in the right half plane outside the images of the disks D_n under the mapping $x \to (2s)^{1/2}$, U satisfies the Cauchy–Riemann equations. Thus, U is an analytic function of $s + \mathrm{i}y$ in this region. Since the complement of $\bigcup_{n=1}^\infty D_n$ is connected we can conclude from (3.85b) and the identity theorem for analytic functions that $U \equiv 0$ in its domain of analyticity. Hence u vanishes on ∂D_n for $n = 1, 2, \ldots$. But from (3.83) and Green's theorem we have

$$\iint_{D_n} f(x, y)\,dx\,dy = \iint_{D_n} (u_x + ixu_y)\,dx\,dy = \int_{\partial D_n} (u\,dy - ixu\,dx) = 0, \tag{3.86}$$

contradicting condition (3) satisfied by f.

The proof of Theorem 3.11 can be easily extended to show that there is also no distributional solution of (3.82) in any neighborhood of the origin (Grushin [79]).

Exercises

(1) Show that when f is linear in u, u_t and u_x the solution of (3.1) exists 'in the large'.

(2) Modify the analysis of Section 3.1 to treat the characteristic initial-value problem

$$u_{\xi\eta} = F(\xi, \eta, u, u_\xi, u_\eta)$$
$$u(\xi, 0) = \phi(\xi)$$
$$u(0, \eta) = \psi(\eta).$$

Show by using the above results that the solution of

$$u_{\xi\eta} + \lambda u = 0$$
$$u(\xi, 0) = 1$$
$$u(0, \eta) = 1$$

where λ is a constant is given by $u = J_0(2(\lambda xy)^{1/2})$ where J_0 denotes a Bessel function of order zero.

(3) Complete the proof of the existence of a unique analytic solution to the elliptic Cauchy problem (3.21).

(4) Show that any solution of the wave equation $u_{tt} = u_{xx} + u_{yy}$, which assumes Cauchy data on the plane $x = 0$ that is independent of t, must be a harmonic function of x and y.

(5) Let u be a solution of Laplace's equation $\Delta_3 u = 0$ in a bounded simply connected domain D. Show that on compact subsets of D, u can be approximated in the maximum norm by a finite linear combi-

nation of functions of the form

$$u_{nm} = r^n P_n^m(\cos\theta) e^{im\phi}; \qquad n \geq 0, \qquad -n \leq m \leq n$$

where P_n^m is an associated Legendre function and (r, θ, ϕ) denote spherical coordinates. Show that if ∂D is analytic then u can be approximated over $\bar{D}$ by a finite linear combination of the functions u_{nm}.

(6) Construct an example of a solution to the heat equation $u_t = u_{xx}$ in the rectangle $0 < x < 1$, $0 < t < T$, that is not analytic with respect to t.

(7) Let u be a solution of the n-dimensional Laplace equation $\Delta_n u = 0$ in a domain D. Use the mean value theorem to show that

$$|D^m u(\mathbf{x})| \leq \frac{n^m e^{m-1} m!}{(r - r_0)^m} \max_{\mathbf{y} \in S} |u(\mathbf{y})|$$

where $S \subset D$ is a closed sphere of radius r and $\mathbf{x} \in S_0$ where S_0 is a closed concentric sphere of radius r_0. Use this result to show that the Taylor series for u about a point in D is convergent and hence conclude that u is an analytic function of its independent variables.

(8) Show that the values of a solution u of (3.41) and its gradient along a short enough arc in the real domain suffice to determine the analytic extension of u at a point in the complex domain where the imaginary characteristics of (3.41) through the endpoints of that arc intersect.

4

Analytic continuation

In this chapter we are concerned with the problem of uniquely continuing solutions of linear partial differential equations beyond their original domain of definition. We are primarily interested in 'global' continuation phenomena as opposed to 'local' continuation (for example, knowing a solution is analytic 'up to the boundary' and hence, by definition, in some neighborhood containing the boundary) and hence our main efforts shall be directed towards obtaining various versions of the reflection principle for partial differential equations with variable coefficients (due to Colton [32], [34] and Lewy [118]; see also Yu [183], [186]). In the case of elliptic equations we assume for the sake of simplicity that the coefficients are entire, which implies that the solutions are analytic, and hence our continuation is in fact an analytic continuation. However, for parabolic equations we no longer assume that the coefficients are analytic, and hence our continuation, although unique, is not in general an analytic continuation with respect to the space and time variables (cf. John [94]). An interesting phenomenon that appears in the course of our investigations is that the reflection property for solutions of partial differential equations is essentially a two-dimensional result and is not in general valid for higher-dimensional equations. Surfaces and equations which allow such a continuation provide an example of *Huygen's principle for reflection* (cf. Garabedian [64]; Hill [85]), the classical example being the Schwarz reflection principle for Laplace's equation across a sphere. In connection with our development of reflection principles for partial differential equations we shall give another example of Huygen's principle for reflection: that for solutions of the Helmholtz equation defined in a ball and vanishing on a portion of the spherical boundary. This result will later be used in our investigation of the inverse scattering problem (Chapter 8). We then continue our discussion of methods for the unique continuation of solutions to partial differential equations by presenting Gilbert's 'envelope method' which gives criteria for locating the singularities of solutions defined in terms of integral representation involving analytic functions (see Gilbert [69], [70], [71]). The 'envelope method' has been widely used by physicists in connection with potential scattering and is usually mistakenly credited by them to either Landau, Bjorken and/or Polkinghorne and Screaton. It was actually first proved by Gilbert in his 1958 thesis (Gilbert [69]) and for a historical discussion of the origin of this theorem we refer the reader to the introduction in Gilbert [70]. Finally, we conclude this chapter by presenting some results on the analytic continuation of solutions to the heat equation arising from melting problems in a homogeneous medium.

4.1 Reflection principles for elliptic equations

We shall be interested here in solutions of the second order elliptic partial differential equation in two independent variables written in canonical form as

$$L[u] \equiv \frac{\partial^2 u}{\partial x^2} + \frac{\partial^2 u}{\partial y^2} + a(x, y)\frac{\partial u}{\partial x} + b(x, y)\frac{\partial u}{\partial y} + c(x, y)u = 0 \tag{4.1}$$

where we assume that a, b and c are entire functions of their independent complex variables x and y. Our primary aim is to develop a reflection principle for solutions of (4.1) that satisfy a first order boundary condition on a portion of the boundary of the domain of definition, and we assume that this portion of the boundary is situated on the line $y = 0$. To this end we first want to construct a special solution of the adjoint equation

$$M[v] \equiv \frac{\partial^2 v}{\partial x^2} + \frac{\partial^2 v}{\partial y^2} - \frac{\partial(av)}{\partial x} - \frac{\partial(bv)}{\partial y} + cv = 0 \tag{4.2}$$

known as the complex Riemann function for $L[u] = 0$ (Vekua [172]). By introducing the conjugate coordinates (see Chapter 3)

$$z = x + iy, \qquad z^* = x - iy \tag{4.3}$$

we can rewrite (4.1) and (4.2) as the formally hyperbolic equations

$$L^*[U] \equiv \frac{\partial^2 U}{\partial z\,\partial z^*} + A(z, z^*)\frac{\partial U}{\partial z} + B(z, z^*)\frac{\partial U}{\partial z^*} + C(z, z^*)U = 0 \tag{4.4}$$

$$M^*[V] \equiv \frac{\partial^2 V}{\partial z\,\partial z^*} - \frac{\partial(AV)}{\partial z} - \frac{\partial(BV)}{\partial z^*} + CV = 0 \tag{4.5}$$

where

$$U(z, z^*) = u\left(\frac{z+z^*}{2}, \frac{z-z^*}{2i}\right), \qquad V(z, z^*) = v\left(\frac{z+z^*}{2}, \frac{z-z^*}{2i}\right),$$
$$A = \tfrac{1}{4}(a + ib), \qquad B = \tfrac{1}{4}(a - ib), \qquad C = \tfrac{1}{4}c. \tag{4.6}$$

Note that from Theorem 3.6 we have that U and V are well defined. The Riemann function for (4.1) is defined to be the (unique) solution $R(z, z^*; \zeta, \zeta^*)$ of (4.5) depending on the complex parameters $\zeta=\xi+i\eta$, $\zeta^*=\xi-i\eta$ (where ξ, η are complex variables) which satisfies the initial conditions

$$\begin{aligned} R(z, \zeta^*; \zeta, \zeta^*) &= \exp\left[\int_{\zeta}^{z} B(\sigma, \zeta^*)\,d\sigma\right] \\ R(\zeta, z^*; \zeta, \zeta^*) &= \exp\left[\int_{\zeta^*}^{z^*} A(\zeta, \tau)\,d\tau\right] \end{aligned} \tag{4.7}$$

on the complex hyperplanes $z^*=\zeta^*$ and $z=\zeta$. Note that by Cauchy's theorem the integrals in (4.7) are independent of the path of integration.

We shall now construct R. By making a change of variables shifting (ζ, ζ^*) to the origin we can restrict ourselves to the problem of solving a characteristic initial value problem of the form

$$\frac{\partial^2 R}{\partial z\,\partial z^*}=\frac{\partial(\tilde{A}R)}{\partial z}+\frac{\partial(\tilde{B}R)}{\partial z^*}-\tilde{C}R \tag{4.8a}$$

$$\begin{aligned} R(z, 0) &= \exp\left[\int_0^z \tilde{B}(\sigma, 0)\,d\sigma\right] \\ R(0, z) &= \exp\left[\int_0^{z^*} \tilde{A}(0, \tau)\,d\tau\right] \end{aligned} \tag{4.8b}$$

where

$$\begin{aligned} &\tilde{A}(z, z^*)=A(z+\zeta, z^*+\zeta^*), \qquad \tilde{B}(z, z^*)=B(z+\zeta, z^*+\zeta^*), \\ &\tilde{C}(z, z^*)=C(z+\zeta, z^*+\zeta^*). \end{aligned} \tag{4.9}$$

But (4.8) can be solved by the methods of Section 1, Chapter 3. In particular, let

$$s(z, z^*)=\frac{\partial^2 R}{\partial z\,\partial z^*}(z, z^*) \tag{4.10}$$

and integrate using (4.8) to obtain

$$\begin{aligned} R(z, z^*) &= \mathbf{B}_1[s] \\ &= \int_0^z\int_0^{z^*} s(\xi, \eta)\,d\eta\,d\xi+\exp\left[\int_0^z \tilde{B}(\sigma, 0)\,d\sigma\right] \\ &\quad +\exp\left[\int_0^{z^*} \tilde{A}(0, \tau)\,d\tau\right]-1 \end{aligned} \tag{4.11}$$

$$R_z(z, z^*) = \mathbf{B}_2[s]$$
$$= \int_0^{z^*} s(z, \eta)\, d\eta + \frac{d}{dz} \exp\left[\int_0^z \tilde{B}(\sigma, 0)\, d\sigma\right] \tag{4.12}$$

$$R_{z^*}(z, z^*) = \mathbf{B}_3[s]$$
$$= \int_0^{z} s(\xi, z^*)\, d\xi + \frac{d}{dz^*} \exp\left[\int_0^{z^*} \tilde{A}(0, \tau)\, d\tau\right]. \tag{4.13}$$

Hence, s is the solution of the operator equation

$$s = \tilde{A}\mathbf{B}_2[s] + \tilde{B}\mathbf{B}_3[s] + [\tilde{A}_z + \tilde{B}_{z^*} - \tilde{C}]\mathbf{B}_1[s]. \tag{4.14}$$

We now want to show the existence of a solution to (4.11) which depends analytically on z, z^*, ζ and ζ^* for $|z|<r$, $|z^*|<r$, $|\zeta|<r$, $|\zeta^*|<r$, r arbitrarily large but fixed. To this end we introduce the norm (3.25) of Chapter 3 and note that if $\mathbf{T}[s]$ is defined to be the right-hand side of (4.14) then

$$\|\mathbf{T}[s_1] - \mathbf{T}[s_2]\|_\lambda \leq M\{\|\mathbf{B}_1[s_1] - \mathbf{B}_1[s_2]\|_\lambda + \|\mathbf{B}_2[s_1] - \mathbf{B}_2[s_2]\|_\lambda$$
$$+ \|\mathbf{B}_3[s_1] - \mathbf{B}_3[s_2]\|_\lambda\} \tag{4.15}$$

where, since $\tilde{A}$, $\tilde{B}$ and $\tilde{C}$ are entire, M can be chosen independently of z, z^*, ζ and ζ^*. The existence of a unique analytic fixed point for the operator $\mathbf{T}$ now follows exactly as in Chapter 3 and, in particular, can be found from the iterative scheme

$$s_{n+1} = \mathbf{T}[s_n], \qquad s_0 = 0. \tag{4.16}$$

Since M is independent of z, z^*, ζ and ζ^*, the convergence is uniform with respect to these variables and hence we can conclude that

$$s = \lim_{n\to\infty} s_n \tag{4.17}$$

exists and is an analytic function of z, z^*, ζ and ζ^* for $|z|<r$, $|z^*|<r$, $|\zeta|<r$, $|\zeta^*|<r$. Since r was arbitrarily large we can conclude that s, and hence R, exist and that s is an entire function of z, z^*, ζ and ζ^*.

Theorem 4.1 *As a function of its last two arguments $R(\zeta, \zeta^*; z, z^*)$ is the Riemann function for $M[v] = 0$.*

Proof Let D be a bounded simply connected domain in the complex plane and let U be an analytic function of z and z^* for $(z, z^*) \in D \times D^*$ where $D^* = \{z^*: \overline{z^*} \in D\}$. Now let $R = R(z, z^*; \zeta, \zeta^*)$ and note that since $M^*[R] = 0$ we have

$$\frac{\partial^2(UR)}{\partial z\, \partial z^*} - RL^*[U] = \frac{\partial}{\partial z}\left\{U\left(\frac{\partial R}{\partial z^*} - AR\right)\right\} + \frac{\partial}{\partial z^*}\left\{U\left(\frac{\partial R}{\partial z} - BR\right)\right\}. \tag{4.18}$$

Interchange z, z^* and ζ, ζ^* in (4.18) and integrate with respect to ζ and ζ^* from z_0 to z and z_0^* to z^* where $(z_0, z_0^*) \in D \times D^*$. Using (4.7) we have

$$
\begin{aligned}
U(z, z^*) = {} & U(z_0, z_0^*)R(z_0, z_0^*; z, z^*) \\
& + \int_{z_0}^{z} R(\zeta, z_0^*; z, z^*)\left\{\frac{\partial U(\zeta, z_0^*)}{\partial \zeta} + B(\zeta, z_0^*)U(\zeta, z_0^*)\right\} d\zeta \\
& + \int_{z_0^*}^{z^*} R(z_0, \zeta^*; z, z^*) \\
& \qquad \times \left\{\frac{\partial U(z_0, \zeta^*)}{\partial \zeta^*} + A(z_0, \zeta^*)U(z_0, \zeta^*)\right\} d\zeta^* \\
& + \int_{z_0}^{z}\int_{z_0^*}^{z^*} R(\zeta, \zeta^*; z, z^*)L^*[U(\zeta, \zeta^*)]\, d\zeta^*\, d\zeta.
\end{aligned}
\tag{4.19}
$$

If we now set $U(z, z^*) = R(z_0, z_0^*; z, z^*)$ and use (4.7) again, we have from (4.19) that

$$
\int_{z_0}^{z}\int_{z_0^*}^{z^*} R(\zeta, \zeta^*; z, z^*)L^*[R(z_0, z_0^*; \zeta, \zeta^*)]\, d\zeta^*\, d\zeta = 0. \tag{4.20}
$$

From (4.20) we can immediately conclude that with respect to its last two arguments $R(z_0, z_0^*; z, z^*)$ is a solution of $L^*[U] = 0$ and (4.7) now shows that as a function of z, z^*, $R(\zeta, \zeta^*; z, z^*)$ is the Riemann function for $M[v] = 0$.

We now want to prove one of the fundamental theorems in the analytic theory of elliptic equations in two independent variables. The following theorem is due to Bergman [12] and Vekua [172] and, in addition to showing that solutions of (4.1) are real analytic in D, gives the region of regularity of the solution in the space of two complex variables. The reader should compare the theorem below with Theorems 3.5 and 3.6.

Theorem 4.2 (*Bergman–Vekua theorem*) *Let $u \in C^2(D)$ be a solution of $L[u] = 0$ in a bounded simply connected domain D of the x, y plane. Then $U(z, \bar{z}) = u(x, y)$ is analytic for $(x, y) \in D$ and $U(z, z^*) = u((z + z^*)/2, (z - z^*)/2i)$ can be analytically continued into the domain $D \times D^*$ where $D^* = \{z^*: \overline{z^*} \in D\}$.*

Proof Without loss of generality we assume that D has a smooth boundary ∂D and let ν be the outer normal and s the arclength along ∂D.

Then from Green's theorem we have for $u, v \in C^2(D) \cap C^1(\bar{D})$

$$\iint_D (vL[u] - uM[v])\, dx\, dy - \int_{\partial D} H[u, v] = 0 \tag{4.21}$$

where

$$H[u, v] = \left\{ v\frac{\partial u}{\partial \nu} - u\frac{\partial v}{\partial \nu} + \left(a\frac{\partial x}{\partial \nu} + b\frac{\partial y}{\partial \nu} \right) uv \right\} ds. \tag{4.22}$$

Now let u be a solution of $L[u] = 0$ in D and without loss of generality assume $u \in C^1(D)$. Let $v = R(z, \bar{z}; \zeta, \bar{\zeta}) \log r$ where $\zeta = \xi - i\eta$, $\bar{\zeta} = \xi - i\eta$, $r^2 = (z - \zeta)(\bar{z} - \bar{\zeta})$. Letting Ω_ε be a small circle of radius ε about the point (ξ, η) and applying (4.21) to the region $D \backslash \Omega_\varepsilon$ instead of D leads in a straightforward fashion, as ε tends to zero, to the formula

$$u(x, y) = -\frac{1}{2\pi} \int_{\partial D} H[u, R \log r] - \frac{1}{2\pi} \iint_D uM[R \log r]\, d\xi\, d\eta \tag{4.23}$$

where we have interchanged the roles of (x, y) and (ξ, η), i.e. $R = R(\zeta, \bar{\zeta}; z, \bar{z})$ and M is a differential operator with respect to the (ξ, η) variables. Since $M[R] = 0$ we have

$$M^*[R \log r] = 2\frac{\partial R/\partial \zeta - BR}{\zeta^* - z^*} + 2\frac{\partial R/\partial \zeta^* - AR}{\zeta - z} \tag{4.24}$$

and hence, from (4.7), we have that $M[R \log r]$ is in fact an entire function of its independent complex variables. Hence the second integral in (4.23) can be continued to an entire function of z and z^*. The first integral in (4.23) can be continued to an analytic function of z and z^* for $z \in D$, $z^* \in D^*$ (i.e. for z and z^* such that r is not zero). Hence (4.23) shows that $U(z, z^*)$ is analytic for $(z, z^*) \in D \times D^*$.

Example $u(x, y) = (1 + x^2 + y^2)^{-1}$ is not a solution of any elliptic equation of the form $L[u] = 0$ in the domain $D = \{(x, y): x^2 + y^2 < 2\}$. This follows from the fact that $u \in C^2(D)$, but $U(z, z^*) = (1 + zz^*)^{-1}$ is singular in $D \times D^*$, for example at $(i, i) \in D \times D^*$.

We are now in a position to prove Lewy's reflection principle for solutions of $L[u] = 0$ satisfying analytic Dirichlet data on a portion σ of the x-axis (Lewy [118]). Let D be a simply connected domain of the x, y plane whose boundary contains a segment σ of the x-axis with the origin as an interior point and such that D contains the portion $y < 0$ of a neighborhood of each point of σ (see Fig. 4.1).

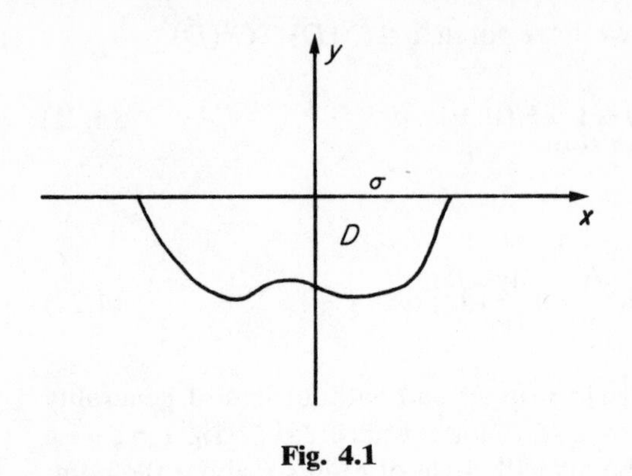

Fig. 4.1

Theorem 4.3 (*Reflection principle for elliptic equations*) *Let* $u \in C^2(D) \cap C(\bar{D})$ *satisfy* $L[u]=0$ *in* D, *where* D *is as described above. Suppose* $u(x,0)=\delta(x)$ *can be continued to an analytic function* $\delta(z)$ *regular in* $D \cup \sigma \cup D^*$ *where* D^* *is the mirror image of* D *reflected across* σ. *Then* u *can be analytically continued into all of* $D \cup \sigma \cup D^*$.

Remark From the regularity theory for elliptic equations (cf. Friedman [58]) we can conclude that $u \in C^1(D \cup \sigma)$.

Proof For analytic functions U, V (of z and z^*) we have the identity

$$VL^*[U]-UM^*[V]=(VU)_{zz^*}-(U[V_{z^*}-AV])_z-(U[V_z-BV])_{z^*} \tag{4.25}$$

where the notation is as in (4.4)–(4.6). Setting $V=R(z,z^*;\zeta,\zeta^*)$, U a solution of $L^*[U]=0$, and using Green's theorem to integrate (4.25) over a plane region S in C^2 (the space of two complex variables) bounded by a piecewise smooth curve ∂S gives

$$0=\int_{\partial S}(UR)_z\,dz-\int_{\partial S}U(R_z-BR)\,dz+\int_{\partial S}U(R_{z^*}-AR)\,dz^*. \tag{4.26}$$

In particular, letting S be the triangle with corners $(\zeta,\bar{\zeta})$, (ζ,ζ^*), $(\overline{\zeta^*},\zeta^*)$ gives (relabelling (ζ,ζ^*) by (z,z^*))

$$\begin{aligned} U(z,z^*)&=U(\overline{z^*},z^*)R(\overline{z^*},z^*,z,z^*)\\ &\quad+\int_d U(t,\bar{t})[R_{\bar{t}}(t,\bar{t};z,z^*)-A(t,\bar{t})R]\,d\bar{t}\\ &\quad+\int_d [(U(t,\bar{t})R(t,\bar{t};z,z^*))_t-U(R_t-BR)]\,dt \end{aligned} \tag{4.27}$$

where d is the diagonal from $(\overline{z^*}, z^*)$ to $(z, \bar{z})$ and use has been made of the initial conditions satisfied by R. Similarly, letting S be the quadrilateral with corners $(0, 0)$, $(\zeta, 0)$, $(\zeta, \bar{\zeta})$, $(0, \bar{\zeta})$, we have

$$U(z, \bar{z}) = -U(0,0)R(0,0; z, \bar{z}) + U(z,0)R(z,0; z, \bar{z}) + U(0, \bar{z})R(0, \bar{z}; z, \bar{z}) - \int_0^z U(t,0)[R_t(t,0; z, \bar{z}) - B(t,0)R]\, dt - \int_0^{\bar{z}} U(0, \bar{t})[R_{\bar{t}}(0, \bar{t}; z, \bar{z}) - A(0, \bar{t})R]\, d\bar{t}. \tag{4.28}$$

Now suppose on σ we have

$$u(x, 0) = U(x, x) = \delta(x) \tag{4.29}$$

where δ is regular for x in $D \cup \sigma \cup D^*$. In (4.28) set $U(z, 0) = f(z)$, $U(0, z) = g(z)$, and obtain for z on σ

$$\delta(z) = -\delta(0)R(0,0; z, z) + f(z)R(z,0; z, z) + g(z)R(0, z; z, z) - \int_0^z f(t)[R_t(t,0; z, z) - B(t,0)R]\, dt - \int_0^z g(t)[R_t(0, t; z, z) - A(0, t)R]\, dt. \tag{4.30}$$

Note that the initial conditions satisfied by R imply that $R(z, 0; z, z)$ and $R(0, z; z, z)$ do not vanish. From (4.27) and the Bergman–Vekua theorem we have that if u is defined in D then f is known for z in $D \cup \sigma$. (Set $z^* = 0$ in (4.27).) Similarly, setting $z = 0$ in (4.27), we have that g is known for z in $D^* \cup \sigma$.

Now for z in $D \cup \sigma$, (4.30) is a Volterra integral equation for g since f and δ are both known in $D \cup \sigma$. Since the kernel and terms not involving g are analytic in D, and continuous in $D \cup \sigma$, so must the solution g. But g is already known to be analytic in D^* and continuous in $D^* \cup \sigma$, which implies that the above construction of g furnishes the analytic continuation of g into $D \cup \sigma \cup D^*$.

Similarly, f can be continued into $D \cup \sigma \cup D^*$. Hence (4.28) now gives $u(x, y) = U(z, \bar{z})$ for arbitrary z in $D \cup \sigma \cup D^*$, and this proves the theorem.

Theorem 4.3 can be generalized to include the situation when u satisfies a first order boundary condition on σ.

Theorem 4.4 (*Generalized reflection principle for elliptic equations*) *Let $u \in C^2(D) \cap C^1(\bar{D})$ satisfy $L[u] = 0$ in D, where D is as in Theorem* 4.3.

Suppose that on σ

$$\alpha(x)\frac{\partial u}{\partial x}(x,0)+\beta(x)\frac{\partial u}{\partial y}(x,0)+\gamma(x)u(x,0)=\delta(x)$$

where α, β, γ *and* δ *can be analytically continued into* $D\cup\sigma\cup D^*$ *where* D^* *is the mirror image of* D *reflected across* σ. *We assume further that*

$$\alpha(z)+i\beta(z)\neq 0, \qquad \alpha(z)-i\beta(z)\neq 0$$

throughout $D\cup\sigma\cup D^*$. *Then* u *can be analytically continued into all of* $D\cup\sigma\cup D^*$.

Proof In conjugate coordinates the boundary condition on σ can be written in the form

$$\tilde{\alpha}(x)\frac{\partial U}{\partial z}(x,x)+\tilde{\beta}(x)\frac{\partial U}{\partial \bar{z}}(x,x)+\tilde{\gamma}(x)U(x,x)=\tilde{\delta}(x) \tag{4.31}$$

where $\tilde{\alpha}=\alpha+i\beta$, $\tilde{\beta}=\alpha-i\beta$, $\tilde{\gamma}=\gamma$, $\delta=\tilde{\delta}$. We now differentiate (4.27) with respect to z, integrate by parts, and set $z^*=\bar{z}$ to arrive at an expression of the form

$$\begin{aligned} U_z(z,\bar{z}) = R(z,0;z,\bar{z})U_z(z,0) &+ \text{Volterra integral of } U_t(t,0) \\ &+ \text{Volterra integral of } U_{\bar{t}}(0,\bar{t}) \\ &+ \text{known function times } U(0,0) \end{aligned}$$

with a similar expression holding for $U_{\bar{z}}(z,\bar{z})$. If we now set $z=\bar{z}=x$ and substitute into (4.31) we arrive at an expression of the form

$$\begin{aligned} &\tilde{\alpha}(z)R(z,0;z,z)f'(z)+\tilde{\beta}(z)R(0,z;z,z)g'(z) \\ &\quad + \text{Volterra integral of } f'(t) + \text{Volterra integral of } g'(t) \\ &\quad = \text{known analytic function of } z \text{ times } U(0,0) \\ &\quad\quad + \text{known analytic function of } z, \end{aligned}$$

where $U(z,0)=f(z)$ and $U(0,z)=g(z)$. From the fact that $\tilde{\alpha}(z)\neq 0$, $\tilde{\beta}(z)\neq 0$, and the initial conditions satisfied by the Riemann function, we can conclude that the coefficients of f' and g' are nonzero in $D\cup\sigma\cup D^*$. We can now proceed exactly as in Theorem 4.3 to conclude that f' and g' are analytic in $D\cup\sigma\cup D^*$ and hence from (4.28) that u can be continued into $D\cup\sigma\cup D^*$.

We now ask the question of whether reflection principles such as those given above are valid for elliptic equations in higher dimensions. In certain special cases an explicit reflection formula can be obtained and in such cases the answer is yes. For example, if u is a solution of Laplace's equation defined in a three-dimensional ball of radius a and vanishes on a portion σ of the spherical boundary, then from the regularity theorems for solutions of elliptic equations (cf. Friedman [58]) we can conclude that u is continuously differentiable up to and including that part of the boundary on which u vanishes. It can then be explicitly verified that

$$u_1(r, \theta, \phi) = -(a/r)u(a^2/r, \theta, \phi) \tag{4.32}$$

is a solution of Laplace's equation for $r > a$ and agrees with u and its first derivatives on σ. Hence

$$w = \begin{cases} u_1 & \text{for} \quad r > a \\ 0 & \text{on} \quad \sigma \\ u & \text{for} \quad r < a \end{cases} \tag{4.33}$$

defines a weak solution of Laplace's equation in $\{r < a\} \cup \sigma \cup \{r > a\}$ and thus, from the regularity theory of solutions to elliptic equations, is a twice continuously differentiable and hence an analytic solution in this region. We have therefore 'reflected' u by the formula (4.32) into the 'mirror image' of $r \leqslant a$ across σ.

However, in general a reflection principle is not true for solutions of elliptic equations in more than two independent variables. As an example, we consider the harmonic function defined by (see Lewy [118])

$$u(x, y, z) = \int_{-\infty}^{z} e^{-z+t}(x^2 + y^2 + t^2)^{-1/2}\, dt \tag{4.34}$$

for all x, y, z not on the positive z-axis. One can verify that

$$\partial u/\partial z + u = (x^2 + y^2 + z^2)^{-1/2}. \tag{4.35}$$

Now let p be a plane through the origin which does not contain the z-axis and let ν be the normal to p. Let D be the half space bounded by p which contains the negative z-axis. Define

$$u_1(x, y, z) = (\partial u/\partial \nu)(x, y, z). \tag{4.36}$$

Then u_1 is harmonic in D, infinitely differentiable in $D \cup p \backslash (0, 0, 0)$, and from (4.35) satisfies

$$\partial u_1/\partial z + u_1 = 0 \quad \text{on} \quad p. \tag{4.37}$$

It is clear that u_1 can be continued across $p \backslash (0, 0, 0)$ as a harmonic function only into a domain which does not contain a portion of the positive z-axis (u_1 becomes unbounded as (x, y, z) tends to the positive z-axis). Thus we have an example of a harmonic function satisfying a first order homogeneous boundary condition along a plane, but the harmonic function cannot in general be continued into the mirror image of its domain of definition (see Fig. 4.2).

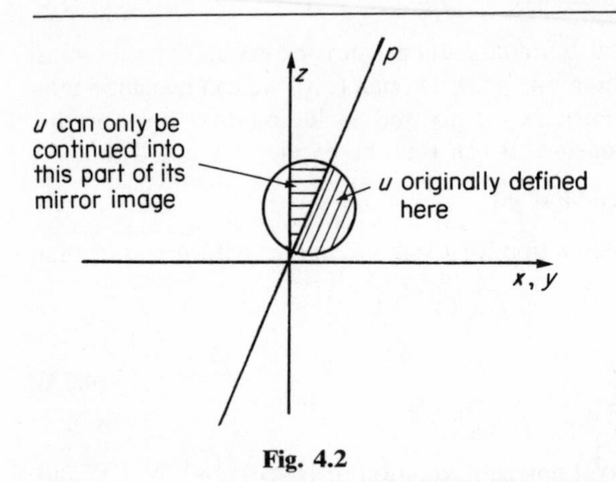

Fig. 4.2

The problem of when solutions of higher dimensional elliptic equations can be reflected across analytic boundaries was taken up by Garabedian in 1960 (Garabedian [64]) who showed that the domain of dependence associated with a solution of an n-dimensional elliptic equation at a point on one side of an analytic surface is in general a whole n-dimensional ball on the other side. Only in exceptional circumstances does some kind of degeneracy occur which causes the domain of dependence to collapse onto a lower dimensional subset, thus allowing a continuation into a larger region than that afforded in general. Such is the situation for example in the case of Lewy's reflection principle for elliptic equations in two independent variables (where the domain of dependence degenerates to a one-dimensional line segment) and the Schwarz reflection principle (4.32) for harmonic functions defined inside or outside a sphere (where the domain of dependence degenerates to a point). Such a collapsing of the domain of dependence can be viewed as a *Huygen's principle for reflection*, analogous to the classical Huygen's principle for hyperbolic equations, and in recent years there have been a number of intriguing examples of when such a degeneracy can occur (cf. Filippenko [55]; John [95]; Lewy [119]). We shall now establish a Huygen's principle for reflection for solutions of the Helmholtz equation defined in the exterior of a ball and vanishing on a portion of the spherical boundary (Colton [33]). We shall show that in this case the domain of dependence degenerates to a one-dimensional line segment, and this result will subsequently be used in Chapter 8 to investigate certain problems arising in the study of the inverse scattering problem for acoustic waves.

We consider solutions of the Helmholtz equation

$$\Delta_n u + \lambda u = 0 \tag{4.38}$$

defined in $D\backslash\bar{S}$ where D is a bounded star-like domain containing the open ball

$$S=\{\mathbf{x}: r=(x_1^2+\cdots+x_n^2)^{1/2}<a\}, \qquad \mathbf{x}=(x_1,\ldots,x_n) \tag{4.39}$$

and λ is a constant. On the surface $r=a$ we assume that $u(r, \theta)=u(\mathbf{x})$ continuously assumes the boundary data

$$u(a, \theta)=0 \tag{4.40}$$

where $(r, \theta)=(r, \theta_1, \ldots, \theta_{n-1})$ are spherical coordinates. We shall obtain a reflection principle for solutions of (4.38), (4.40) through the use of an integral operator which maps harmonic functions defined in $D\backslash\bar{S}$ and vanishing on $r=a$ onto solutions of (4.38), (4.40). In this connection our approach resembles in some ways our use of transformation operators to investigate the wave propagation problem of Section 2.3, except that we are now concerned with solutions defined in a multiply connected domain instead of a simply connected domain.

We look for a solution of (4.38) in the form

$$u(r, \theta)=h(r, \theta)+\int_a^r s^{n-3}K(r, s; \lambda)h(s, \theta)\, ds \tag{4.41}$$

where $h\in C^2(D\backslash\bar{S})\cap C(D\backslash S)$ is a solution of

$$\Delta_n h=0 \tag{4.42}$$

such that

$$h(a, \theta)=0. \tag{4.43}$$

Substituting (4.41) into (4.38) and integrating by parts shows that (4.41) will be a solution of (4.38) provided K satisfies

$$r^2\left[K_{rr}+\frac{(n-1)}{r}K_r+\lambda K\right]=s^2\left[K_{ss}+\frac{(n-1)}{s}K_s\right] \tag{4.44}$$

and the initial data

$$K(r, r; \lambda)=-\tfrac{1}{4}\lambda r^{2-n}(r^2-a^2) \tag{4.45a}$$

$$K(r, a; \lambda)=0. \tag{4.45b}$$

Setting

$$\xi=\log r, \qquad \eta=\log s \tag{4.46}$$

we transform (4.44), (4.45) into the initial value problem

$$M_{\xi\xi}-M_{\eta\eta}+\lambda e^{2\xi}M=0 \tag{4.47a}$$

$$M(\xi, \log a; \lambda)=0 \tag{4.47b}$$

$$M(\xi, \xi; \lambda)=-\tfrac{1}{4}\lambda(e^{2\xi}-a^2) \tag{4.47c}$$

for the function

$$M(\xi, \eta; \lambda) = \exp\{\tfrac{1}{2}(n-2)(\xi+\eta)\}K(e^{\xi}, e^{\eta}; \lambda) \tag{4.48}$$

defined in the cone $\{(\xi, \eta): \xi \leqslant \eta, \eta \leqslant \log a, \text{ or } \xi \geqslant \eta, \eta \geqslant \log a\}$. (4.47) is a Goursat problem for a hyperbolic equations and by using the methods of Chapter 3, Section 1, we can show that it has a unique (analytic) solution in this cone (see the Exercises). Since (4.41) is a Volterra operator, it is easy to show that if $u \in C^2(D\setminus\bar{S}) \cap C(D\setminus S)$ is any solution of (4.38), (4.40), then u can be represented in the form (4.41) for some harmonic function satisfying (4.43) (see the Exercises).

Before turning to the proof of the reflection principle for solutions of (4.38) satisfying (4.40), we take the opportunity to construct another integral operator which in a sense is complementary to (4.41) and which we shall use in Chapter 8 to investigate the inverse scattering problem for acoustic waves. This operator is of the form

$$u(r, \theta) = h(r, \theta) + \int_a^r s^{n-3}\tilde{K}(r, s; \lambda)h(s, \theta)\, ds, \tag{4.49}$$

where $h \in C^2(D\setminus\bar{S}) \cap C^1(D\setminus S)$ is a solution of (4.42) such that

$$h_r(a, \theta) + \frac{(n-2)}{2a} h(a, \theta) = 0. \tag{4.50}$$

In order for u as defined by (4.49) to be a solution of (4.38) we must have that $\tilde{K}$ is a solution of (4.44) satisfying (4.45a) and the initial data

$$\tilde{K}_s(r, a; \lambda) + \frac{(n-2)}{2a}\tilde{K}(r, a; \lambda) = 0. \tag{4.51}$$

This can be verified directly by substituting (4.49) into (4.38) and integrating by parts using (4.50). Using the change of variables (4.46) and setting

$$\tilde{M}(\xi, \eta; \lambda) = \exp\{\tfrac{1}{2}(n-2)(\xi+\eta)\}\tilde{K}(e^{\xi}, e^{\eta}; \lambda) \tag{4.52}$$

we obtain the initial value problem

$$\tilde{M}_{\xi\xi} - \tilde{M}_{\eta\eta} + \lambda e^{2\xi}\tilde{M} = 0 \tag{4.53a}$$

$$\tilde{M}_{\eta}(\xi, \log a; \lambda) = 0 \tag{4.53b}$$

$$\tilde{M}(\xi, \xi; \lambda) = -\tfrac{1}{4}\lambda(e^{2\xi} - a^2). \tag{4.53c}$$

To solve (4.53) we introduce the function E defined as the (unique) solution of the characteristic initial value problem

$$E_{\xi\xi} - E_{\eta\eta} + \lambda e^{2\xi}E = 0 \tag{4.54a}$$

$$E(\xi, \xi; \lambda) = -\tfrac{1}{4}\lambda(e^{2\xi} - a^2) \tag{4.54b}$$

$$E(\xi, -\xi + 2\log a; \lambda) = -\tfrac{1}{4}\lambda(e^{2\xi} - a^2). \tag{4.54c}$$

The existence of a unique (analytic) solution to (4.54) in the cone $\{(\xi, \eta): \xi \leqslant \eta,\ \eta + \xi \leqslant 2 \log a, \text{ or } \xi \geqslant \eta,\ \eta + \xi \geqslant 2 \log a\}$ again follows from the methods of Chapter 3, Section 1, and a solution of (4.53) is now given by

$$\tilde{M}(\xi, \eta; \lambda) = \tfrac{1}{2}[E(\xi, \eta; \lambda) + E(\xi, -\eta + 2 \log a; \lambda)]. \tag{4.55}$$

We have thus established the existence of the operator defined by (4.49). It is again easy to show that if $u \in C^2(D \backslash \bar{S}) \cap C^1(D \backslash S)$ is any solution of (4.38) satisfying

$$u_r(a, \theta) + \frac{(n-2)}{2a} u(a, \theta) = 0, \tag{4.56}$$

then u can be represented in the form (4.49) for some harmonic function satisfying (4.50).

We are now in a position to prove the following reflection principle for solutions of (4.38), (4.40).

Theorem 4.5 Let $u \in C^2(D \backslash \bar{S}) \cap C(D \backslash S)$ be a solution of (4.38), (4.40) and let D^* denote the set obtained by inverting $D \backslash S$ across ∂S, i.e. $(r, \theta) \in D^*$ if and only if $(a^2/r, \theta) \in D \backslash S$. Then u can be analytically continued as a solution of (4.38) in $(D \backslash S) \cup D^*$.

Proof From the results of Chapter 3 it follows that the continuation, if it exists, is analytic. Represent u in the form (4.41) where h satisfies (4.42), (4.43). Then $u \in C^2(D \backslash \bar{S}) \cap C(D \backslash S)$ implies that $h \in C^2(D \backslash \bar{S}) \cap C(D \backslash S)$. Hence from the reflection principle (4.32) for harmonic functions, h is harmonic in $(D \backslash S) \cup D^*$, and so by (4.41) u can be continued as a solution of (4.38) into $(D \backslash S) \cup D^*$.

Note that the domain of dependence of u at a point on one side of ∂S is a one-dimensional line segment on the other side of ∂S. We further note that Theorem 4.5 is also valid if u only vanishes on a portion of the boundary ∂S, in which case D^* is replaced by a cone of the form $\{(r, \theta): (a^2/r, \theta) \in D \backslash S,\ (a, \theta) \in \sigma\}$, where σ is that part of ∂S on which u vanishes.

4.2 Reflection principles for parabolic equations

Our aim in this section is to derive reflection principles for parabolic equations of the form

$$L[u] \equiv \frac{\partial^2 u}{\partial x^2} + a(x, t) \frac{\partial u}{\partial x} + b(x, t) u - \frac{\partial u}{\partial t} = 0 \tag{4.57}$$

defined in the rectangle $R^+ = \{(x, t): 0 < x < x_0, 0 < t < T\}$ and satisfying analytic boundary data on the side $x=0$. By making a preliminary change of variables of the form

$$\xi = x - s(t) \tag{4.58}$$

we could also treat the case when (4.57) is defined in a domain with moving boundaries (see the Exercises). We make the assumption that the coefficients of L are continuously differentiable for $-x_0 \leq x \leq x_0$, $0 \leq t \leq T$, and are real analytic as a function of t on the interval $0 \leq t \leq T$. By making the change of variables

$$u(x, t) = v(x, t) \exp\left\{-\frac{1}{2}\int_0^x a(\xi, t)\, d\xi\right\}. \tag{4.59}$$

We can arrive at an equation for v of the form

$$\partial^2 u/\partial x^2 + q(x, t)u = \partial u/\partial t \tag{4.60}$$

where q enjoys the same regularity properties as the coefficients of L.

In order to derive our reflection principles we first need to construct a class of transformation operators for (4.60) that map solutions of the heat equation

$$\partial^2 h/\partial x^2 = \partial h/\partial t \tag{4.61}$$

onto solutions of (4.60). These operators will also be used in Chapter 5 to construct a complete family of solutions to (4.60). We first look for solutions of (4.60) in the form

$$u(x, t) = h(x, t) + \int_0^x K(s, x, t)h(s, t)\, ds \tag{4.62}$$

where h is a solution of (4.61) in R^+ and satisfies the Dirichlet data

$$h(0, t) = 0. \tag{4.63}$$

Note that from the regularity properties of solutions to parabolic equations we can conclude that h is continuously differentiable for $0 \leq x < x_0$, $0 < t < T$. Furthermore, (4.62) and (4.63) imply that $u(0, t) = 0$. Substituting (4.62) into (4.60) and integrating by parts shows that (4.62) is a solution of (4.60) provided K is a solution of

$$K_{xx} - K_{ss} + q(x, t)K = K_t \tag{4.64a}$$

for $0 < s \leq x < x_0$ which satisfies the initial data

$$K(x, x, t) = -\frac{1}{2}\int_0^x q(s, t)\, ds \tag{4.64b}$$

$$K(0, x, t) = 0. \tag{4.64c}$$

Now suppose that instead of satisfying $h(0, t)=0$, h satisfies

$$h_x(0, t)=0. \tag{4.65}$$

We again look for a solution of (4.60) in the form

$$u(x, t)=h(x, t)+\int_0^x M(s, x, t)h(s, t)\,ds. \tag{4.66}$$

Then it can again be shown that (4.66) will be a solution of (4.60) provided M is a solution of

$$M_{xx}-M_{ss}+q(x, t)M=M_t \tag{4.67a}$$

for $0<s\leqslant x<x_0$ which satisfies the initial data

$$M(x, x, t)=-\frac{1}{2}\int_0^x q(s, t)\,ds \tag{4.67b}$$

$$M_s(0, x, t)=0. \tag{4.67c}$$

If the functions K and M exist, we can now define two operators $\mathbf{I}+\mathbf{T}_1$ and $\mathbf{I}+\mathbf{T}_2$ mapping solutions of the heat equation onto solutions of (4.60) by

$$(\mathbf{I}+\mathbf{T}_1)[h]=h(x, t)+\int_0^x K(s, x, t)h(s, t)\,ds \tag{4.68}$$

$$(\mathbf{I}+\mathbf{T}_2)[h]=h(x, t)+\int_0^x M(s, x, t)h(x, t)\,ds \tag{4.69}$$

where the domain of $\mathbf{I}+\mathbf{T}_1$ is the class of solutions to the heat equation in R^+ satisfying $h(0, t)=0$ and the domain of $\mathbf{I}+\mathbf{T}_2$ is the class of solutions to the heat equation in R^+ satisfying $h_x(0, t)=0$.

We shall now show the existence of the functions K and M. Due to the regularity assumptions on q we shall in fact show that K and M are twice continuously differentiable solutions of (4.64) and (4.67) for $-x_0\leqslant x\leqslant x_0$, $-x_0\leqslant s\leqslant x_0$, $0\leqslant t\leqslant T$. Suppose E satisfies

$$E_{xx}-E_{ss}+q(x, t)E=E_t \tag{4.70}$$

for $-x_0\leqslant x\leqslant x_0$, $-x_0\leqslant s\leqslant x_0$, $0\leqslant t\leqslant T$, and assumes the initial data

$$E(x, x, t)=-\frac{1}{2}\int_0^x q(s, t)\,ds \tag{4.71a}$$

$$E(-x, x, t)=\frac{1}{2}\int_0^x q(s, t)\,ds. \tag{4.71b}$$

Then

$$K(s, x, t)=\tfrac{1}{2}[E(s, x, t)-E(-s, x, t)] \tag{4.72}$$

satisfies (4.64). Similarly, if G satisfies (4.70) for $-x_0 \leq x \leq x_0$, $-x_0 \leq s \leq x_0$, $0 \leq t \leq T$, and assumes the initial data

$$G(x, x, t) = -\frac{1}{2}\int_0^x q(s, t)\, ds \tag{4.73a}$$

$$G(-x, x, t) = -\frac{1}{2}\int_0^x q(s, t)\, ds \tag{4.73b}$$

then

$$M(s, x, t) = \tfrac{1}{2}[G(s, x, t) + G(-s, x, t)] \tag{4.74}$$

satisfies (4.67). Hence it suffices to show the existence of the functions E and G. We shall now do this for E; the existence of G follows in an identical fashion. Let

$$x = \xi + \eta, \qquad s = \xi - \eta \tag{4.75}$$

and define $\tilde{E}$ by

$$\tilde{E}(\xi, \eta, t) = E(\xi - \eta, \xi + \eta, t). \tag{4.76}$$

Then (4.70), (4.71) become

$$\tilde{E}_{\xi\eta} + q(\xi + \eta, t)\tilde{E} = \tilde{E}_t \tag{4.77a}$$

$$\tilde{E}(\xi, 0, t) = -\frac{1}{2}\int_0^{\xi} q(s, t)\, ds \tag{4.77b}$$

$$\tilde{E}(0, \eta, t) = \frac{1}{2}\int_0^{\eta} q(s, t)\, ds \tag{4.77c}$$

and hence $\tilde{E}$ satisfies the Volterra integrodifferential equation

$$\tilde{E}(\xi, \eta, t) = -\frac{1}{2}\int_0^{\xi} q(s, t)\, ds + \frac{1}{2}\int_0^{\eta} q(s, t)\, ds + \int_0^{\eta}\int_0^{\xi} (\tilde{E}_t(\xi, \eta, t) - q(\xi + \eta, t)\tilde{E}(\xi, \eta, t))\, d\xi\, d\eta. \tag{4.78}$$

We shall establish the existence of a solution to (4.78) by using the *method of dominants.* If we are given two series

$$S = \sum_{n=0}^{\infty} a_n t^n, \qquad \tilde{S} = \sum_{n=0}^{\infty} \tilde{a}_n t^n; \qquad |t| < t_0, \tag{4.79}$$

where $\tilde{a}_n \geq 0$, then we say $\tilde{S}$ *dominates* S if $|a_n| \leq \tilde{a}_n$, $n = 0, 1, 2, \ldots$, and write $S \ll \tilde{S}$. It is easily verified that dominants can be multiplied and if $S \ll \tilde{S}$ then

$$\partial S/\partial t \ll \partial \tilde{S}/\partial t \tag{4.80a}$$

and

$$S \ll \tilde{S}(1-t/t_0)^{-1}. \tag{4.80b}$$

Furthermore, if f is an analytic function for $|t| \leq t_0$ then, since the terms of the Taylor series for f are uniformly bounded, there exists a positive constant C such that

$$f \ll C(1-t/t_0)^{-1}. \tag{4.81}$$

We now return to the problem of establishing the existence of a solution to the integrodifferential equation (4.78). It suffices to establish the existence of a solution for t in a (complex) neighborhood of the origin since the solution about any other point $t \in [0, T]$ can be reduced to this case by a preliminary translation. The solution of (4.78) in a neighborhood of the origin can formally be obtained by iteration in the form

$$\tilde{E}(\xi, \eta, t) = \sum_{n=1}^{\infty} \tilde{E}_n(\xi, \eta, t) \tag{4.82}$$

where

$$\begin{aligned} \tilde{E}_1(\xi, \eta, t) &= -\frac{1}{2}\int_0^{\xi} q(s, t)\, ds + \frac{1}{2}\int_0^{\eta} q(s, t)\, ds \\ \tilde{E}_{n+1}(\xi, \eta, t) &= \int_0^{\eta}\int_0^{\xi} (\tilde{E}_{nt}(\xi, \eta, t) - q(\xi+\eta, t)\tilde{E}_n(\xi, \eta, t))\, d\xi\, d\eta. \end{aligned} \tag{4.83}$$

We shall now show that the series (4.82) converges absolutely and uniformly for (ξ, η, t) on $\Omega = \{(\xi, \eta, t)\colon -x_0 \leq \xi \leq x_0,\ -x_0 \leq \eta \leq x_0,\ |t| \leq t_0\}$ where t_0 is such that q is an analytic function of t for $|t| \leq t_0$. To this end let C be a positive constant such that for $(\xi, \eta, t) \in \Omega$ we have with respect to t

$$q(\xi+\eta, t) \ll C(1-t/t_0)^{-1}. \tag{4.84}$$

Without loss of generality assume $C \geq 1$, $x_0 \leq 1$. Then from (4.83) and the properties of dominants it follows by induction that

$$\tilde{E}_n \ll \frac{2^n C^n\, |\xi|^{n-1}\, |\eta|^{n-1}}{(n-1)!}\left(1-\frac{t}{t_0}\right)^{-n} \tag{4.85}$$

and hence

$$|\tilde{E}_n| \leq \frac{2^n C^n\, |\xi|^{n-1}\, |\eta|^{n-1}}{(n-1)!}\left(1-\frac{|t|}{t_0}\right)^{-n} \tag{4.86}$$

for $(\xi, \eta, t) \in \Omega$. Hence the series (4.82) converges absolutely and uniformly for $(\xi, \eta, t) \in \Omega$. In a similar manner it is easily seen that $\tilde{E}$ is twice continuously differentiable in Ω, and we can now conclude the existence of the function $\tilde{E}=E$. Similarly, the function G exists and is twice continuously differentiable in Ω, and we have therefore now established the existence of the operators $\mathbf{I}+\mathbf{T}_1$ and $\mathbf{I}+\mathbf{T}_2$.

We now want to show that if u is a solution of (4.60) in R^+ and satisfies $u(0, t)=0$, then u can be represented in the form $u=(\mathbf{I}+\mathbf{T}_1)[h]$, where h is a solution of the heat equation in R^+ satisfying $h(0, t)=0$. A similar point was rather lightly passed over in our discussion of the operators (4.41) and (4.49), and here we wish to be a little more precise. From the regularity theorems for parabolic equations (cf. Friedman [57]) we can conclude that u is continuously differentiable for $0 \leqslant x < x_0$, $0<t<T$, and since $u=(\mathbf{I}+\mathbf{T}_1)[h]$ is a Volterra integral equation of the second kind for h, there exists a solution which has the same regularity properties as u and satisfies $h(0, t)=u(0, t)=0$. To show that this solution of $u=(\mathbf{I}+\mathbf{T}_1)[h]$ is in fact a solution of the heat equation, we substitute $u=(\mathbf{I}+\mathbf{T}_1)[h]$ into the differential equation (4.60) and integrate by parts to obtain

$$\begin{aligned} 0 &= u_{xx}+q(x, t)u-u_t \\ &= (h_{xx}-h_t)+\int_0^x K(s, x, t)(h_{ss}(x, t)-h_t(x, t))\, ds. \end{aligned} \tag{4.87}$$

Since solutions of Volterra integral equations of the second kind are unique, we must have

$$h_{xx}-h_t=0, \tag{4.88}$$

i.e. h is a solution of the heat equation in R^+. In a similar manner we can show that if u is a solution of (4.60) in R^+ satisfying $u_x(0, t)=0$, then u can be represented in the form $u=(\mathbf{I}+\mathbf{T}_2)[h]$ where h is a solution of the heat equation in R^+ such that $h_x(0, t)=0$.

We now want to combine the results obtained above to construct an integral operator whose domain and range are independent of the boundary data at $x=0$. This operator will be of use in connection with our discussion of Runge's theorem in Chapter 5. Let u be a solution of (4.60) in $R=\{(x, t)\colon -x_0<x<x_0,\ 0<t<T\}$. We shall show that there exists a solution h of (4.61) in R such that u can be represented in the form

$$u(x, t)=(\mathbf{I}+\mathbf{T}_3)[h]=h(x, t)+\frac{1}{2}\int_{-x}^{x}[K(s, x, t)+M(s, x, t)]h(s, t)\, ds. \tag{4.89}$$

(4.89) is a Volterra equation of the second kind for h and hence can be uniquely solved for h where h is defined in R. It remains to be shown that h is a solution of (4.61). From (4.72) and (4.74) we have that $K(s, x, t)=-K(-s, x, t)$ and $M(s, x, t)=M(-s, x, t)$ and hence we can rewrite (4.89) in the form

$$u(x,t)=\frac{1}{2}[h(x,t)-h(-x,t)]+\frac{1}{2}\int_0^x K(s,x,t)[h(s,t)-h(-s,t)]\,ds$$
$$+\frac{1}{2}[h(x,t)+h(-x,t)]+\frac{1}{2}\int_0^x M(s,x,t)$$
$$\times[h(s,t)+h(-s,t)]\,ds. \tag{4.90}$$

Substituting (4.90) into (4.60) and integrating by parts, gives

$$0=(h_{xx}-h_t)+\frac{1}{2}\int_{-x}^x [K(s,x,t)+M(s,x,t)](h_{ss}(s,t)-h_t(s,t))\,ds, \tag{4.91}$$

and by the invertibility of Volterra integral equations of the second kind we can conclude that h is a solution of the heat equation in R.

We are now in a position to begin proving our reflection principles for solutions of $L[u]=0$ defined in R^+. However, we first need two results on the continuation of solutions to the heat equation. The first of these is the well-known reflection principle for solutions of the heat equation, and can be proved either by the method used to prove the reflection formula (4.32) for Laplace's equation, or by making use of the Green or Neumann function for the heat equation defined in a rectangle (Widder [179]).

Theorem 4.6 (1) *Let $h\in C^2(R^+)\cap C(R^+\cup\sigma)$ be a solution of the heat equation in R^+ such that $h(0,t)=0$ on σ: $x=0$. Then h can be uniquely continued into $R^+\cup R^-\cup\sigma$ as a solution of the heat equation by the rule $h(x,t)=-h(-x,t)$, where $R^-=\{(x,t): -x_0<x<0, 0<t<T\}$.*

(2) *Let $h\in C^2(R^+)\cap C^1(R^+\cup\sigma)$ be a solution of the heat equation in R^+ such that $h_x(0,t)=0$. Then h can be uniquely continued into $R^+\cup R^-\cup\sigma$ as a solution of the heat equation by the rule $h(x,t)=h(-x,t)$.*

Our next result is concerned with the analytic continuation of solutions to the heat equation satisfying analytic Cauchy data on $x=0$ (Colton [28]; Hill [84]; Widder [179]).

Theorem 4.7 *Let $h\in C^2(R^+)\cap C^1(R^+\cup\sigma)$ be a solution of the heat equation in R^+ such that $h(0,t)$ and $h_x(0,t)$ are real analytic for $0<t<T$. Then h can be uniquely continued into $-\infty<x<\infty$, $0<t<T$, as a solution of the heat equation that is an analytic function of x and t for $|x|<\infty$, $0<t<T$.*

Proof By Holmgren's uniqueness theorem and direct calculation we can represent h in the form

$$h(x, t) = -\frac{1}{2\pi i} \oint_{|t-\tau|=\delta} E_x(x, t-\tau) h(0, \tau)\, d\tau$$
$$-\frac{1}{2\pi i} \oint_{|t-\tau|=\delta} E(x, t-\tau) h_x(0, \tau)\, d\tau \tag{4.92}$$

where $\delta > 0$ and

$$E(x, t) = \sum_{j=0}^{\infty} \frac{x^{2j+1}(-1)^j j!}{(2j+1)!\, t^{j+1}}. \tag{4.93}$$

Since E is analytic for $|x| < \infty$, $|t| > 0$, the statement of the theorem now follows.

Theorem 4.8 (*Reflection principle for parabolic equations*) *Let* $u \in C^2(R^+) \cap C(R^+ \cup \sigma)$ *satisfy* $L[u] = 0$ *in* R^+. *Suppose* $u(0, t) = f(t)$ *is analytic for* $0 < t < T$. *Then* u *can be uniquely continued into* $R^+ \cup R^- \cup \sigma$ *as a solution of* $L[u] = 0$.

Remark From the regularity theory of solutions to parabolic equations we have that $u \in C^1(R^+ \cup \sigma)$. Note that since q is not in general an analytic function of x, the continuation of u is not in general an analytic continuation with respect to x.

Proof By making the change of variables (4.59) we can assume without loss of generality that u satisfies an equation of the form (4.60). Let h be the analytic solution of the heat equation given by (4.92) where $h(0, t) = f(t)$ and $h_x(0, t) = 0$, and define the solution v of (4.60) by $v = (\mathbf{I} + \mathbf{T}_2)[h]$. By construction, the operator $\mathbf{I} + \mathbf{T}_2$ preserves Cauchy data on $\sigma: x = 0$ and hence from Theorem 4.7 and the regularity of the kernel M of $\mathbf{I} + \mathbf{T}_2$ we can conclude that v is a solution of (4.60) in $R^+ \cup R^- \cup \sigma$ such that $v(0, t) = f(t)$. Hence, by considering $u - v$ instead of u we can assume without loss of generality that $f \equiv 0$, i.e. $u(0, t) = 0$. But now we can represent u in the form $u = (\mathbf{I} + \mathbf{T}_1)[h]$ where h is a solution of the heat equation satisfying $h(0, t) = 0$, and from Theorem 4.6 h can be uniquely continued into $R^+ \cup R^- \cup \sigma$ as a solution of the heat equation. This implies that u can be continued as a solution of (4.60) into $R^+ \cup R^- \cup \sigma$. The uniqueness of the continuation follows from the invertibility of the operator $\mathbf{I} + \mathbf{T}_1$ and the uniqueness of the continuation of h.

Theorem 4.8 can be generalized to include the situation when u satisfies a first order boundary condition on σ.

Theorem 4.9 *(Generalized reflection principle for parabolic equations) Let $u \in C^2(R^+) \cap C^1(R^+ \cup \sigma)$ satisfy $L[u]=0$ in R^+. Suppose that on σ*

$$\alpha(t)u(0, t)+\beta(t)u_x(0, t)+\gamma(t)u_t(0, t)=f(t)$$

where α, β, γ and f are real analytic for $0<t<T$. Assume that $\vec{\mu}(t)=(\beta(t), \gamma(t))$ is either never parallel to the x-axis or always parallel to the x-axis, and that $\beta(t)\neq 0$ for $0<t<T$. Then u can be uniquely continued into $R^+ \cup R^- \cup \sigma$ as a solution of $L[u]=0$.

Remark If $\beta \equiv 0$ then Theorem 4.9 can be reduced to Theorem 4.8.

Proof We can again assume without loss of generality that u is the solution of an equation of the form (4.60) and by considering the Cauchy problem

$$\begin{aligned} h_{xx} &= h_t \\ h(0, t) &= 0 \\ h_x(0, t) &= f(t)/\beta(t) \end{aligned} \tag{4.94}$$

we can proceed as in Theorem 4.8 to assume without loss of generality that $f\equiv 0$. Now assume $\vec{\mu}(t)$ is always parallel to the x-axis, i.e. $\gamma\equiv 0$. Then we can rewrite the boundary condition on σ as

$$u_x(0, t)+\frac{\alpha(t)}{\beta(t)}u(0, t)=0. \tag{4.95}$$

Then from our previous analysis we have that

$$u(x, t)=h(x, t)+\int_0^x M(s, x, t)h(s, t)\,ds \tag{4.96}$$

will be a solution of (4.60) in R^+ provided h is a solution of the heat equation in R^+ satisfying $h_x(0, t)=0$ and

$$M_{xx}-M_{ss}+q(x, t)M=M_t \tag{4.97a}$$

$$M_s(0, x, t)=0 \tag{4.97b}$$

$$\frac{\partial}{\partial x}M(x, x, t)=-\tfrac{1}{2}q(x, t). \tag{4.97c}$$

This is a slightly more general result than (4.68), (4.69) where we arbitrarily imposed the condition that $M(0, 0, t)=0$ and integrated (4.97c). In the present case we integrate (4.97c) to obtain

$$M(x, x, t)=-\frac{1}{2}\int_0^x q(s, t)\,ds-\frac{\alpha(t)}{\beta(t)}. \tag{4.98}$$

Due to the regularity assumptions on α and β we can conclude that M exists in $R^+ \cup R^- \cup \sigma$ and from (4.96), (4.98) we have

$$u_x(0, t) = h_x(0, t) + M(0, 0, t)h(0, t) = -\frac{\alpha(t)}{\beta(t)} u(0, t), \tag{4.99}$$

i.e. u satisfies (4.95). Conversely, it follows from an analysis similar to that leading to (4.87) that every solution of (4.60), (4.95) in R^+ that is continuously differentiable in $R^+ \cup \sigma$ can be represented in the form (4.96). Since h is a solution of the heat equation satisfying $h_x(0, t) = 0$, the reflection principle now follows from (4.96) and Theorem 4.6. The uniqueness of the continuation follows from the same reasons as in Theorem 4.8.

Now assume that $\vec{\mu}(t)$ is never parallel to the x-axis, i.e. $\gamma(t) \neq 0$ for $0 < t < T$. By setting

$$u(x, t) = v(x, t) \exp\left\{-\int_0^t \alpha(\tau)\, d\tau\right\} \tag{4.100}$$

we arrive at an equation for v of the form (4.60) where v satisfies a boundary condition on σ of the form stated in the theorem, but with $\alpha \equiv 0$. Hence without loss of generality we consider u to be a solution of (4.60) in R^+ satisfying

$$u_x(0, t) + \eta(t) u_t(0, t) = 0 \tag{4.101}$$

where $\eta(t) = \alpha(t)/\beta(t)$. Let $h^{(1)}$ be a solution of the heat equation in R^+ such that $h_x^{(1)}(0, t) = 0$ and define $h^{(2)}$ by

$$h^{(2)}(x, t) = -\eta(t) h_x^{(1)}(x, t). \tag{4.102}$$

Then $h^{(2)}$ satisfies

$$h_{xx}^{(2)} + \frac{\eta'(t)}{\eta(t)} h^{(2)} = h_t^{(2)}. \tag{4.103}$$

Note that by the reflection principle for the heat equation $h^{(1)}$ and $h^{(2)}$ are solutions of the heat equation and (4.103), respectively, in $R^+ \cup R^- \cup \sigma$. We now look for a solution of (4.60), (4.101) in the form

$$u(x, t) = h^{(1)}(x, t) + h^{(2)}(x, t) + \int_0^x K^{(1)}(s, x, t) h^{(1)}(s, t)\, ds$$
$$+ \int_0^x K^{(2)}(s, x, t) h^{(2)}(s, t)\, ds. \tag{4.104}$$

Substituting (4.104) into (4.60) shows that (4.104) will be a solution of

(4.60) provided

$$K^{(1)}_{xx} - K^{(1)}_{ss} + q(x, t)K^{(1)} = K^{(1)}_t$$
$$K^{(1)}(x, x, t) = -\frac{1}{2}\int_0^x q(s, t)\, ds \tag{4.105}$$
$$K^{(1)}_s(0, x, t) = 0$$

and

$$K^{(2)}_{xx} - K^{(2)}_{ss} + \left[q(x, t) - \frac{\eta'(t)}{\eta(t)}\right]K^{(2)} = K^{(2)}_t$$
$$K^{(2)}(x, x, t) = -\frac{1}{2}\int_0^x \left[q(s, t) - \frac{\eta'(t)}{\eta(t)}\right] ds \tag{4.106}$$
$$K^{(2)}(0, x, t) = 0.$$

The existence of the functions $K^{(1)}$ and $K^{(2)}$ in $R^+ \cup R^- \cup \sigma$ follows from our previous analysis. From (4.104) we have

$$\begin{aligned} u_x(0, t) + \eta(t)u_t(0, t) &= h^{(1)}_x(0, t) + h^{(2)}_x(0, t) + \eta(t)[h^{(1)}_t(0, t) + h^{(2)}_t(0, t)] \\ &= \eta(t)[h^{(1)}_t(0, t) - h^{(1)}_{xx}(0, t)] \\ &= 0, \end{aligned} \tag{4.107}$$

i.e. u satisfies the boundary data (4.101). The fact that every solution u of (4.60), (4.101) can be represented in the form (4.104) follows along the same lines as that leading to (4.87), although in the present case the analysis is somewhat more involved (see the Exercises). The reflection principle for (4.60), (4.101) now follows from (4.104) and the fact that from Theorem 4.6 $h^{(1)}$ (and hence $h^{(2)}$) can be continued into $R^+ \cup R^- \cup \sigma$. The uniqueness of the continuation follows as in Theorem 4.8. This completes the proof of the theorem.

To conclude this section we turn to the case when the solution of $L[u] = 0$ is known to be analytic in some neighborhood of the origin and make the assumption that the coefficients of L are real analytic for $-\infty < x < \infty$, $0 < t < T$. In this case we have that $u(0, t)$ and $u_x(0, t)$ are analytic functions of t for $0 < t < t_0$ for some constant $t_0 < T$, and if $L[u] = 0$ is reduced to the canonical form (4.60) and u is represented in the form $u = (\mathbf{I} + \mathbf{T}_3)[h]$ we have that $h(0, t) = u(0, t)$ and $h_x(0, t) = u_x(0, t)$ are analytic functions of t for $0 < t < t_0$. Hence by Theorem 4.7 h is analytic for $|x| < \infty$, $0 < t < t_0$ and hence from the analyticity of the kernel of the operator $\mathbf{T}_3$, u is analytic for $-\infty < x < \infty$, $0 < t < t_0$. Hence we have the following theorem (note that due to analyticity we do not need to make any assumption at all on the boundary data satisfied by u).

Theorem 4.10 *Let u be an analytic solution of $L[u]=0$ in a neighborhood of the origin, where the coefficients of L are assumed to be analytic for $-\infty<x<\infty$, $0<t<T$, and the neighborhood of analyticity of u contains $S_\varepsilon=\{(x,t)\colon -\varepsilon<x<\varepsilon,\ 0<t<t_0\}$, $t_0<T$. Then u can be analytically continued into S_∞ as a solution of $L[u]=0$.*

In contrast to the case of parabolic equations in one space variable, the results on the global continuation of solutions to parabolic equations in several space variables are rather limited. Even in the case of parabolic equations in two space variables with analytic coefficients it is not yet known whether or not solutions of such equations can be reflected across a plane boundary on which the Dirichlet data vanish. However, it has been shown by Hill [85] that the domain of dependence associated with an *analytic* solution of a parabolic equation in two space variables with analytic coefficients at a point on one side of a plane on which the solution vanishes is a one-dimensional line segment on the other side, and in this case reflection principles and explicit reflecting formulae can be derived (Colton [37]; Hill [85]).

4.3 The envelope method

The 'envelope method' (Gilbert [69], [70], [71]; Marsden [124]) gives a procedure for the analytic continuation of analytic functions of several complex variables defined in terms of a contour integral. Its applicability to the analytic continuation of solutions to partial differential equations rests on being able to represent such a solution by means of a contour integral involving analytic functions having known singularity manifold. In this section we shall give a proof of the envelope method and show its applicability to locating the singularities of the axially symmetric potential equation (cf. Gilbert [70]; Weinstein [174]). In what follows C^{n+1} denotes the space of $n+1$ complex variables.

Theorem 4.11 (*The envelope method*) *Let $F(z)=F(z_1,\ldots,z_n)$ be defined by the integral representation*

$$F(z)=\int_{\mathsf{L}} K(z;\zeta)\,\mathrm{d}\zeta$$

where K is a holomorphic function of $n+1$ complex variables in C^{n+1} except for possible singularities lying on the analytic set $G_0=\{(z,\zeta)\colon S(z;\zeta)=0,\ \zeta\in C^1\}$. Let the integration path L be a closed rectifiable contour in C^1. Then F is regular for all points $z\notin G_0\cap G_1$ where $G_1=\{(z,\zeta)\colon \partial S(z;\zeta)/\partial\zeta=0,\ \zeta\in C^1\}$.

Proof Let F be regular at $z=z_0$ and hence in a neighborhood $N(z_0)$ of z_0. Now analytically continue F along a path γ with one endpoint in $N(z_0)$. This can be done as long as no point of γ corresponds to a singularity of the integrand on L. Even when this happens we can keep on continuing F along γ by deforming the path of integration to avoid the singularity $\zeta=\alpha(z)$ threatening to cross it. In particular, suppose we have continued F along γ to a point $z=z_1$ and at that point there exists a singularity $\zeta=\alpha$ on L. Suppose, however, $S(z;\zeta)$ has a simple zero at $\zeta=\alpha$, i.e. in a sufficiently small neighborhood $N(\alpha)=\{\zeta: |\zeta-\alpha|<\varepsilon\}$ we have

$$S(z_1;\zeta)\approx(\zeta-\alpha)\partial S(z_1;\zeta)/\partial\zeta \tag{4.108}$$

where $\partial S(z_1;\zeta)/\partial\zeta\neq 0$. Then we can deform L about the point $\zeta=\alpha$ by letting it follow a portion of the circle $|\zeta-\alpha|=\varepsilon/2$, which implies that F is regular at z_1. This proves the theorem.

An immediate corollary of Theorem 4.11 is Hadamard's multiplication of singularities theorem (Titchmarsh [167]).

Corollary 4.1 (Hadamard's multiplication of singularities theorem) Let $z\in C^1$ and suppose f is an analytic function of z except for isolated singularities at $\alpha_1, \alpha_2, \ldots, \alpha_i\neq 0$, $i=1,2,\ldots$, and g is an analytic function of z except for isolated singularities at $\beta_1, \beta_2, \ldots, \beta_i\neq 0$, $i=1,2,\ldots$. Then if L is a closed contour lying in a neighborhood of the origin where f is analytic,

$$F(z)=\frac{1}{2\pi i}\int_{\mathsf{L}} f(\zeta)g\left(\frac{z}{\zeta}\right)\frac{d\zeta}{\zeta}$$

is regular for $z\neq\alpha_m\beta_n$, $m, n=1,2,\ldots$.

Proof Without loss of generality suppose that α and β are the only finite singularities of f and g, respectively. Then $S(z;\zeta)=\zeta(\zeta-\alpha)(z-\beta\zeta)$ and $\partial S(z;\zeta)/\partial\zeta=\zeta(z-2\beta\zeta+\alpha\beta)+(\zeta-\alpha)(z-\beta\zeta)$. Setting $S(z;\zeta)=\partial S(z;\zeta)/\partial\zeta=0$ and eliminating ζ gives $(z-\alpha\beta)^2=0$ if $\zeta\neq 0$ and $z=0$ if $\zeta=0$. But since we know *a priori* that F is analytic at $z=0$, the corollary follows.

Remark If L is not a closed contour but an open contour between two fixed points ζ_1 and ζ_2, then we cannot deform L away from these points and hence F may be singular on the set $G_2=\{(z,\zeta): S(z;\zeta)=0, \zeta=\zeta_1$ and $\zeta=\zeta_2\}$. Such singularities are called 'endpoint-pinch' singularities.

We now apply the envelope method to locate the possible singular points of solutions to the axially symmetric potential equation

$$\frac{\partial^2 u}{\partial z^2}+\frac{\partial^2 u}{\partial r^2}+\frac{1}{r}\frac{\partial u}{\partial r}=0 \tag{4.109}$$

in terms of the location of the singularities of $u(z, 0)$ in the complex z plane. It is easy to show that if u is an analytic solution of (4.109) in some neighborhood of the origin, then u is an even function of r and is uniquely determined by $u(z, 0)=f(z)$ (see the Exercises). In order to apply the envelope method we need to have an integral representation of the solution to (4.109) satisfying $u(z, 0)=f(z)$. It can be easily verified that such a representation is given by

$$u(z, r)=\mathbf{A}[f]=\frac{1}{2\pi i}\int_{\mathsf{L}} f(\sigma)\frac{d\zeta}{\zeta} \tag{4.110}$$

where L is a closed contour surrounding the origin in a counterclockwise direction and $\sigma=z+i(r/2)(\zeta+\zeta^{-1})$. We can now prove the following theorem (cf. Erdélyi [52]; Gilbert [73]; Henrici [83]).

Theorem 4.12 *If the only finite singularities of f are at $z=\alpha$, then the only possible singularities of u on its first Riemann sheet are at $z+ir=\alpha$ and $z-ir=\alpha$.*

Proof We represent u by the operator $\mathbf{A}$ and apply the envelope method. The envelope singularities are given by $G=G_0\cap G_1$ where

$$\begin{aligned} G_0&=\{(z, r)\colon (z-\alpha)\zeta+i(r/2)(\zeta^2+1)=0\}\\ G_1&=\{(z, r)\colon (z-\alpha)+ir\zeta=0\}. \end{aligned} \tag{4.111}$$

Eliminating ζ gives

$$G=G_0\cap G_1=\{(z, r)\colon (z-\alpha)^2+r^2=0\} \tag{4.112}$$

which implies $z+ir=\alpha$ or $z-ir=\alpha$.

It should be noted that singularities of u can arise on other sheets of the Riemann surface for u if u has branch points; for an example of this we refer the reader to Gilbert [70]. It is also possible to show by constructing the operator $\mathbf{A}^{-1}$ that $z+ir=\alpha$ and $z-ir=\alpha$ are indeed singular points of u if f is singular at $z=\alpha$. For this result the reader is again referred to Gilbert [70], [71].

4.4 Analytic continuation of analytic solutions to the heat equation

In the case of one-space variable the topic of this section has already been considered in Section 4.2. However, here we want to relate this problem to a particular physical problem arising in the melting of solids and to present some results for the case of two-space variables. We first consider the case of one-space variable. Consider a thin block of ice at zero degrees Centigrade occupying the interval $0 \leqslant x < \infty$ and suppose at $x=0$ the temperature is given by a prescribed function $\phi(t)$ where $t \geqslant 0$ denotes time. Then the ice will begin to melt and for $t>0$ the water will occupy an interval $0 \leqslant x < s(t)$. If u is the temperature of the water then we have, after a suitable normalization and scaling, that

$$\partial^2 u/\partial x^2 = \partial u/\partial t \quad \text{for} \quad 0<x<s(t) \tag{4.113a}$$

$$u(0, t) = \phi(t) \quad \text{for} \quad t>0 \tag{4.113b}$$

$$u(s(t), t) = 0 \quad \text{for} \quad t>0 \tag{4.114a}$$

and, from the law of conservation of energy,

$$\frac{\partial u}{\partial x}(s(t), t) = -\frac{\mathrm{d}s(t)}{\mathrm{d}t}. \tag{4.114b}$$

This problem is known as the (single-phase) *Stefan problem* (cf. Rubinstein [150]) and the problem is to determine the temperature u and free boundary s from (4.113a)–(4.114b). It is easily seen that this problem is nonlinear in s. The *inverse Stefan problem* is, given s, to determine u and ϕ, i.e. how must we heat the water to melt the ice along a prescribed curve? The idea is to construct a 'catalog' of solutions u corresponding to a large class of 'free' boundaries s and then to be in a position to solve the Stefan problem by looking in the catalog for a solution whose boundary data at $x=0$ are close to ϕ. Although the inverse problem is linear, it is improperly posed in the real domain in the sense that u does not depend continuously on the initial data on the curve $x=s(t)$ (see the Exercises). However, assuming that s is analytic, we can easily construct a solution to the inverse Stefan problem described above by using Theorem 4.7 (Hill [84]). Indeed, if in (4.92) we place the cycle $|t-\tau|=\delta$ on the two-dimensional manifold $x=s(t)$ in the space of two complex variables, and note that since $u(s(t), t)=0$ the first integral in (4.92) vanishes, we are led to the following solution of the inverse Stefan problem:

$$u(x, t) = \frac{1}{2\pi \mathrm{i}} \oint_{|t-\tau|=\delta} E(x-s(\tau), t-\tau)\dot{s}(\tau)\, \mathrm{d}\tau. \tag{4.115}$$

Computing the residue in (4.115) now gives

$$u(x, t) = \sum_{n=1}^{\infty} \frac{1}{(2n)!} \frac{\partial^n}{\partial t^n} [x-s(t)]^{2n}. \tag{4.116}$$

As an example of the use of (4.116) we see that in order for the ice to melt along the curve $x=\sqrt{t}$ we must supply heat at $x=0$ given by

$$\phi(t)=\sum_{n=1}^{\infty}\frac{n!}{(2n)!}=\text{constant}. \tag{4.117}$$

We now want to extend (4.115), (4.116) to the case of the two-dimensional inverse Stefan problem (Colton [28], [31]). In this case (4.113a)–(4.114b) is replaced by

$$\frac{\partial^2 u}{\partial x^2}+\frac{\partial^2 u}{\partial y^2}=\frac{\partial u}{\partial t} \quad \text{in} \quad \Phi(x, y, t)<0 \tag{4.118a}$$

$$u|_{\partial D\times[0,T]}=\gamma(x, y, t) \tag{4.118b}$$

$$u|_{\Phi=0}=0 \tag{4.118c}$$

$$\left.\frac{\partial u}{\partial \nu}\right|_{\Phi=0}=\left.\frac{1}{|\nabla\Phi|}\frac{\partial \Phi}{\partial t}\right|_{\Phi=0} \tag{4.118d}$$

where u is the temperature in the water, $\Phi=0$ is the interphase boundary, D is the region originally filled with ice, γ is the temperature applied to the boundary ∂D of D, ν is the normal with respect to the space variables that point into the water region $\Phi<0$, and ∇ denotes the gradient with respect to the space variables. The inverse Stefan problem is to find u and γ given Φ. We note that a solution to the two-dimensional inverse Stefan problem will exist only if Φ lies in a class of functions such that the solution of the noncharacteristic Cauchy problem (4.118a), (4.118c), (4.118d) exists in a region containing $\bar{D}\times[0, T]\cap\{(x, y, t): \Phi(x, y, t)<0\}$, i.e. no singularities appear in the water region. We now proceed to define a suitable class of admissible functions Φ. Let D_t, $0\leq t\leq T$, be a family of simply connected domains which depend analytically on a parameter t such that $\bigcup_{0\leq t\leq T} D_t$ contains $\bar{D}\times[0, T]\cap\{(x, y, t): \Phi(x, y, t)<0\}$. Let $z=\phi(\zeta, t)$ conformally map the unit disk Ω onto D_t such that the image of $(-1, 1)$ intersects D and for $\zeta^*\in\Omega$, $0\leq t\leq T$, define $\bar{\phi}(\zeta^*, t)$ by $\bar{\phi}(\zeta^*, t)=\overline{\phi(\bar{\zeta}^*, t)}$ where the bars denote conjugation. Now set $z^*=\bar{\phi}(\zeta^*, t)$ and note that $z^*=\bar{z}$ if and only if $\zeta^*=\bar{\zeta}$. We now define the function Φ for (possibly) complex values of x, y and t by

$$\Phi(x, y, t)=(2i)^{-1}[\phi^{-1}(z, t)-\bar{\phi}^{-1}(z^*, t)] \tag{4.119}$$

where $z=x+iy$, $z^*=x-iy$. Noting that $z^*=\bar{z}$ if and only if x and y are real it is seen that $\Phi=0$ corresponds to $\operatorname{Im}\zeta=0$, i.e. the interval $(-1, 1)$ in the complex ζ plane. Similarly, the region $\Phi<0$ corresponds to $\operatorname{Im}\zeta<0$, i.e. the part of Ω which lies in the lower half plane. We finally note that on $\Phi=0$ we have that $\partial\Phi/\partial t$ can be analytically continued for each fixed t to an analytic function of z for z in D_t.

We shall now construct a solution to (4.118a)–(4.118d) with Φ (as given by (4.119)) as a free boundary. Let x and y be considered as independent complex variables and define the transformation of C^2, the space of two complex variables, into itself by

$$z = x + iy, \qquad z^* = x - iy. \tag{4.120}$$

By the Cauchy–Kowalewski theorem we know that in a neighborhood of $\Phi = 0$ there exists a unique analytic solution of (4.118a), (4.118b), (4.118c) and hence in this neighborhood we can rewrite (4.118a) as

$$L[U] = \frac{\partial^2 U}{\partial z\, \partial z^*} - \frac{1}{4}\frac{\partial U}{\partial t} = 0 \tag{4.121}$$

where

$$U(z, z^*, t) = u\left(\frac{z+z^*}{2}, \frac{z-z^*}{2i}, t\right). \tag{4.122}$$

Now let V be the fundamental solution of the adjoint equation

$$M[V] = \frac{\partial^2 V}{\partial z\, \partial z^*} + \frac{1}{4}\frac{\partial V}{\partial t} = 0 \tag{4.123}$$

defined by

$$V(z, z^*, t; \xi, \bar{\xi}, \tau) = \frac{1}{t-\tau} \exp\left\{\frac{(z-\xi)(z^*-\bar{\xi})}{4(t-\tau)}\right\} \tag{4.124}$$

where $\xi = \xi_1 + i\xi_2$, $\bar{\xi} = \xi_1 - i\xi_2$. Note that V satisfies the Goursat data

$$\begin{aligned} V(z, \bar{\xi}, t; \xi, \bar{\xi}, \tau) &= 1/(t-\tau) \\ V(\xi, z^*, t; \xi, \bar{\xi}, \tau) &= 1/(t-\tau). \end{aligned} \tag{4.125}$$

We shall obtain the solution of the inverse Stefan problem by using Stokes' theorem to integrate $VL[U] - UM[V]$ over a torus lying in the space of three complex variables. First assume $(\xi, \bar{\xi}, \tau)$ lies in the known region of analyticity of U and for t on the circle $|t-\tau| = \delta$ let $G(t)$ be a cell whose boundary consists of a curve $C(t)$ lying on the surface $\phi^{-1}(z, t) = \bar{\phi}^{-1}(z^*, t)$ and line segments lying on the characteristic planes $z = \xi$ and $z^* = \bar{\xi}$, respectively, which join the point $(\xi, \bar{\xi})$ to $C(t)$. For $(\xi, \bar{\xi})$ sufficiently near $\Phi = 0$, τ real, and δ sufficiently small, $G(t) \times |t-\tau| = \delta$ lies in the region of analyticity of U. Now use Stokes' theorem to integrate the identity

$$VL[U] - UM[V] = (\tfrac{1}{2}VU_{z^*} - \tfrac{1}{2}V_{z^*}U)_z + (\tfrac{1}{2}VU_z - \tfrac{1}{2}V_zU)_{z^*} - (\tfrac{1}{4}VU)_t \tag{4.126}$$

over the torus $G(t) \times |t-\tau| = \delta$, making use of (4.125) and (4.118c) to obtain

$$
\begin{aligned}
u(\xi_1, \xi_2, \tau) &= U(\xi, \bar{\xi}, \tau) \\
&= \frac{1}{4\pi i} \oint_{|t-\tau|=\delta} \int_{C(t)} [VU_z \, dz - VU_{z^*} \, dz^*] \, dt \\
&= \frac{1}{4\pi} \oint_{|t-\tau|=\delta} \int_{C(t)} \frac{1}{t-\tau} \exp\left[\frac{(z-\xi)(z^*-\bar{\xi})}{4(t-\tau)}\right] \\
&\quad \cdot \frac{1}{|\nabla \Phi|} \frac{\partial \Phi}{\partial t} |dz| \, dt. \qquad (4.127)
\end{aligned}
$$

From the definition of Φ in (4.119) it is seen that (4.127) is valid for (ξ_1, ξ_2, τ) in $\bar{D} \times [0, T]$, i.e. (4.127) provides an analytic continuation of u from its original domain of definition in a neighborhood of $\Phi = 0$ to all of $\bar{D} \times [0, T]$. Computing the residue in (4.127) now gives

$$
u(\xi_1, \xi_2, \tau) = \frac{i}{2} \sum_{n=0}^{\infty} \frac{1}{4^n (n!)^2} \frac{\partial^n}{\partial \tau^n} \left\{ \int_{C(\tau)} \frac{(z-\xi)^n (z^*-\bar{\xi})^n}{|\nabla \Phi|} \frac{\partial \Phi}{\partial \tau} |dz| \right\}.
$$

(4.128)

Equation (4.128) gives the desired generalization to two (space) dimensions of the series solution (4.116) for the one-dimensional inverse Stefan problem. We note that the integral in (4.128) is pure imaginary.

Exercises

(1) Show that if the coefficients of (4.4) are analytic for $(z, z^*) \in D \times D^*$, where D and D^* are simply connected domains in the z and z^* planes, respectively, then the Riemann function for (4.1) exists and is an analytic function of its independent complex variables in $D \times D^* \times D \times D^*$. Use this result in conjunction with a conformal mapping to establish a reflection principle for solutions of (4.4) satisfying Dirichlet or a first order boundary condition along an analytic arc.

(2) Establish a reflection principle for solutions u of the biharmonic equation $\Delta_n^2 u = 0$ across a plane on which the Dirichlet and Neumann data vanish.

(3) Let u be a solution of the two-dimensional Laplace's equation which satisfies the free boundary conditions $u = 0$, $\partial u / \partial \nu = 1$ along a sufficiently smooth arc C, where ν denotes the positive normal to C. Show that C must be analytic.

(4) Show that (4.47) has a unique analytic solution in the cone $\{(\xi, \eta): \xi \leq \eta, \eta \leq \log a \text{ or } \xi \geq \eta, \eta \geq \log a\}$.

(5) Show that if $u \in C^2(D \setminus \bar{S}) \cap C(D \setminus S)$ is any solution of (4.38), (4.40)

then u can be represented in the form (4.41) for some harmonic function satisfying (4.43).

(6) Use the change of variables (4.58) to establish a reflection principle and generalized reflection principle for solutions of (4.57) satisfying an analytic boundary condition on an analytic arc $x = s(t)$.

(7) Show that every solution $u \in C^2(R^+) \cap C^1(R^+ \cup \sigma)$ of (4.60), (4.101) can be represented in the form (4.106).

(8) Show that every solution of (4.109) that is analytic in some neighborhood of the origin is an even function of r and is uniquely determined by its values on $r = 0$.

(9) Show by constructing an example that the solution of the inverse Stefan problem is improperly posed in the real domain.

5

Runge's theorem

We have already encountered Runge's theorem for elliptic partial differential equations in Chapter 3, where it was shown that the uniqueness of the solution to Cauchy's problem is equivalent to the Runge approximation property for elliptic equations. As a consequence of this result, a complete family of solutions can be obtained in many cases by separation of variables (see the Exercises to Chapter 3) and an approximate solution to the Dirichlet problem can be found by matching a linear combination of functions taken from this family to the boundary data. In this chapter we shall continue this line of investigation and first obtain a constructive method for obtaining a complete family of solutions to elliptic equations in two independent variables with analytic coefficients. This is followed by a brief description of the Bergman kernel function (Bergman [13]; Bergman and Schiffer [15]) and the development of Runge approximation theorems for certain classes of parabolic equations. We conclude by considering the problem of constructing a complete family of solutions for the Helmholtz equation in an exterior domain. Numerical examples of the use of a complete family of solutions to solve boundary value problems for partial differential equations can be found in Bergman and Herriot [14], Chang and Colton [25], Eisenstat [50], Gilbert and Linz [76], Schryer [157], Wilton and Mittra [181], and Yasuura [182].

5.1 Elliptic equations and the kernel function

Let D be a bounded simply connected domain in the plane containing the origin and let $u \in C^2(D)$ be a solution of the elliptic equation

$$L[u] \equiv \frac{\partial^2 u}{\partial x^2} + \frac{\partial^2 u}{\partial y^2} + a(x, y)\frac{\partial u}{\partial x} + b(x, y)\frac{\partial u}{\partial y} + c(x, y)u = 0 \tag{5.1}$$

defined in D where for the sake of simplicity we assume that the coefficients of L are real-valued entire functions of their independent complex variables. Let D_0 be a compact subset of D. We want to construct a set of solutions $\{u_n\}$ to $L[u]=0$ such that for any $\varepsilon>0$ there exists an integer N and constants $a_0, \ldots, a_N$ such that

$$\max_{D_0} \left| u - \sum_{n=0}^{N} a_n u_n \right| < \varepsilon. \tag{5.2}$$

The set $\{u_n\}$ is then said to be complete in the maximum norm over compact subsets of D. The problem of when the set $\{u_n\}$ is complete over D (i.e. we can replace D_0 by D in (5.2)) is more difficult (Colton [28]; Vekua [172]). However, if ∂D is smooth enough and u is continuous in $\bar{D}$, it is possible to prove that the set $\{u_n\}$ is complete over D. For example, if ∂D is analytic then u can be approximated by a solution u_1 assuming analytic boundary data, and from the results of Chapter 3 we can continue u_1 across ∂D into a domain $D_1 \supset D$. The completeness of $\{u_n\}$ over D now follows from the completeness of $\{u_n\}$ over compact subsets of D_1. If ∂D is not analytic, but sufficiently smooth, other methods are available to arrive at the same result (cf. Section 2 of this chapter). In what follows we shall only construct a set $\{u_n\}$ that is complete over compact subsets of D and assume without explicitly mentioning it that when we want $\{u_n\}$ to be complete over D that ∂D is smooth enough such that this is the case.

To construct our complete family $\{u_n\}$ we shall first need a few preliminary results concerning (5.1) or its complex form

$$L^*[U] \equiv \frac{\partial^2 U}{\partial z\, \partial z^*} + A(z, z^*)\frac{\partial U}{\partial z} + B(z, z^*)\frac{\partial U}{\partial z^*} + C(z, z^*)U = 0 \tag{5.3}$$

where A, B and C are defined as in (4.6). Let

$$\Lambda(z, z^*) = \exp\left\{ -\int_0^z B(\zeta, 0)\, d\zeta - \int_0^{z^*} A(0, \zeta^*)\, d\zeta^* + \int_0^z \int_0^{z^*} [A(\zeta, \zeta^*)B(\zeta, \zeta^*) - C(\zeta, \zeta^*)]\, d\zeta^*\, d\zeta \right\} \tag{5.4}$$

and set $U = \Lambda \tilde{U}$. Then $\tilde{U}$ satisfies

$$\frac{\partial^2 \tilde{U}}{\partial z\, \partial z^*} + \tilde{A}(z, z^*)\frac{\partial \tilde{U}}{\partial z} + \tilde{B}(z, z^*)\frac{\partial \tilde{U}}{\partial z^*} + \tilde{A}(z, z^*)\tilde{B}(z, z^*)\tilde{U} = 0 \tag{5.5}$$

where

$$\tilde{A}(z, z^*) = \int_0^z h(\zeta, z^*)\, d\zeta, \qquad \tilde{B}(z, z^*) = \int_0^z k(z, \zeta^*)\, d\zeta^* \tag{5.6}$$

with

$$h = \partial A/\partial z^* + AB - C, \qquad k = \partial B/\partial z^* + AB - C, \tag{5.7}$$

i.e. $\tilde{A}(0, z^*) = \tilde{B}(z, 0) = 0$. Hence we can assume that the coefficients of (5.3) satisfy this condition to begin with. The second result we shall need is that the Riemann function for $L[u] = 0$ takes real values when $z^* = \bar{z}$, $\zeta^* = \bar{\zeta}$. To see this note that since the coefficients of L are real we have

$$\operatorname{Im} R(\zeta, \bar{\zeta}; z, \bar{z}) = (2i)^{-1}[R(\zeta, \bar{\zeta}; z, \bar{z}) - \bar{R}(\bar{\zeta}, \zeta; \bar{z}, z)] \tag{5.8}$$

where $\bar{R}(\zeta, \bar{\zeta}; z, \bar{z}) = \overline{R(\bar{\zeta}, \zeta; \bar{z}, z)}$ is a solution of $L[u] = 0$. Extending (5.8) into the complex domain and evaluating along the characteristic $z = 0$ gives

$$\operatorname{Im} R(\zeta, \bar{\zeta}; 0, z^*) = \frac{1}{2i}\left[\exp\left(-\int_{\bar{\zeta}}^{z^*} A(\zeta, \tau)\, d\tau\right) - \exp\left(-\int_{\bar{\zeta}}^{z^*} \bar{B}(\sigma, \zeta)\, d\sigma\right)\right] = 0 \tag{5.9}$$

since $A(z, \bar{z}) = \overline{B(z, \bar{z})} = \bar{B}(\bar{z}, z)$. Similarly, $\operatorname{Im} R(\zeta, \bar{\zeta}; z, 0) = 0$, and since an elementary power series analysis shows that solutions of $L^*[U] = 0$ are uniquely determined by their data on $z^* = 0$ and $z = 0$, we have $\operatorname{Im} R = 0$ for $z^* = \bar{z}$, $\zeta^* = \bar{\zeta}$.

We can now prove the following result.

Theorem 5.1 *Let u be a real-valued solution of $L[u] = 0$ in D. Then*

$$u(x, y) = \operatorname{Re}\left[H_0(z)\phi(z) + \int_0^z H(z, \zeta)\phi(\zeta)\, d\zeta\right]$$

where

$$H_0(z) = R(z, 0; z, \bar{z}), \qquad \phi(z) = 2U(z, 0) - U(0, 0),$$

$$H(z, \zeta) = -\frac{\partial}{\partial \zeta} R(\zeta, 0; z, \bar{z}), \qquad U(z, z^*) = u\left(\frac{z + z^*}{2}, \frac{z - z^*}{2i}\right).$$

Proof From (4.19) and the Bergman–Vekua theorem we have

$$U(z, \bar{z}) = \alpha_0 R(0, 0; z, z^*) + \int_0^z f(\zeta) R(\zeta, 0; z, z^*)\, d\zeta + \int_0^{z^*} g(\zeta^*) R(0, \zeta^*; z, z^*)\, d\zeta^* \tag{5.10}$$

where

$$\alpha_0 = U(0, 0), \qquad f(z) = \frac{\partial U(z, 0)}{\partial z}, \qquad g(z^*) = \frac{\partial U(0, z^*)}{\partial z^*}, \tag{5.11}$$

and we have made use of the fact that $A(0, z^*) = B(z, 0) = 0$. Integrating

(5.10) by parts now gives

$$\begin{aligned}U(z, z^*) &= -U(0,0)R(0,0; z, z^*) + U(z,0)R(z,0; z, z^*)\\&\quad - \int_0^z U(\zeta, 0)\frac{\partial R}{\partial \zeta}(\zeta, 0; z, z^*)\,d\zeta + U(0, z^*)R(0, z^*; z, z^*)\\&\quad - \int_0^z U(0, \zeta^*)\frac{\partial R}{\partial \zeta}(0, \zeta^*; z, z^*)\,d\zeta^*\\&= R(z, 0; z, z^*)\phi(z) - \int_0^z \phi(\zeta)\frac{\partial R}{\partial \zeta}(\zeta, 0; z, z^*)\,d\zeta\\&\quad + R(0, z^*; z, z^*)\phi(z^*) - \int_0^{z^*} \phi(\zeta^*)\frac{\partial R}{\partial \zeta^*}(0, \zeta^*; z, z^*)\,d\zeta^*\end{aligned} \tag{5.12}$$

where

$$\begin{aligned}\phi(z) &= U(z,0) - \tfrac{1}{2}U(0,0)\\\phi(z^*) &= U(0,z^*) - \tfrac{1}{2}U(0,0).\end{aligned} \tag{5.13}$$

Since u is real-valued we have $U(z, \bar{z}) = \overline{U(z, \bar{z})}$ and hence $\phi(z) = \overline{\phi(\bar{z})}$. But from (5.8) we have $R(\zeta, 0; z, \bar{z}) = \bar{R}(0, \zeta; \bar{z}, z) = \overline{R(0, \bar{\zeta}; z, \bar{z})}$ and hence the theorem follows from (5.12).

From Theorem 5.1 and using Runge's theorem in analytic function theory we now have the following corollary.

Corollary 5.1 *Let u be a real-valued solution of $L[u]=0$ in D and let*

$$u_{2n}(x, y) = \text{Re}\left[H_0(z)z^n + \int_0^z H(z, \zeta)\zeta^n\,d\zeta\right]$$

$$u_{2n+1}(x, y) = \text{Im}\left[H_0(z)z^n + \int_0^z H(z, \zeta)\zeta^n\,d\zeta\right].$$

Then the set $\{u_n\}$ is complete in the maximum norm over compact subsets of D.

Having constructed our complete set $\{u_n\}$ there is now a variety of methods that can be used to approximate solutions of the Dirichlet problem for $L[u]=0$ (as well as other boundary value problems for this equation). Assuming that ∂D is sufficiently smooth, for example, we can orthonormalize the set $\{u_n\}$ in the L_2 norm over ∂D to obtain the complete set $\{\phi_n\}$. Then if $u = f$ on ∂D we can approximate u by a sum of the form

$$u_N = \sum_{n=0}^{N} c_n\phi_n \tag{5.14}$$

where

$$c_n = \int_{\partial D} f\phi_n \, ds. \tag{5.15}$$

Since each ϕ_n is a solution of $L[u]=0$, error estimates can be found by finding the maximum of $|f-u_N|$ on ∂D and then applying the maximum principle (assuming $c(x, y) \leq 0$).

Instead of using the $L_2(\partial D)$ inner product to arrive at the approximate solution u_N, one can use a variety of other inner products, for example the Dirichlet inner product defined by

$$(f, u)_D = \int_{\partial D} f(\partial u/\partial \nu) \, ds \tag{5.16}$$

where ν is the unit outward normal to ∂D. This choice has certain attractive properties and leads to the concept of the Bergman kernel function for elliptic equations (Bergman and Schiffer [15]). We shall illustrate this development by considering the elliptic equation

$$\partial^2 u/\partial x^2 + \partial^2 u/\partial y^2 + q(x, y)u = 0 \tag{5.17}$$

where $q(x, y) < 0$ is a real-valued entire function of x and y and $u \in C^2(D) \cap C^1(\bar{D})$. We assume that ∂D is smooth enough such that our complete set $\{u_n\}$ can approximate u and its first derivatives along ∂D (see Colton [28] and Vekua [172] for sufficient conditions for this to be valid). Let N and G be the Neumann and Green functions, respectively, for (5.17) in D. Then the *kernel function* of (5.17) in D is defined by

$$K(\mathbf{x}; \boldsymbol{\xi}) = N(\mathbf{x}; \boldsymbol{\xi}) - G(\mathbf{x}; \boldsymbol{\xi}). \tag{5.18}$$

Note that since the singularities of the singular parts of N and G cancel we have that K is regular in D both as a function of $\mathbf{x} = (x, y)$ and $\boldsymbol{\xi} = (\xi, \eta)$. Furthermore, due to the symmetry of the Neumann and Green functions we have

$$K(\mathbf{x}; \boldsymbol{\xi}) = K(\boldsymbol{\xi}; \mathbf{x}). \tag{5.19}$$

From the boundary conditions satisfied by the Neumann and Green functions we have

$$\begin{aligned} K(\mathbf{x}; \boldsymbol{\xi}) &= N(\mathbf{x}; \boldsymbol{\xi}); \qquad \mathbf{x} \in \partial D \\ \frac{\partial K}{\partial \nu}(\mathbf{x}; \boldsymbol{\xi}) &= -\frac{\partial G}{\partial \nu}(\mathbf{x}; \boldsymbol{\xi}); \qquad \mathbf{x} \in \partial D. \end{aligned} \tag{5.20}$$

Hence, if $u \in C^2(D) \cap C^1(\bar{D})$ is a solution of (5.17) we have from Green's formulae that

$$u(\boldsymbol{\xi}) = \int_{\partial D} u(\mathbf{x}) \frac{\partial K(\mathbf{x}; \boldsymbol{\xi})}{\partial \nu} \, ds, \qquad u(\boldsymbol{\xi}) = \int_{\partial D} K(\mathbf{x}; \boldsymbol{\xi}) \frac{\partial u(\mathbf{x})}{\partial \nu} \, ds. \tag{5.21}$$

If we define the inner product $(\cdot, \cdot)_D$ by

$$(u, v)_D = \iint_D [u_x v_x + u_y v_y + quv]\, dx\, dy \tag{5.22}$$

where $u, v \in C^1(\bar{D})$ we see that $(\cdot, \cdot)_D$ satisfies all the conditions of an inner product. In particular, since $q(x, y) < 0$ for $(x, y) \in \bar{D}$ we have $\|u\|_D^2 = (u, u)_D = 0$ if and only if $u \equiv 0$ in D. From Green's formula we have the fact that if v is a solution of (5.17) then

$$(u, v)_D = \int_{\partial D} u(\partial v/\partial \nu)\, ds, \tag{5.23}$$

and in particular if u is also a solution of (5.17), then (5.21) can be written as the single relation

$$u = (u, K)_D \tag{5.24}$$

since both u and K are solutions of (5.17). (5.24) is known as the *reproducing property* of the kernel function. In particular, from Schwarz's inequality we have

$$\begin{aligned} |u(\boldsymbol{\xi})|^2 &= |(u, K)_D|^2 \\ &\leq \|u\|_D^2 \|K\|_D^2 \\ &= K(\boldsymbol{\xi}; \boldsymbol{\xi}) \|u\|_D^2. \end{aligned} \tag{5.25}$$

Now let $\{u_n\}$ be the complete set constructed in Corollary 5.1 and orthonormalize this set with respect to $(\cdot, \cdot)_D$ to obtain the set $\{\phi_n\}$. Assuming that ∂D is sufficiently smooth we have that for any $\varepsilon > 0$ there exists an integer N and constants $a_0, \ldots, a_N$ such that

$$\left\| u - \sum_{n=0}^{N} a_n \phi_n \right\|_D^2 < \varepsilon. \tag{5.26}$$

In particular, since the set $\{\phi_n\}$ is an orthonormal set, the optimum choice of the constants $a_0, \ldots, a_N$ is given by

$$a_n = (u, \phi_n)_D = \int_{\partial D} u \frac{\partial \phi_n}{\partial \nu}\, ds. \tag{5.27}$$

From (5.25) and (5.26) we have

$$\left| u(\boldsymbol{\xi}) - \sum_{n=0}^{N} a_n \phi_n(\boldsymbol{\xi}) \right|^2 < \varepsilon K(\boldsymbol{\xi}; \boldsymbol{\xi}) \tag{5.28}$$

and hence the series

$$u(\mathbf{x}) = \sum_{n=0}^{\infty} a_n \phi_n(\mathbf{x}), \qquad a_n = (u, \phi_n)_D \tag{5.29}$$

converges uniformly to u in every closed subdomain D_0 of D. In particular, setting $u = K$ we have from the reproducing property that

$$(\phi_n, K) = \phi_n \tag{5.30}$$

and hence for $\mathbf{x}$ and $\boldsymbol{\xi}$ on compact subsets of D we have the remarkable representation

$$K(\mathbf{x}; \boldsymbol{\xi}) = \sum_{n=0}^{\infty} \phi_n(\mathbf{x})\phi_n(\boldsymbol{\xi}). \tag{5.31}$$

Note that the representation (5.31) is in fact independent of the particular orthonormal system $\{\phi_n\}$ we started out with.

The kernel function can of course also be constructed in more than two space variables, provided either that a complete family of solutions is available from which we can construct the orthonormal system $\{\phi_n\}$, or that the Green and Neumann functions are known. In the case of second order elliptic equations with constant coefficients a complete family of solutions can be obtained by separation of variables and an application of Theorem 3.4 (see the Exercises in Chapter 3).

5.2 Parabolic equations

We now want to construct a complete family of solutions for certain classes of parabolic equations. We first consider parabolic equations in one space variable

$$\partial^2 u/\partial x^2 + a(x, t)(\partial u/\partial x) + b(x, t)u = \partial u/\partial t \tag{5.32}$$

defined in a domain with moving boundaries of the form $D = \{(x, t): s_1(t) < x < s_2(t), 0 < t < t_0\}$ where s_1 and s_2 are analytic functions of t for $0 \leq t \leq t_0$. We assume that the coefficients a and b of (5.32) are continuously differentiable in a rectangle containing D and are analytic functions of t for $0 \leq t \leq t_0$. By making a change of independent variable we can assume without loss of generality that $a \equiv 0$, i.e. we can consider the equation

$$L[u] \equiv \frac{\partial^2 u}{\partial x^2} + q(x, t)u - \frac{\partial u}{\partial t} = 0 \tag{5.33}$$

where q has the same regularity properties as a and b. The construction of a complete family of solutions for (5.33) is accomplished through the use of the transformation operators obtained in Chapter 4 and the application of the reflection principle for parabolic equations.

We begin by considering the set $\{h_n\}$ of polynomial solutions to the heat equation

$$\partial^2 h/\partial x^2 = \partial h/\partial t \tag{5.34}$$

defined by

$$h_n(x, t) = n! \sum_{k=0}^{[n/2]} \frac{x^{n-2k} t^k}{(n-2k)!\, k!} = (-t)^{n/2} H_n(x/(-4t)^{1/2}) \tag{5.35}$$

where H_n denotes the Hermite polynomial of degree n (Rosenbloom and Widder [149]). Let x_0 be a positive constant, $R = \{(x, t): -x_0 < x < x_0, 0 < t < t_0\}$, and $\bar{R}$ denote the closure of R. Then we have the following result (Colton [28]).

Theorem 5.2 *Let $h \in C^2(R) \cap C(\bar{R})$ be a solution of the heat equation in R. Then given $\varepsilon > 0$ there exist an integer N and constants $a_0, a_1, \ldots, a_N$ such that*

$$\max_{\bar{R}} \left| h(x, t) - \sum_{n=0}^{N} a_n h_n(x, t) \right| < \varepsilon.$$

Proof By the Weierstrass approximation theorem and the maximum principle for the heat equation there exists a solution w_1 of (5.34) in R which assumes polynomial initial and boundary data such that

$$\max_{\bar{R}} |h(x, t) - w_1(x, t)| < \varepsilon/3. \tag{5.36}$$

Let $w_1(-x_0, t) = \sum_{m=0}^{M} b_m t^m$ and $w_1(x_0, t) = \sum_{m=0}^{M} c_m t^m$ and look for a solution of (5.34) in the form

$$v(x, t) = \sum_{m=0}^{M} v_m(x) t^m \tag{5.37}$$

where $v(-x_0, t) = w_1(-x_0, t)$, $v(x_0, t) = w_1(x_0, t)$. Substituting (5.37) into (5.34) leads to the following recursion scheme for the v_m:

$$\begin{aligned} v_M'' &= 0; & v_M(-x_0) &= b_M, & v_M(x_0) &= c_M \\ v_{M-1}'' &= m v_m; & v_{m-1}(-x_0) &= b_{m-1}, & v_{m-1}(x_0) &= c_{m-1}, \end{aligned} \tag{5.38}$$

for $m = 1, 2, \ldots, M$. Hence, each v_m is a uniquely determined polynomial. Now consider $w_2 = w_1 - v$. By the method of separation of variables it is seen that there exist constants $d_1, \ldots, d_L$ such that

$$\max_{\bar{R}} \left| w_2(x, t) - \sum_{l=1}^{L} d_l \sin \frac{l\pi}{2x_0}(x + x_0) \exp\left(-\frac{l^2 \pi^2 t}{4x_0^2}\right) \right| < \frac{\varepsilon}{3}. \tag{5.39}$$

Hence there exists a solution w_3 of the heat equation which is an entire function of the complex variables x and t such that

$$\max_{\bar{R}} |h(x, t) - w_3(x, t)| < \frac{2\varepsilon}{3}. \tag{5.40}$$

We can now represent w_3 in the form (4.92), and by truncating the Taylor series expansions of the Cauchy data for w_3 we can conclude that there exists an entire solution w_4 of the heat equation satisfying polynomial Cauchy data on $x=0$ such that

$$\max_{\bar{R}} |w_3(x, t) - w_4(x, t)| < \frac{\varepsilon}{3}. \tag{5.41}$$

But from (5.35) and Holmgren's uniqueness theorem it is seen that there exist positive constants $a_0, \ldots, a_N$ such that

$$w_4(x, t) = \sum_{k=0}^{N} a_n h_n(x, t), \tag{5.42}$$

and the proof of the theorem now follows.

From the above theorem and the boundness of the transformation operator $\mathbf{I}+\mathbf{T}_3$ of Chapter 4 we can now immediately arrive at the following corollary.

Corollary 5.2 *Let $u \in C^2(R) \cap C(\bar{R})$ be a solution of (5.33) in R. Then given $\varepsilon > 0$ there exist an integer N and constants $a_1, \ldots, a_N$ such that*

$$\max_{\bar{R}} \left| u(x, t) - \sum_{n=0}^{N} a_n u_n(x, t) \right| < \varepsilon$$

where $u_n = (\mathbf{I}+\mathbf{T}_3)[h_n]$.

We now want to use the reflection principle for parabolic equations (as extended by Exercise 6 of Chapter 4) to show that the set $\{u_n\}$ defined in Corollary 5.1 is complete in $\bar{D} = \{(x, t): s_1(t) \leq x \leq s_2(t),\ 0 \leq t \leq t_0\}$ under the assumption that s_1 and s_2 are analytic functions of t for $0 \leq t \leq t_0$ and $s_1(t) < s_2(t)$ for $0 \leq t \leq t_0$ (Colton [28]).

Theorem 5.3 *Let $u \in C^2(D) \cap C(\bar{D})$ be a solution of (5.33) in D. Then given $\varepsilon > 0$ there exist an integer N and constants $a_0, \ldots, a_N$ such that*

$$\max_{\bar{D}} \left| u(x, t) - \sum_{n=0}^{N} a_n u_n(x, t) \right| < \varepsilon.$$

Proof It suffices to show that there exists a solution w of (5.33) defined in a rectangle $R \supset \bar{D}$ such that

$$\max_{\bar{D}} |u(x, t) - w(x, t)| < \varepsilon, \tag{5.43}$$

since in this case we can apply Corollary 5.2 to w. From the existence theory for solutions of initial-boundary value problems for parabolic equations and the maximum principle we can approximate the boundary data for u by analytic functions and construct a solution w having these boundary data and satisfying (5.43). From the reflection principle for parabolic equations we can continue w across the arc $x = s_1(t)$ into the region bounded by the characteristics $t = t_0$, $t = 0$, and the analytic curves $x = 2s_1(t) - s_2(t)$, $x = s_2(t)$ (assuming the coefficient q of the differential equation is regular in this region). Applying the reflection principle a second time, but this time continuing w across the arc $x = s_2(t)$, shows that w can be continued into the region bounded by $t = t_0$, $t = 0$, $x = 2s_1(t) - s_2(t)$, and $x = 3s_2(t) - 2s_1(t)$. Due to the fact that $s_1(t) < s_2(t)$ for $0 \leq t \leq t_0$, it is seen that by repeating the above procedure we can continue w into a rectangle R containing $\bar{D}$, and this completes the proof of the theorem.

As a special case of the above theorem, we have the following corollary which generalizes Theorem 5.2 and is the exact analogue of Runge's theorem in analytic function theory for solutions of the heat equation in the sense that in both cases we have a polynomial approximation.

Corollary 5.3 *Let $h \in C^2(D) \cap C(\bar{D})$ be a solution of the heat equation in D. Then given $\varepsilon > 0$ there exist an integer N and constants $a_0, \ldots, a_N$ such that*

$$\max_{\bar{D}} \left| h(x, t) - \sum_{n=0}^{N} a_n h_n(x, t) \right| < \varepsilon.$$

As a second corollary to Theorem 5.3 we want to give conditions on the separation constants λ_n such that the solutions of the heat equation

$$h_n^{\pm}(x, t) = \exp(\pm i\lambda_n x - \lambda_n^2 t) \tag{5.44}$$

form a complete family of solutions to the heat equation (in the sense of Corollary 5.3). From Corollary 5.3 it suffices to show that every heat polynomial (defined by (5.35)) can be approximated by a linear combination of the functions (5.44), subject to certain restrictions on the constants λ_n. From the representation (5.92) we see that it suffices to show that the set $\{\exp(-\lambda_n^2 t)\}$ is complete for analytic functions defined in some ellipse containing $[0, t_0]$. The type of restriction necessary is indicated by the following theorem in entire functions of a complex variable (Levin [115], p. 219).

Theorem 5.4 *If $\{\mu_n\}$ is a sequence of distinct complex numbers for which the limit*

$$d=\lim_{n\to\infty} n/\mu_n>0$$

exists, the system $\{\exp(\mu_n z)\}$ is complete in the space of analytic functions defined in any simply connected domain G for which every straight line parallel to the imaginary axis cuts out a segment of length less than $2\pi d$, and the system is not complete in any region which contains a segment of length $2\pi d$ parallel to the imaginary axis.

From the above theorem we now see that the set (5.44) is complete for solutions of the heat equation defined in D provided

$$\lim_{n\to\infty} n/\lambda_n^2>0. \tag{5.45}$$

Corollary 5.4 *Let $h\in C^2(D)\cap C(\bar{D})$ be a solution of the heat equation in D. Then given $\varepsilon>0$ there exist an integer N and constants $a_0^\pm,\ldots,a_N^\pm$ such that*

$$\max_{\bar{D}}\left|h(x,t)-\sum_{n=0}^{N} a_n^\pm h_n^\pm(x,t)\right|<\varepsilon$$

where $h_n^\pm$ are defined by (5.44), (5.45).

Given a complete family of solutions we can approximate the solution of initial-boundary value problems for parabolic equations in the same manner as for elliptic equations by first orthonormalizing the set with respect to the L_2 norm over the base and sides of the domain of definition and then approximating the solution by a truncated Fourier series of these orthonormal functions (Chang and Colton [25]).

We now want to extend Corollary 5.3 to the case where h is a solution of the n-dimensional heat equation

$$\Delta_n h=h_t \tag{5.46}$$

defined in a cylindrical domain $D\times[0,T]$ where D is a bounded simply connected domain in R^n with ∂D in class C^{2i+2} where $i=1+[\frac{1}{4}n+\frac{1}{2}]$. In particular, we shall show that the set $\{h_m\}$ where

$$\begin{gathered} h_m(\mathbf{x},t)=h_{m_1}(x_1,t)h_{m_2}(x_2,t)\cdots h_{m_n}(x_n,t)\\ m=(m_1,\ldots,m_n),\qquad \mathbf{x}=(x_1,\ldots,x_n)\end{gathered} \tag{5.47}$$

is a complete family of solutions with respect to the maximum norm for solutions $h\in C^2(D\times(0,T))\times C(\bar{D}\times[0,T])$ of (5.46), where h_{m_i} denotes a heat polynomial defined by (5.35) (Colton and Watzlawek [41]). Corollary 5.4 can also be extended to higher dimensions and we leave this as an exercise for the reader (see the Exercises). The set (5.47) is particularly useful for approximating solutions of initial-boundary value problems for (5.46) since the set is simple and easily calculated. Furthermore, since each h_m is a solution of (5.46) we need only minimize a linear combination of these functions over the base and sides of $D\times(0, T)$, thus reducing the dimensionality of the approximation scheme by one, an important consideration in the numerical solution of higher dimensional initial-boundary value problems. For other versions of Runge's theorem for (5.46) we refer the reader to Jones [97] and Magenes [120], [121].

We first note that without loss of generality we can assume that $h(\mathbf{x}, 0)=0$. This follows from the fact that by the maximum principle and the Weierstrass approximation theorem we can approximate h in the maximum norm over $\bar{D}\times[0, T]$ by a solution $h^{(1)}$ of (5.46) such that $h^{(1)}(\mathbf{x}, 0)$ is a polynomial. From the definition of h_m it is seen that there exist constants a_m and an integer M such that

$$h^{(1)}(\mathbf{x}, 0)=\sum_{|m|\leq M} a_m h_m(\mathbf{x}, 0), \qquad |m|=m_1+m_2+\cdots+m_n. \tag{5.48}$$

To approximate h by a linear combination of the h_m it therefore suffices to approximate

$$h^{(2)}(\mathbf{x}, t)=h^{(1)}(\mathbf{x}, t)-\sum_{|m|\leq M} a_m h_m(\mathbf{x}, t) \tag{5.49}$$

where $h^{(2)}(\mathbf{x}, 0)=0$. Hence without loss of generality we can assume $h(\mathbf{x}, 0)=0$ to begin with.

Lemma 5.1 *Let $h(\mathbf{x}, 0)=0$ and $\varepsilon>0$. Then there exists a bounded simply connected domain $D_1\supset\bar{D}$ with ∂D_1 in class C^{2i+2} with $i=1+[\frac{1}{4}n+\frac{1}{2}]$, and a solution $h^{(0)}\in C^2(D_1\times(-1, T))\cap C(\bar{D}_1\times[-1, T])$ of (5.46) in $D_1\times(-1, T)$ such that*

$$\max_{\bar{D}\times[0,T]} |h(\mathbf{x}, t)-h^{(0)}(\mathbf{x}, t)|<\varepsilon.$$

Proof We first extend h to $D\times[-1, T]$ by constructing a solution $h^{(1)}\in C^2(D\times(-1, T)\cap C(\bar{D}\times[-1, T])$ of (5.46) such that

$$\begin{aligned} h^{(1)}(\mathbf{x}, t)&=\begin{cases} h(\mathbf{x}, t) & \text{for } \mathbf{x}\in\partial D, \quad t\geq 0\\ 0 & \text{for } \mathbf{x}\in\partial D, \quad t\leq 0\end{cases}\\ h^{(1)}(\mathbf{x}, -1)&=0.\end{aligned} \tag{5.50}$$

The existence of $h^{(1)}$ follows from the standard existence theorems for solutions of parabolic initial-boundary value problems (cf. Friedman [57]). Note that from the maximum principle for the heat equation we have $h^{(1)}=h$ on $\bar{D}\times[0, T]$ and $h^{(1)}=0$ on $\bar{D}\times[-1, 0]$. Now let ∂D_1 be the surface obtained from ∂D by moving ∂D a distance $d>0$ in the direction of the outer normal, where d is small enough such that ∂D_1 is not self-intersecting. Let D_1 be the region bounded by ∂D_1. For $\mathbf{x}'\in\partial D_1$ define $f(\mathbf{x}', t)$ by $f(\mathbf{x}', t)=h^{(1)}(\mathbf{x}, t)$ where $\mathbf{x}$ is the point on ∂D associated with $\mathbf{x}'\in\partial D_1$ under the above deformation, and let $h^{(2)}\in C^2(D_1\times(-1, T))\cap C(\bar{D}_1\times[-1, T])$ be the solution of (5.46) satisfying the initial-boundary data

$$\begin{aligned} h^{(2)}(\mathbf{x}, t) &= f(\mathbf{x}, t), \qquad \mathbf{x}\in\partial D_1 \\ h^{(2)}(\mathbf{x}, -1) &= 0. \end{aligned} \tag{5.51}$$

From the Weierstrass approximation theorem, the maximum principle for the heat equation, and the existence of solutions to the first initial-boundary value problem for the heat equation we can construct a solution $h^{(0)}\in C^2(D_1\times(-1, T))\cap C(\bar{D}_1\times[-1, T])$ of (5.46) such that on $\partial D_1\times[-1, T]$ $h^{(0)}$ is the restriction of an analytic function of $\mathbf{x}$ and t, $h^{(0)}(\mathbf{x}, -1)=0$, and

$$\max_{\bar{D}_1\times[-1,T]} |h^{(0)}(\mathbf{x}, t)-h^{(2)}(\mathbf{x}, t)|<\varepsilon_1 \tag{5.52}$$

for $\varepsilon_1>0$ arbitrarily small. From the smoothness of ∂D_1 and the regularity theorems for solutions of initial-boundary value problems for parabolic equations (cf. Friedman [57]) we can conclude that there exists a positive constant C which is independent of d for $d\leqslant d_0$ such that

$$|\nabla_x h^{(0)}(\mathbf{x}, t)|\leqslant C \tag{5.53}$$

for $(\mathbf{x}, t)\in\bar{D}_1\times[-1+\delta, T]$ where $\delta>0$ is arbitrarily small. In particular, the constant C depends only on δ, d_0, D, and the boundary data of $h^{(0)}$. Hence from the mean value theorem, for $t\in[-1+\delta, T]$

$$|h^{(0)}(\mathbf{x}', t)-h^{(0)}(\mathbf{x}, t)|\leqslant Cd \tag{5.54}$$

where $|\mathbf{x}'-\mathbf{x}|=d$ and hence from (5.52) we have

$$|h^{(2)}(\mathbf{x}', t)-h^{(0)}(\mathbf{x}, t)|<\varepsilon_1+Cd \tag{5.55}$$

for $t\in[-1+\delta, T]$. But for $\mathbf{x}'\in\partial D_1$ we have $h^{(2)}(\mathbf{x}', t)=h^{(1)}(\mathbf{x}, t)$, $\mathbf{x}\in\partial D$, and hence

$$|h^{(1)}(\mathbf{x}, t)-h^{(0)}(\mathbf{x}, t)|\leqslant\varepsilon_1+Cd \tag{5.56}$$

for $\mathbf{x}\in\partial D$, $t\in[-1+\delta, T]$. We now note that from the maximum principle $h^{(1)}(\mathbf{x}, t)=h^{(2)}(\mathbf{x}, t)=0$ for $(\mathbf{x}, t)\in\bar{D}_1\times[-1, 0]$ and hence (5.52) and

(5.56) imply

$$\max_{\bar{D}\times[0,T]} |h^{(1)}(\mathbf{x}, t) - h^{(0)}(\mathbf{x}, t)| \leq \varepsilon_1 + Cd, \tag{5.57}$$

again by the maximum principle. Since d can be made arbitrarily small and C is independent of d for $d \leq d_0$, the lemma now follows, recalling that $h^{(1)}(\mathbf{x}, t) = h(\mathbf{x}, t)$ for $(\mathbf{x}, t) \in \bar{D} \times [0, T]$.

Lemma 5.2 *Let $h(\mathbf{x}, 0) = 0$ and $\varepsilon > 0$. Then there exists a solution $h^{(1)}$ of* (5.46) *which is an entire function of its independent complex variables such that*

$$\max_{\bar{D}\times[0,T]} |h(\mathbf{x}, t) - h^{(1)}(\mathbf{x}, t)| < \varepsilon.$$

Proof We consider the function $h^{(0)}$ constructed in Lemma 5.1. We can assume that $h^{(0)}$ is a polynomial in t for $\mathbf{x} \in \partial D_1$, i.e.

$$h^{(0)}(\mathbf{x}, t) = \sum_{k=0}^{k_0} \psi_k(\mathbf{x}) t^k; \qquad (\mathbf{x}, t) \in \partial D_1 \times [-1, T] \tag{5.58}$$

where the ψ_k are analytic functions of $\mathbf{x}$ defined in a neighborhood of ∂D_1. We now look for a solution $w \in C^2(D_1 \times (-1, T) \cap C(\bar{D}_1 \times [-1, T])$ of (5.46) in the form

$$w(\mathbf{x}, t) = \sum_{k=0}^{k_0} w_k(\mathbf{x}) t^k \tag{5.59}$$

such that

$$w(\mathbf{x}, t) = h^{(0)}(\mathbf{x}, t); \qquad (x, t) \in \partial D_1 \times [-1, T]. \tag{5.60}$$

Substituting (5.60) into (5.46) and using (5.59) shows that

$$\begin{aligned} \Delta_n w_{k_0} &= 0; & \mathbf{x} &\in D_1 \\ w_{k_0}(\mathbf{x}) &= \psi_{k_0}(\mathbf{x}); & \mathbf{x} &\in \partial D_1 \end{aligned} \tag{5.61a}$$

$$\begin{aligned} \Delta_n w_k &= (k+1) w_{k+1}; & \mathbf{x} &\in D_1 \\ w_k(\mathbf{x}) &= \psi_k(\mathbf{x}); & \mathbf{x} &\in \partial D_1 \end{aligned} \tag{5.61b}$$

for $k = 0, 1, \ldots, k_0 - 1$. The existence of the functions w_k follows from the analyticity of the ψ_k, the regularity of ∂D_1, and the existence of solutions to the Dirichlet problem for the Poisson equation. Hence w as defined by (5.59), (5.60) exists. Now w_k is a solution of

$$\Delta_n^{k_0 - k + 1} w_k = 0 \tag{5.62}$$

for $k = 0, 1, \ldots, k_0$, $\mathbf{x} \in D_1$, and hence we can write (cf. Bergman and

Schiffer [15])

$$w_k(\mathbf{x}) = \sum_{j=1}^{k_0-k+1} r^{2(j-1)} u_j(\mathbf{x}) \tag{5.63}$$

where $r = |\mathbf{x}|$ and the u_j are harmonic functions in D_1. From the Runge approximation property for elliptic equations (Theorem 3.4) we have that each u_j can be approximated in the maximum norm on compact subsets of D_1 by a finite linear combination of harmonic polynomials. From (5.59) and (5.63) we can now conclude that there exists a solution v_0 of (5.46) that is an entire function of its independent complex variables such that

$$\max_{\bar{D}\times[-1,T]} |w(\mathbf{x}, t) - v_0(\mathbf{x}, t)| < \varepsilon \tag{5.64}$$

for $\varepsilon > 0$ arbitrarily small.

Now let $v_1 = h^{(0)} - w$ and let λ_j and ϕ_j be the eigenvalues and eigenfunctions, respectively, of the eigenvalue problem

$$\begin{aligned} \Delta_n u + \lambda u &= 0, \qquad \mathbf{x} \in D_1 \\ u(\mathbf{x}) &= 0, \qquad \mathbf{x} \in \partial D_1. \end{aligned} \tag{5.65}$$

From the expansion theorem for the above eigenvalue problem (cf. Hellwig [82]) we have that there exists an integer $j_0 = j_0(\varepsilon)$ and constants a_j, $j = 0, 1, \ldots, j_0$, such that for $\varepsilon > 0$ we have

$$\left\| v_1(\mathbf{x}, \mathbf{t}) - \sum_{j=0}^{j_0} a_j \phi_j(\mathbf{x}) e^{-\lambda_j(t+1)} \right\| < \varepsilon, \tag{5.66}$$

where $\|\cdot\|$ denotes the L_2 norm over $\partial D_1 \times [-1, T] \cup D_1$. But from an application of Schwarz's inequality to the representation of solutions to the heat equation in terms of the Green's function, we can conclude from (5.66) that

$$\max_{\bar{D}\times[0,T]} \left| v_1(\mathbf{x}, t) - \sum_{j=1}^{j_0} a_j \phi_j(\mathbf{x}) e^{-\lambda_j(t+1)} \right| < C_1 \varepsilon, \tag{5.67}$$

where C_1 is a positive constant. Applying the Runge approximation property to each of the eigenfunctions ϕ_j and approximating each ϕ_j by a finite linear combination taken from the family

$$\{r^{-(n-2)/2} J_{(n-2+2l)/2}(\lambda_j r) S_{lm}(\theta; \phi)\}$$

where $k = 0, 1, \ldots, J_\nu$ is Bessel's function and S_{lm} is a spherical harmonic, now shows that there exists a solution v_2 of (5.46) that is an entire function of its independent complex variables such that

$$\max_{\bar{D}\times[0,T]} |v_1(\mathbf{x}, t) - v_2(\mathbf{x}, t)| < \varepsilon \tag{5.68}$$

for $\varepsilon > 0$ arbitrarily small. If we now set $h^{(1)} = v_0 + v_2$ it is seen from (5.64) and (5.68) that

$$\max |h^{(0)}(\mathbf{x}, t) - h^{(1)}(\mathbf{x}, t)| < 2\varepsilon. \tag{5.69}$$

The function $h^{(0)}$ is the same $h^{(0)}$ appearing in the statement of Lemma 5.1, and hence from this lemma and (5.69) we have completed the proof of Lemma 5.2.

We are now in a position to prove the following theorem.

Theorem 5.5 *Let $h \in C^2(D \times (0, T)) \cap C(\bar{D} \times [0, T])$ be a solution of (5.46) in $D \times (0, T)$. Then for every $\varepsilon > 0$ there exist an integer M and constants a_m, $|m| \leq M$, such that*

$$\max_{\bar{D} \times [0,T]} \left| h(\mathbf{x}, t) - \sum_{|m| \leq M} a_m h_m(\mathbf{x}, t) \right| < \varepsilon$$

where the h_m are defined in (5.47).

Proof Without loss of generality we can assume $h(\mathbf{x}, 0) = 0$ (see the discussion before Lemma 5.1). From Lemma 5.2 we can also assume that h is an entire function of its independent complex variables, and in particular as a function of t is a finite linear combination of exponentials and powers of t. We now note that one can verify directly (see the Exercises) that if u is a harmonic function depending analytically on the parameter t, then

$$h(\mathbf{x}, t) = u(\mathbf{x}, t) + \frac{1}{2\pi i} \oint_{|t-\tau|=\delta} \int_0^1 \sigma^{n-1} G(r, 1-\sigma^2, \tau - t) u(\mathbf{x}\sigma^2, \tau) \, d\sigma \, d\tau \tag{5.70}$$

defines a solution of (5.46), where $\delta > 0$ and

$$G(r, \xi, t) = (r^2/2t^2) \exp(\xi r^2/4t). \tag{5.71}$$

(For a generalization of the operator (5.70) see Rundell and Stecher [153].) In particular, u can be chosen such that h as defined by (5.70) is the (entire) solution h we are trying to approximate over $\bar{D} \times [0, T]$ (represent u as a linear combination of exponentials and powers of t). Now let Ω be a sphere in R^n such that $\Omega \supset \bar{D}$ and let $\{u_j\}$ denote the set of harmonic polynomials. Then from the Runge approximation property for elliptic equations we can approximate u on $\Omega \times \{t: |t| \leq T + \delta\}$ by a finite sum of the form

$$\sum_{j=0}^{j_0} \sum_{k=0}^{k_0} a_{jk} u_j(\mathbf{x}) t^k, \tag{5.72}$$

where the a_{jk} are constants, and hence from (5.70) we have that for every $\varepsilon>0$ there exist integers j_0 and k_0 and constants a_{jk} such that

$$\max_{\bar{D}\times[0,T]}\left|h(\mathbf{x},t)-\sum_{j=0}^{j_0}\sum_{k=0}^{k_0}a_{jk}h_{jk}(\mathbf{x},t)\right|<\varepsilon \tag{5.73}$$

where the h_{jk} are polynomial solutions of (5.46) defined by

$$h_{jk}(\mathbf{x},t)=u_j(\mathbf{x})t^k + \frac{1}{2\pi i}\oint_{|t-\tau|=\delta}\int_0^1 \sigma^{n-1}G(r,1-\sigma^2,\tau-t)u_j(\mathbf{x}\sigma^2)\tau^k\,d\sigma\,d\tau. \tag{5.74}$$

Since the h_{jk} are polynomials in $x_1,\ldots,x_n$ and t, there exist an integer $M=M(j,k)$ and constants $b_m=b_m(j,k)$, $|m|\leq M$, such that

$$h_{jk}(\mathbf{x},0)=\sum_{|m|\leq M}b_m h_m(\mathbf{x},0). \tag{5.75}$$

From the uniqueness theorem for Cauchy's problem for the heat equation we have that

$$h_{jk}(\mathbf{x},t)=\sum_{|m|\leq M}b_m h_m(\mathbf{x},t) \tag{5.76}$$

for all $\mathbf{x}$ and t, and the conclusion of the theorem now follows from (5.73) and (5.76).

5.3 The Helmholtz equation in an exterior domain

Until now we have only been considering the problem of constructing a complete family of solutions for partial differential equations defined in interior domains. However, as was seen in Chapter 2, in scattering theory we are interested in solutions of the Helmholtz equation

$$\Delta_3 u+k^2u=0 \tag{5.77}$$

defined in exterior domains and satisfying the Sommerfeld radiation condition

$$\lim_{r\to\infty} r(\partial u/\partial r-iku)=0 \tag{5.78}$$

where k is assumed to be real and the limit in (5.78) is assumed to hold uniformly with respect to the angular coordinates θ and ϕ. If one separates variables in (5.77) and makes use of (5.78) it is seen that for $n=0,1,2,\ldots$, $m=0,\pm1,\pm2,\ldots,\pm n$, the spherical wave functions

$$u_{nm}(r,\theta,\phi)=h_n^{(1)}(kr)P_n^m(\cos\theta)\exp(im\phi) \tag{5.79}$$

are solutions of (5.77), (5.78), where $h_n^{(1)}$ is a spherical Hankel function and P_n^m is an associated Legendre function. Now suppose u is a solution of (5.77) in the exterior of a bounded domain D with C^2 boundary ∂D and satisfies (5.78). Without loss of generality assume D contains the origin. Then our aim in this section is to show that in the exterior of D, u can be approximated by a finite linear combination of functions taken from the set $\{u_{nm}\}$. More specifically, we shall show that $\{u_{nm}\}$ is complete in $L_2(\partial D)$ (Millar [127]; Vekua [173]). Suppose for the moment that this is true, and let G be the Green's function for (5.77) in $R^3\backslash D$ satisfying (5.78). Then u can be represented in the form

$$u(\mathbf{x}) = \int_{\partial D} u(\boldsymbol{\xi}) \frac{\partial G}{\partial \nu}(\mathbf{x}; \boldsymbol{\xi})\, dS_\xi \tag{5.80}$$

where ν is the outward normal to ∂D at ξ. Let D_0 be a compact set such that $D_0 \supset D$. Then for $\mathbf{x} \in R^3\backslash D_0$, $\|\partial G/\partial \nu\|_{L_2(\partial D)} \leq M$ where M is a positive constant independent of $\mathbf{x}$. Now suppose constants a_{nm} are chosen such that

$$\|u - u_N\|_{L_2(\partial D)} < \varepsilon \tag{5.81}$$

where

$$u_N(\boldsymbol{\xi}) = \sum_{n=0}^{N} \sum_{m=-n}^{n} a_{nm} u_{nm}(\boldsymbol{\xi}). \tag{5.82}$$

Then for $\mathbf{x} \in R^3\backslash D$

$$u_N(\mathbf{x}) = \int_{\partial D} u_N(\boldsymbol{\xi}) \frac{\partial G}{\partial \nu}(\mathbf{x}; \boldsymbol{\xi})\, dS_\xi \tag{5.83}$$

and hence for $\mathbf{x} \in R^3\backslash D_0$

$$|u(\mathbf{x}) - u_N(\mathbf{x})| = \left| \int_{\partial D} (u(\boldsymbol{\xi}) - u_N(\boldsymbol{\xi})) \frac{\partial G}{\partial \nu}(\mathbf{x}; \boldsymbol{\xi})\, dS_\xi \right|$$

$$\leq \|u - u_N\| \left\| \frac{\partial G}{\partial \nu} \right\| \leq \varepsilon M \tag{5.84}$$

i.e. u can be uniformly approximated in $R^3\backslash D_0$ by a finite linear combination of the functions u_{nm}.

In order to show that the set $\{u_{nm}\}$ is complete in $L_2(\partial D)$ we first need the following uniqueness theorem (cf. Rellich [147]; Vekua [172]).

Theorem 5.6 *Let $u \in C^2(R^3\backslash \bar{D}) \cap C^1(R^3\backslash D)$ be a solution of (5.77) in the exterior of D satisfying the Sommerfeld radiation condition (5.78) at infinity and the boundary condition $u = 0$ on ∂D. Then $u(\mathbf{x}) = 0$ for $\mathbf{x} \in R^3\backslash D$.*

Proof Let Ω be a ball centered at the origin containing D in its interior. Then from Green's formula we have

$$\iint_{\Omega\setminus D} (u\Delta_3\bar{u} - \bar{u}\Delta_3 u)\,\mathrm{d}V = \int_{\partial D}\left(\bar{u}\frac{\partial u}{\partial \nu} - u\frac{\partial \bar{u}}{\partial \nu}\right)\mathrm{d}S - \int_{\partial\Omega}\left(\bar{u}\frac{\partial u}{\partial r} - u\frac{\partial \bar{u}}{\partial r}\right)\mathrm{d}S. \quad (5.85)$$

Since k is real and $u = \bar{u} = 0$ on ∂D, we have from (5.85) that

$$\int_{\partial\Omega}\left(\bar{u}\frac{\partial u}{\partial r} - u\frac{\partial \bar{u}}{\partial r}\right)\mathrm{d}S = 0. \quad (5.86)$$

But from the results of Chapter 2, in $R^3\setminus\Omega$ we can expand u in the form

$$u(\mathbf{x}) = \sum_{n=0}^{\infty}\sum_{m=-n}^{n} a_{nm}h_n^{(1)}(kr)P_n^m(\cos\theta)\mathrm{e}^{im\phi}, \quad (5.87)$$

where the series (5.87) converges absolutely and uniformly. By the orthogonality of the spherical harmonics $P_n^m(\cos\theta)\mathrm{e}^{im\phi}$ and the formula

$$\overline{h_n^{(1)}(kr)}\frac{\mathrm{d}}{\mathrm{d}r}h_n^{(1)}(kr) - h_n^{(1)}(kr)\frac{\mathrm{d}}{\mathrm{d}r}\overline{h_n^{(1)}(kr)} = \frac{2\mathrm{i}}{kr^2} \quad (5.88)$$

we can conclude from (5.86) and (5.87) that

$$\sum_{n=0}^{\infty}\sum_{m=-n}^{n}|a_{nm}|^2 = 0, \quad (5.89)$$

which implies that $u(\mathbf{x}) = 0$ for $\mathbf{x}\in R^3\setminus\Omega$. But by the analyticity of solutions to (5.77) we can now conclude that $u(\mathbf{x}) = 0$ for $\mathbf{x}\in R^3\setminus D$.

We now turn to the proof of the fact that the set $\{u_{nm}\}$ is complete in $L_2(\partial D)$. To prove this result we shall need the following two facts concerning the continuity properties of the single-layer potential

$$u(\mathbf{x}) = \int_{\partial D} g(\boldsymbol{\xi})\frac{\mathrm{e}^{ikR}}{R}\,\mathrm{d}S, \qquad R = |\mathbf{x}-\boldsymbol{\xi}|, \quad (5.90)$$

where g is assumed to be continuous on ∂D

(1) u is continuous across ∂D.
(2) Let $\partial u^+/\partial\nu$ denote the limit of $\partial u(\mathbf{x})/\partial\nu$ as $\mathbf{x}$ tends to ∂D from inside D and $\partial u^-/\partial\nu$ the limit as $\mathbf{x}$ tends to ∂D from outside D. Then

$$\partial u^+/\partial\nu - \partial u^-/\partial\nu = 4\pi g(\mathbf{x}). \quad (5.91)$$

A proof of these facts is left to the Exercises. (For the analogous result in two dimensions see Chapter 8.)

Theorem 5.7 *The set $\{u_{nm}\}$ is complete in $L_2(\partial D)$.*

Proof Since the set of continuous functions on ∂D is dense in $L_2(\partial D)$, it suffices to show that if $g \in C(\partial D)$ and

$$\int_{\partial D} g(\boldsymbol{\xi}) u_{nm}(\boldsymbol{\xi})\, dS = 0 \tag{5.92}$$

for $n = 0, 1, 2, \ldots, -n \leq m \leq n$, then $g = 0$ on ∂D. Hence suppose (5.92) is true, and let Ω be a ball centered at the origin and contained in D. From (5.92) and the expansions (2.62) and (2.66) of Chapter 2 we can conclude that

$$u(\mathbf{x}) = \int_{\partial D} g(\boldsymbol{\xi}) \frac{e^{ikR}}{R}\, dS = 0; \qquad \mathbf{x} \in \Omega. \tag{5.93}$$

Since u, as defined by (5.93), is a solution of (5.77) inside D, we can conclude by the analyticity of u that $u(\mathbf{x}) = 0$ for $\mathbf{x} \in D$. Now let $\mathbf{x} \in R^3 \backslash D$. Then using the fact that u is continuous across ∂D we can conclude that for $\mathbf{x} \in R^3 \backslash \bar{D}$, u is a solution of (5.77), satisfies the Sommerfeld radiation condition, and vanishes on ∂D. Furthermore, $u \in C^1(R^3 \backslash D)$. Hence by Theorem 5.6, $u(\mathbf{x}) = 0$ for $\mathbf{x} \in R^3 \backslash D$. But now we can conclude from (5.91) that $g = 0$ on ∂D and the proof is complete.

Exercises

(1) Show that the Riemann function for

$$\Delta_2 u + \frac{4n(n+1)}{(1+x^2+y^2)^2} u = 0$$

is given by

$$R(z, z^*; \zeta, \zeta^*) = P_n\left(\frac{(1-\zeta\zeta^*)(1-zz^*) + 2z^*\zeta + 2z\zeta^*}{(1+\zeta\zeta^*)(1+zz^*)}\right)$$

where P_n is a Legendre polynomial. Use this result to construct a complete set of polynomial solutions to the above equation and express each member of this set in terms of known special functions.

(2) Define a kernel function for Laplace's equation $\Delta_2 u = 0$ and give an explicit construction of this function for the special case when the domain D is a circle of radius R about the origin.

(3) A fundamental solution of the biharmonic equation $\Delta_2^2 u = 0$ is given by $\Gamma(\mathbf{x}, \boldsymbol{\xi}) = (1/8\pi) r^2 \log r + v$ where $r = |\mathbf{x} - \boldsymbol{\xi}|$ and v is a regular biharmonic function of $\mathbf{x}$. The biharmonic Green's function of a plane

domain D is defined to be the fundamental solution defined in D and satisfying $\Gamma = \partial\Gamma/\partial\nu = 0$ for $\mathbf{x} \in \partial D$ and ν the outward normal to ∂D. Using two applications of Green's identity, interpret $K(\mathbf{x}; \boldsymbol{\xi}) = -\Delta_{\mathbf{x}}\Delta_{\boldsymbol{\xi}}\Gamma(\mathbf{x}; \boldsymbol{\xi})$ (where $\Delta_{\mathbf{x}} = \partial^2/\partial x_1^2 + \partial^2/\partial x_2^2$, $\Delta_{\boldsymbol{\xi}} = \partial^2/\partial \xi_1^2 + \partial^2/\partial \xi_2^2$) as a kernel function for the biharmonic equation and show that if $\{h_n\}$ is a complete set of harmonic functions in D, orthonormal with respect to the L_2 scalar product over D, then for $\mathbf{x}, \boldsymbol{\xi} \in D$, $K(\mathbf{x}; \boldsymbol{\xi}) = \sum_n^\infty h_n(\mathbf{x})h_n(\boldsymbol{\xi})$. Show how this result can be used to solve the boundary value problem

$$\Delta_2^2 u = 0 \quad \text{in} \quad D$$
$$u = f \quad \text{on} \quad \partial D$$
$$\partial u/\partial\nu = g \quad \text{on} \quad \partial D.$$

(4) Show that if u is a harmonic function of $\mathbf{x}$ depending analytically on the parameter t, then (5.70) defines a solution of (5.46). Show that in the case $n = 1$, (5.70) reduces to the representation (4.92).

(5) Let h be a solution of the heat equation (5.46) in a cylindrical domain $D \times (0, T)$ where D and h have the regularity properties as in Theorem 5.5. Let λ_k be a sequence of real numbers satisfying (5.45). Show that h can be approximated arbitrarily closely over $\bar{D} \times [0, T]$ by a linear combination of the functions

$$r^{-(n-2)/2} J_{(n-2)/2+l}(\lambda_k r) S_{lm}(\theta; \phi) \exp(-\lambda_k^2 t)$$

where J_ν is a Bessel function and S_{lm} a spherical harmonic.

(6) Let u_{nm} be defined by (5.79). Using the fact that for $g \in C(\partial D)$ the double layer potential

$$u(\mathbf{x}) = \int_{\partial D} g(\boldsymbol{\xi}) \frac{\partial}{\partial\nu} \frac{e^{ikR}}{R} \, dS$$

has a jump discontinuity of $4\pi g(\mathbf{x})$ as $\mathbf{x}$ crosses ∂D, but has continuous normal derivative as $\mathbf{x}$ crosses ∂D, show that the set $\{\partial u_{nm}/\partial\nu\}$ is complete in $L_2(\partial D)$.

(7) Derive the continuity properties of the single- and double-layer potentials as defined in (5.90) and Exercise 6.

(8) Prove a result analogous to Theorem 5.7 for bounded simply connected domains lying in the plane.

6

Pseudoanalytic functions

Until this point we have restricted our attention to the analytic theory of a single second order partial differential equation (although, as we have mentioned, many of our results can be extended to higher order equations). However, in recent years a considerable amount of attention has been devoted to systems of elliptic equations in the plane, and we refer the reader to the monograph of Wendland [175] for a systematic and well-written account of the recent developments in this area. The reader should also consult the monograph of Tutschke [168] and the book by Gilbert [72] on this subject. In this chapter we shall only give the rudiments of this theory and restrict ourselves to function theoretic aspects of elliptic systems in two unknown functions defined in the plane. The theory of such systems was originally developed by Bers (Bers [16]; Bers, John and Schechter [17]); Vekua [171], [172] and Haack (Haack and Wendland [80]), and in view of the thorough discussions available in the above-cited references we have chosen to present only the salient features of this subject as discussed, for example, in Courant and Hilbert [43].

The systems we are about to consider are of the form

$$\begin{aligned} u_x - v_y &= a_{11}(x, y)u + a_{12}(x, y)v \\ u_y + v_x &= a_{21}(x, y)u + a_{22}(x, y)v \end{aligned} \tag{6.1}$$

where the coefficients a_{ij}, $i = 1, 2$, $j = 1, 2$, are assumed to be everywhere defined and to satisfy a Hölder condition

$$|a_{ij}(P) - a_{ij}(Q)| \leq K\overline{PQ}^{\alpha} \tag{6.2}$$

where P, Q are points in the plane, $\overline{PQ}$ is the distance from P to Q, and $K = K(i, j)$, $\alpha = \alpha(i, j)$ are positive constants, $0 < \alpha < 1$. We assume further that the coefficients vanish identically outside a large disk. These assumptions are only made for the sake of simplicity and can be considerably weakened (cf. Bers [16]; Vekua [171]). By introducing the conjugate coordinates

$$z = x + \mathrm{i}y, \qquad \bar{z} = x - \mathrm{i}y \tag{6.3}$$

we can rewrite (6.1) in the form

$$w_{\bar{z}} = a(z)w + b(z)\bar{w} \tag{6.4}$$

where

$$\begin{aligned} &w = u + \mathrm{i}v, \\ &a = \tfrac{1}{4}(a_{11} + a_{22} + \mathrm{i}a_{21} - \mathrm{i}a_{12}), \qquad b = \tfrac{1}{4}(a_{11} - a_{22} + \mathrm{i}a_{21} + \mathrm{i}a_{12}). \end{aligned} \tag{6.5}$$

Note that if $a = b = 0$ then (6.1) or (6.4) reduce to the Cauchy–Riemann equations, and for this reason continuously differentiable solutions of (6.4) are called *pseudoanalytic functions* (Bers [16]) or *generalized analytic functions* (Vekua [171]). We shall show that pseudoanalytic functions have many properties that are similar to analytic functions of a single complex variable. However, in order to do this we shall first have to investigate the behavior of integrals of the form

$$q(z) = -\frac{1}{\pi} \iint_D \frac{\rho(\zeta)}{\zeta - z} \, d\xi \, d\eta \tag{6.6}$$

where D is a bounded, simply connected domain, $\zeta = \xi + \mathrm{i}\eta$, and ρ is a complex-valued Hölder continuous function defined everywhere and vanishing outside a disk of radius R. This will be done in the following section, and the applications of these results to solutions of (6.1) (equivalently (6.4)) will be done in Section 6.2. Finally, following the approach of Yu [184], [185], we shall derive a reflection principle for solutions of (6.1) in the case when the coefficients a_{ij} are analytic functions of x and y.

6.1 Singular integrals

We now proceed to investigate the continuity properties of the singular integral (6.6). Our first result is the following theorem.

Theorem 6.1 *Let ρ and q be defined as in (6.6). Then for every ε, $0 < \varepsilon < 1$, there exists a constant K depending only on ε and R such that*

$$|q(z)| \leq \frac{KM}{1 + |z|^{\varepsilon}}$$

$$|q(z_1) - q(z_2)| \leq KM\, |z_1 - z_2|^{\varepsilon}$$

where $|\rho| \leq M$.

Proof We can write

$$|q(z)| \leqslant \frac{1}{\pi} \iint_{|\zeta| \leqslant R} \frac{M}{|\zeta - z|} \, d\xi \, d\eta \tag{6.7}$$

and hence for any z

$$|q(z)| \leqslant MR^2 + \frac{M}{\pi} \iint_{|\zeta - z| \leqslant 1} \frac{1}{|\zeta - z|} \, d\xi \, d\eta = MR^2 + 2M. \tag{6.8}$$

Since from (6.7) we have

$$\lim_{|z| \to \infty} |z|^{\varepsilon} \, |q(z)| = 0 \tag{6.9}$$

we can conclude that the first statement in the theorem is true. In order to prove the second inequality we write

$$q(z_1) - q(z_2) = -\frac{1}{\pi} \iint_D \rho(\zeta) \left[\frac{1}{\zeta - z_1} - \frac{1}{\zeta - z_2} \right] d\xi \, d\eta$$

$$= -\frac{z_1 - z_2}{\pi} \iint_D \frac{\rho(\zeta)}{(\zeta - z_1)(\zeta - z_2)} \, d\xi \, d\eta. \tag{6.10}$$

Now let $2\delta = |z_1 - z_2|$ and choose R_1 such that the disk $|\zeta - z_1| \leqslant R_1$ contains the disk of radius R about the origin for all $z_1, |z_1| \leqslant R$ (note that if $|z_1| > R$ the desired inequality follows easily from (6.10)). Then for $|z_1| \leqslant R$ (see Fig. 6.1)

$$|q(z_1) - q(z_2)| \leqslant \frac{2\delta M}{\pi} \iint_D \frac{d\xi \, d\eta}{|\zeta - z_1| \, |\zeta - z_2|}$$

$$\leqslant \frac{2\delta M}{\pi} \iint_{4\delta \leqslant |\zeta - z_1| \leqslant R_1} \frac{d\xi \, d\eta}{|\zeta - z_1| \, |\zeta - z_2|} + \frac{2\delta M}{\pi} \iint_{|\zeta - z_1| \leqslant \delta} \frac{d\xi \, d\eta}{|\zeta - z_1| \, |\zeta - z_2|}$$

$$+ \frac{2\delta M}{\pi} \iint_{|\zeta - z_2| \leqslant \delta} \frac{d\xi \, d\eta}{|\zeta - z_1| \, |\zeta - z_2|} + \frac{2\delta M}{\pi} \iint_{\substack{\delta \leqslant |\zeta - z_1| \leqslant 4\delta \\ |\zeta - z_2| \geqslant \delta}} \frac{d\xi \, d\eta}{|\zeta - z_1| \, |\zeta - z_2|}$$

$$\leqslant 4\delta M \int_{4\delta}^{R_1} \frac{dr}{r - 2\delta} + 36\delta M$$

$$\leqslant 4\delta M \log \frac{R_1 - 2\delta}{2\delta} + 36\delta M$$

$$\leqslant MK\delta^{\varepsilon} \tag{6.11}$$

for $0 < \varepsilon < 1$, and this completes the proof of the theorem.

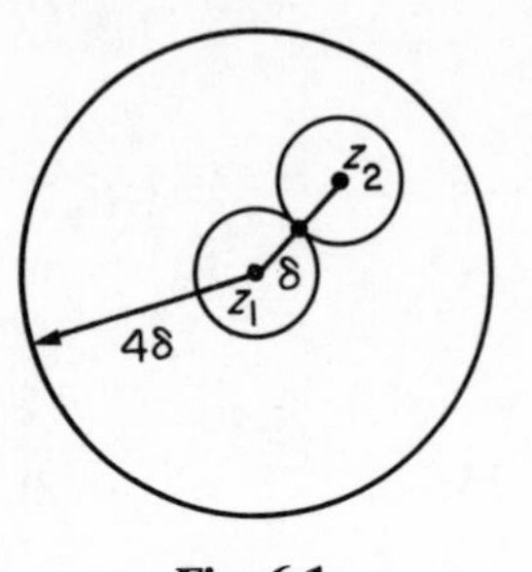

Fig. 6.1

Theorem 6.2 *Let ρ and q be defined as in (6.6). Then q has Hölder continuous partial derivatives and for $z \in D$, $q_{\bar{z}} = \rho$.*

Proof Let $\Omega = \{\zeta : |\zeta| < R_1, R_1 > R\}$ and for z in Ω consider

$$-\frac{1}{\pi}\iint_{\Omega} \frac{\rho(\zeta)}{(\zeta - z)^2}\, d\xi\, d\eta = -\lim_{\varepsilon \to 0} \frac{1}{\pi} \iint_{\Omega \backslash \Omega_\varepsilon} \frac{\rho(\zeta)}{(\zeta - z)^2}\, d\xi\, d\eta \tag{6.12}$$

where $\Omega_\varepsilon = \{\zeta : |\zeta - z| \leqslant \varepsilon\}$. Then

$$-\frac{1}{\pi}\iint_{\Omega} \frac{\rho(\zeta)}{(\zeta - z)^2}\, d\xi\, d\eta = \lim_{\varepsilon \to 0} \frac{1}{\pi} \iint_{\Omega \backslash \Omega_\varepsilon} \frac{\rho(z) - \rho(\zeta)}{(\zeta - z)^2}\, d\xi\, d\eta$$

$$-\rho(z) \lim_{\varepsilon \to 0} \frac{1}{\pi} \iint_{\Omega \backslash \Omega_\varepsilon} \frac{d\xi\, d\eta}{(\zeta - z)^2}. \tag{6.13}$$

On the other hand, from Green's formula we have that for $w \in C^1(\bar{\Omega})$

$$\iint_{\Omega} \frac{\partial w}{\partial \bar{z}}\, dx\, dy = \frac{1}{2i} \int_{\partial\Omega} w(z)\, dz \quad \text{and} \quad \iint_{\Omega} \frac{\partial w}{\partial z}\, dx\, dy = -\frac{1}{2i} \int_{\partial\Omega} w(z)\, d\bar{z} \tag{6.14}$$

and therefore

$$\frac{1}{\pi} \iint_{\Omega \backslash \Omega_\varepsilon} \frac{d\xi\, d\eta}{(\zeta - z)^2} = -\frac{1}{\pi} \iint_{\Omega \backslash \Omega_\varepsilon} \frac{\partial}{\partial \zeta}\left(\frac{1}{\zeta - z}\right) d\xi\, d\eta$$

$$= \frac{1}{2\pi i} \int_{\partial\Omega} \frac{d\bar{\zeta}}{\zeta - z} - \frac{1}{2\pi i} \int_{\partial\Omega_\varepsilon} \frac{d\bar{\zeta}}{\zeta - z}$$

$$= 0 \tag{6.15}$$

since each of the line integrals vanishes. Hence from (6.13), (6.15)

$$-\frac{1}{\pi}\iint_{\Omega}\frac{\rho(\zeta)}{(\zeta-z)^2}\,d\xi\,d\eta=-\frac{1}{\pi}\iint_{\Omega}\frac{\rho(\zeta)-\rho(z)}{(\zeta-z)^2}\,d\xi\,d\eta \tag{6.16}$$

where, since ρ is Hölder continuous, the integral on the right-hand side of (6.16) is an ordinary improper integral. In particular, the singular integral on the left-hand side of (6.16) exists in the sense of the Cauchy principal value.

We shall now show that provided ρ is Hölder continuous the integral on the left-hand side of (6.16) defines a Hölder continuous function of z. To this end let z_1 and z_2 be in Ω, $z_1 \neq z_2$, and define

$$g(z)=-\frac{1}{\pi}\iint_{\Omega}\frac{\rho(\zeta)}{(\zeta-z)^2}\,d\xi\,d\eta. \tag{6.17}$$

Then

$$\begin{aligned} g(z_1)-g(z_2)&=\frac{z_1-z_2}{\pi}\iint_{\Omega}\frac{\rho(\zeta)-\rho(z_2)}{(\zeta-z_2)^2(\zeta-z_1)}\,d\xi\,d\eta\\ &\quad+\frac{z_1-z_2}{\pi}\iint_{\Omega}\frac{\rho(\zeta)-\rho(z_1)}{(\zeta-z_2)(\zeta-z_1)^2}\,d\xi\,d\eta\\ &\quad+\frac{\rho(z_2)(z_1-z_2)}{\pi}\iint_{\Omega}\frac{d\xi\,d\eta}{(\zeta-z_2)^2(\zeta-z_1)}\\ &\quad+\frac{\rho(z_1)(z_1-z_2)}{\pi}\iint_{\Omega}\frac{d\xi\,d\eta}{(\zeta-z_2)(\zeta-z_1)^2}. \end{aligned} \tag{6.18}$$

Since

$$\begin{aligned} \frac{1}{\pi}\iint_{\Omega}\frac{d\xi\,d\eta}{(\zeta-z_2)^2(\zeta-z_1)}&=\frac{1}{\pi(z_2-z_1)}\iint_{\Omega}\frac{d\xi\,d\eta}{(\zeta-z_2)^2}\\ &\quad-\frac{1}{\pi(z_2-z_1)^2}\left(\iint_{\Omega}\frac{d\xi\,d\eta}{\zeta-z_2}-\iint_{\Omega}\frac{d\xi\,d\eta}{\zeta-z_1}\right) \end{aligned} \tag{6.19}$$

and from Green's formula

$$\frac{1}{\pi}\iint_{\Omega}\frac{d\xi\,d\eta}{\zeta-z}=-\bar{z}+\frac{1}{2\pi i}\int_{\partial\Omega}\frac{\bar{\zeta}\,d\zeta}{\zeta-z}=-\bar{z}, \tag{6.20}$$

we have

$$\frac{1}{\pi}\iint_{\Omega}\frac{d\xi\,d\eta}{(\zeta-z_2)^2(\zeta-z_1)}=\frac{\bar{z}_2-\bar{z}_1}{(z_2-z_1)^2}, \tag{6.21}$$

and (6.18) can be rewritten as

$$\begin{aligned} g(z_1)-g(z_2)&=\frac{z_1-z_2}{\pi}\iint_{\Omega}\frac{\rho(\zeta)-\rho(z_2)}{(\zeta-z_2)^2(\zeta-z_1)}\,d\xi\,d\eta \\ &\quad+\frac{z_1-z_2}{\pi}\iint_{\Omega}\frac{\rho(\zeta)-\rho(z_1)}{(\zeta-z_2)(\zeta-z_1)^2}\,d\xi\,d\eta \\ &\quad-[\rho(z_2)-\rho(z_1)]\left(\frac{\bar{z}_2-\bar{z}_1}{z_2-z_1}\right). \end{aligned} \tag{6.22}$$

From the same method of calculation as that used to derive (6.11) we now have from (6.22) that if ρ has Hölder exponent α, then there exists a constant K such that

$$|g(z_1)-g(z_2)|\leq K\,|z_2-z_1|^{\alpha}. \tag{6.23}$$

We are now (finally!) in a position to prove our theorem. It suffices to consider (6.6) defined over Ω instead of D since the integral over Ω can be written as the sum of an integral over D and an integral over $\Omega\backslash D$ where the second integral is an analytic function of z for z in D. Let z_1 and z_2 be in D, $z_1\neq z_2$. Then, using (6.21), we have

$$\begin{aligned} \frac{q(z_1)-q(z_2)}{z_1-z_2}-g(z_2)&=\frac{z_2-z_1}{\pi}\iint_{\Omega}\frac{\rho(\zeta)-\rho(z_2)}{(\zeta-z_2)^2(\zeta-z_1)}\,d\xi\,d\eta \\ &\quad+\frac{\rho(z_2)(z_2-z_1)}{\pi}\iint_{\Omega}\frac{d\xi\,d\eta}{(\zeta-z_2)^2(\zeta-z_1)} \\ &=\frac{z_2-z_1}{\pi}\iint_{\Omega}\frac{\rho(\zeta)-\rho(z_2)}{(\zeta-z_2)^2(\zeta-z_1)}\,d\xi\,d\eta+\left(\frac{\bar{z}_2-\bar{z}_1}{z_2-z_1}\right)\rho(z_2) \end{aligned} \tag{6.24}$$

and since ρ is Hölder continuous

$$\left|\frac{q(z_1)-q(z_2)}{z_1-z_2}-g(z_2)-\left(\frac{\bar{z}_2-\bar{z}_1}{z_2-z_1}\right)\rho(z_2)\right|\leq K\,|z_2-z_1|^{\alpha} \tag{6.25}$$

for some positive constant K. Now let z_1 tend to z_2 along a line making an angle θ with the real axis, i.e.

$$z_1 - z_2 = |z_1 - z_2|\, e^{i\theta}. \tag{6.26}$$

Then from (6.25) we have

$$\lim_{z_1 \to z_2} \frac{q(z_1) - q(z_2)}{z_1 - z_2} = g(z_2) + e^{-2i\theta}\rho(z_2). \tag{6.27}$$

Letting $\theta = 0$ and $\theta = \pi/2$ now gives (setting $z_2 = z$)

$$\partial q(z)/\partial x = g(z) + \rho(z), \qquad \partial q(z)/\partial y = ig(z) - i\rho(z), \tag{6.28}$$

and since

$$\partial q/\partial z = \tfrac{1}{2}(\partial q/\partial x - i(\partial q/\partial y)), \qquad \partial q/\partial \bar{z} = \tfrac{1}{2}(\partial q/\partial z + i(\partial q/\partial y)) \tag{6.29}$$

we have

$$\partial q/\partial \bar{z} = \rho. \tag{6.30}$$

Since we have just shown that g is Hölder continuous provided ρ is, the theorem now follows from (6.28) and (6.30).

Note: A slight modification of the above analysis shows that if ρ is Hölder continuous in a neighborhood of a point $z_0 \in D$ then q has Hölder continuous partial derivative in this neighborhood and $q_{\bar{z}} = \rho$ there.

6.2 The similarity principle and its applications

We shall now use the results obtained in the previous section to deduce the function theoretic properties of solutions to (6.4). We first note that a continuous bounded function w defined in D is a solution of (6.4) if and only if

$$f(z) = w(z) + \frac{1}{\pi} \iint_D \frac{a(\zeta)w(\zeta) + b(\zeta)\overline{w(\zeta)}}{\zeta - z}\, d\xi\, d\eta \tag{6.31}$$

is analytic in D. To see this we note that the integral in (6.31) is a Hölder continuous function of z (note that in Theorem 6.1 the Hölder continuity of ρ was not used) and hence w is Hölder continuous if and only if f is. Similarly, if w is Hölder continuous, then w is continuously differentiable if and only if f is. Hence from Theorem 6.2 we have that if either f is analytic or w is pseudoanalytic then

$$f_{\bar{z}} = w_{\bar{z}} - aw - b\bar{w} = 0 \tag{6.32}$$

in D and the result follows. By the removable singularity theorem in analytic function theory and the fact that the integral in (6.31) is unchanged if a point is removed from its region of integration, we thus have the following theorem.

Theorem 6.3 (*Removable singularity theorem*) *Let w be pseudoanalytic and bounded for $0<|z-z_0|<r$ for some number $r>0$. Then w can be defined at z_0 in such a way that it is pseudoanalytic in the whole disk $|z-z_0|<r$.*

The name of the next theorem is motivated by the fact that two complex-valued functions w and f, defined in a domain D, are called *similar* if w/f is bounded away from zero and continuous in $\bar{D}$.

Theorem 6.4 (*Similarity principle*) *Let w be pseudoanalytic in a domain D. Then there exist an analytic function f defined in D and a Hölder continuous function s defined in $\bar{D}$ such that*

$$w(z)=f(z)e^{s(z)}.$$

Proof We shall explicitly construct the functions f and s and show that they satisfy the stated conditions. If $w\equiv 0$ there is nothing to prove. Hence assume w is not identically zero and let D_0 be the open subset of D in which $w(z)\neq 0$, $z\in D_0$. Define

$$s(z)=-\frac{1}{\pi}\iint_{D_0}\left[a(\zeta)+b(\zeta)\frac{\overline{w(\zeta)}}{w(\zeta)}\right]\frac{d\xi\, d\eta}{\zeta-z} \tag{6.33}$$

$$f(z)=e^{-s(z)}w(z).$$

Since $s(z)$ is (1) continuous everywhere, (2) continuously differentiable in D_0, and (3) satisfies

$$s_{\bar{z}}=a(z)+b(z)\frac{\overline{w(z)}}{w(z)} \tag{6.34}$$

in D_0 we can conclude from (6.33) that $f_{\bar{z}}\equiv 0$ in D_0 and hence f is analytic in D_0. From the removable singularities theorem for analytic functions it follows that f is also analytic at every isolated point of $D\backslash D_0$. Now let z_0 be a nonisolated point of $D\backslash D_0$ and $\{z_n\}_{n=1}^{\infty}$ a sequence of points such that $\lim\limits_{n\to\infty} z_n=z_0$. By selecting a subsequence we may assume that $\lim\limits_{n\to\infty} \arg(z_n-z_0)=\theta$ and from the mean value theorem we have

$$w_x\cos\theta+w_y\sin\theta=0 \tag{6.35}$$

at $z=z_0$. On the other hand $w(z_0)=0$ implies that $2w_{\bar{z}}=w_x+iw_y=0$ at z_0, and this together with (6.35) implies that $w_x(z_0)=w_y(z_0)=0$. Hence from the mean value theorem again we have (using $w(z_0)=f(z_0)=0$)

$$\lim_{z\to z_0}\frac{w(z)}{z-z_0}=0 \tag{6.36}$$

and therefore

$$\lim_{z\to z_0}\frac{f(z)}{z-z_0}=0. \tag{6.37}$$

We can thus conclude that f is analytic at every nonisolated point of $D\backslash D_0$ and thus in all of D. Since if f is not identically zero we have that the zeros of f are isolated, we can conclude that $D\backslash D_0$ consists only of isolated points and hence we can write

$$s(z)=-\frac{1}{\pi}\iint_D\left[a(\zeta)+b(\zeta)\frac{\overline{w(\zeta)}}{w(\zeta)}\right]\frac{d\xi\,d\eta}{\zeta-z}. \tag{6.38}$$

We now show that for every analytic function f defined in D we can define a pseudoanalytic function in D by $w(z)=f(z)e^{s(z)}$ where s is Hölder continuous in $\bar{D}$.

Theorem 6.5 *Let f be an analytic function defined in a domain D. Then there exists a function s, Hölder continuous in $\bar{D}$ and vanishing at a point $z_0\in D$, such that $w(z)=f(z)e^{s(z)}$ is pseudoanalytic in D.*

Proof From Theorem 6.3 it is seen that there is no loss in generality if we remove all the zeros of f from the domain D, i.e. assume $f(z)\neq 0$ for z in D. Then if we can find a function s satisfying

$$s_{\bar{z}}=a(z)+b(z)\frac{\overline{f(z)}}{f(z)}e^{\overline{s(z)}-s(z)}, \tag{6.39}$$

$f(z)e^{s(z)}$ will be pseudoanalytic. To find s we first define the operator $\mathbf{T}$ by

$$\mathbf{T}[s]=\sigma(z)-\sigma(z_0) \tag{6.40}$$

where

$$\sigma(z)=-\frac{1}{\pi}\iint_D\left[a(\zeta)+b(\zeta)\frac{\overline{f(\zeta)}}{f(\zeta)}e^{\overline{s(\zeta)}-s(\zeta)}\right]\frac{d\xi\,d\eta}{\zeta-z}, \tag{6.41}$$

and note that $\mathbf{T}$ maps the Banach space $C(D)$ of bounded continuous functions defined on D (equipped with the norm $\|s\|=\sup_D|s(z)|$)

onto a subset of $C(D)$ consisting of Hölder continuous functions defined in D and, if D is unbounded, tending uniformly to zero like $|z|^{-\varepsilon}$, $0<\varepsilon<1$, as z tends to infinity (Theorem 6.1). We call this subset Λ. Since elements of Λ have the same bound and Hölder exponent, it follows from Arzela's theorem that Λ is compact and since $\mathbf{T}$ is a continuous mapping of $C(D)$ onto Λ, it follows from Shauder's fixed point theorem (cf. Hochstadt [87]) that there exists a function s in Λ such that $s=\mathbf{T}[s]$, i.e.

$$s(z)=-\frac{1}{\pi}\iint_D\left[a(\zeta)+b(\zeta)\frac{\overline{f(\zeta)}}{f(\zeta)}\,e^{\overline{s(\zeta)}-s(\zeta)}\right]\left[\frac{1}{\zeta-z}-\frac{1}{\zeta-z_0}\right]d\xi\,d\eta. \tag{6.42}$$

This function satisfies (6.39) and the condition $s(z_0)=0$, and the theorem is now proved.

From Theorems 6.4 and 6.5 we can now deduce a variety of function theoretic results for pseudoanalytic functions, and we list three of these as corollaries.

Corollary 6.1 (*Carleman's theorem*) *A pseudoanalytic function which does not vanish identically has only isolated zeros and vanishes at most of finite order at any point.*

Proof This follows immediately from Theorem 6.4 and the corresponding classical results from analytic function theory.

Corollary 6.2 (*Liouville's theorem*) *A bounded pseudoanalytic function defined in the entire plane and vanishing at a fixed point z_0 of the plane (z_0 may be infinity) is identically zero.*

Proof By Theorem 6.5 $w(z)=f(z)e^{s(z)}$ where f is an entire function and s is given by (6.42) with $D=R^2$. Since s is bounded (due to the fact that a and b have compact support) we have that f is bounded and hence, by Liouville's theorem in analytic function theory, is a constant. Since $w(z_0)=0$, this constant must be zero, and the corollary follows.

Corollary 6.3 (*Maximum modulus theorem*) *Let w be a pseudoanalytic function defined in D and continuous in $\bar{D}$. Then there exists a positive constant $M\geq 1$ depending only on a and b such that for z in D*

$$|w(z)|\leq M\max_{t\in\partial D}|w(t)|.$$

Proof This follows from Theorem 6.4 (using (6.38)) and the maximum modulus principle for analytic functions.

6.3 Reflection principles

We now consider solutions of (6.4) in the special case where the coefficients a and b are analytic functions of x and y. In order to avoid confusion between analyticity with respect to x and y and with respect to the complex variable z we rewrite (6.4) in the form

$$\partial w/\partial \bar{z} = A(z, \bar{z})w + B(z, \bar{z})\bar{w} \tag{6.43}$$

and for the sake of simplicity assume that A and B are in fact entire functions of the two independent complex variables z and $\bar{z}$ (i.e. x and y are allowed to be complex variables). Assume now that the solution $w(z) = W(z, \bar{z})$ of (6.43) is defined for z in a bounded simply connected domain D, $\bar{z}$ in $D^* = \{z: \bar{z} \in D\}$, and that W can be continued as an analytic function of z and $\bar{z}$ into the product domain $D \times D^*$. Since for x and y complex $z = x + iy$, $\bar{z} = x - iy$ are independent complex variables, we relabel $\bar{z}$ by z^*, noting that $\bar{z} = z^*$ if and only if x and y are real, and rewrite (6.43) in the form

$$\partial W/\partial z^* = A(z, z^*)W(z, z^*) + B(z, z^*)W^*(z^*, z) \tag{6.44}$$

where W^* is the analytic continuation of $\bar{w}$ into the domain of complex arguments. In order to derive a reflection principle for solutions of (6.43) we must first derive a formula giving all solutions of (6.44) analytic for $z \in D$, $z^* \in D^*$, and then show that any solution $w \in C^1(D)$ of (6.43) can be continued as an analytic function of z and z^* into $D \times D^*$, i.e. that every continuously differentiable solution of (6.43) in D can be represented by our formula. We now proceed to do this, following Vekua [172].

We shall consider the nonhomogeneous form of (6.44)

$$\partial W/\partial z^* = A(z, z^*)W + B(z, z^*)W^* + F(z, z^*) \tag{6.45}$$

where F is analytic in $D \times D^*$, and make the change of variables

$$W(z, z^*) = V(z, z^*) \exp\left[\int_{z_0^*}^{z^*} A(z, \zeta^*)\, d\zeta^*\right] \tag{6.46}$$

where $z_0^* \in D^*$ to reduce (6.45) to the canonical form

$$\partial V/\partial z^* = C(z, z^*)V^*(z, z^*) + F_0(z, z^*) \tag{6.47}$$

where

$$\begin{aligned} C(z, z^*) &= B(z, z^*) \exp\left[\int_{z_0}^{z} A^*(z^*, \zeta)\, d\zeta - \int_{z_0^*}^{z^*} A(z, \zeta^*)\, d\zeta^*\right] \\ F_0(z, z^*) &= F(z, z^*) \exp\left[-\int_{z_0^*}^{z^*} A(z, \zeta^*)\, d\zeta^*\right] \end{aligned} \tag{6.48}$$

and $z_0^* = \bar{z}_0$. Suppose that V is a solution of (6.47) analytic in $D \times D^*$.

Then we can write (6.47) as

$$(\partial/\partial z^*)\left[V(z, z^*) - \int_{z_0^*}^{z^*} C(z, \zeta)V^*(\zeta, z)\, d\zeta\right] = F_0(z, z^*) \tag{6.49}$$

or

$$V(z, z^*) = \phi(z) + \int_{z_0^*}^{z^*} C(z, \zeta)V^*(\zeta, z)\, d\zeta + \int_{z_0^*}^{z^*} F_0(z, \zeta)\, d\zeta \tag{6.50}$$

where ϕ is an analytic function of z in D. If we substitute the conjugate of (6.50)

$$V^*(z^*, z) = \phi^*(z^*) + \int_{z_0}^{z} C^*(z^*, \zeta)V(\zeta, z^*)\, d\zeta + \int_{z_0}^{z} F_0^*(z^*, \zeta)\, d\zeta \tag{6.51}$$

into (6.50) we obtain

$$V(z, z^*) - \int_{z_0}^{z}\int_{z_0^*}^{z^*} C(z, \eta)C^*(\eta, \xi)V(\xi, \eta)\, d\eta\, d\xi = \Phi(z, z^*) \tag{6.52}$$

where

$$\Phi(z, z^*) = \phi(z) + \int_{z_0^*}^{z^*} C(z, \zeta)\phi^*(\zeta)\, d\zeta + \int_{z_0^*}^{z^*} F_0(z, \zeta)\, d\zeta + \int_{z_0}^{z}\int_{z_0^*}^{z^*} C(z, \eta)F_0^*(\eta, \xi)\, d\eta\, d\xi. \tag{6.53}$$

Hence we have rewritten (6.47) as the Volterra integral equation (6.52), where from (6.50)

$$\phi(z) = V(z, z_0^*). \tag{6.54}$$

By successive approximations we can write the unique solution of (6.52) in the form

$$V(z, z^*) = \Phi(z, z^*) + \int_{z_0}^{z}\int_{z_0^*}^{z^*} \Gamma(z, z^*, \zeta, \zeta^*)\, d\zeta^*\, d\zeta \tag{6.55}$$

where the resolvent Γ is a solution of the integral equation (see the Exercises)

$$\begin{aligned}\Gamma(z, z^*, \zeta, \zeta^*) &= C(z, \zeta^*)C^*(\zeta^*, \zeta) \\ &\quad + \int_{\zeta^*}^{z^*}\int_{\zeta}^{z} C(z, \eta)C^*(\eta, \xi)\Gamma(\xi, \eta, \zeta, \zeta^*)\, d\xi\, d\eta \\ &= C(z, \zeta^*)C^*(\zeta^*, \zeta) \\ &\quad + \int_{\zeta^*}^{z^*}\int_{\zeta}^{z} C(\xi, \zeta^*)C^*(\zeta^*, \zeta)\Gamma(z, z^*, \xi, \eta)\, d\xi\, d\eta.\end{aligned} \tag{6.56}$$

From (6.56) and the methods of Chapter 4 we can establish the existence of Γ and the fact that it is an entire function of its four independent complex variables. If we now substitute (6.53) into (6.55) we obtain

$$V(z, z^*) = \phi(z) + \int_{z_0}^{z} \Gamma_1(z, z^*, \zeta, z_0^*)\phi(\zeta)\, d\zeta + \int_{z_0^*}^{z^*} \Gamma_2(z, z^*, z_0, \zeta^*)\phi^*(\zeta^*)\, d\zeta^* + U_0(z, z^*) \tag{6.57}$$

where

$$U_0(z, z^*) = \int_{z_0^*}^{z^*} F_0(z, \zeta^*)\, d\zeta^* + \int_{z_0^*}^{z^*}\int_{z_0}^{z} \Gamma_1(z, z^*, \zeta, \zeta^*)F_0(\zeta, \zeta^*)\, d\zeta\, d\zeta^* + \int_{z_0^*}^{z^*}\int_{z_0}^{z} \Gamma_2(z, z^*, \zeta, \zeta^*)F_0^*(\zeta^*, \zeta)\, d\zeta\, d\zeta^*$$

$$\Gamma_1(z, z^*, \zeta, \zeta^*) = \int_{\zeta^*}^{z^*} \Gamma(z, z^*, \zeta, \eta)\, d\eta \tag{6.58}$$

$$\Gamma_2(z, z^*, \zeta, \zeta^*) = C(z, \zeta^*) + \int_{\zeta}^{z} C(\xi, \zeta^*)\Gamma_1(z, z^*, \xi, \zeta^*)\, d\xi = \frac{\Gamma(z, z^*, \zeta, \zeta^*)}{C^*(\zeta^*, \zeta)}.$$

By reversing the steps leading to (6.57) we can easily establish the fact that every solution of (6.47) analytic in $D \times D^*$ can be represented in the form (6.57) where ϕ is given by (6.54).

Theorem 6.6 *Let V be a solution of (6.47) that is analytic in $D \times D^*$. Then V can be represented in the form (6.57) where $\phi(z) = V(z, z_0^*)$ is analytic in D.*

We now want to establish that every solution of (6.43) can be analytically continued as a function of z and $\bar{z}$ into $D \times D^*$. By using the transformation (6.46) we can, without loss of generality, consider the case when $A \equiv 0$, i.e.

$$v_{\bar{z}} = C(z, \bar{z})\bar{v}. \tag{6.59}$$

To accomplish our aim we shall construct two singular solutions of the adjoint equation

$$v_{\bar{z}} + \overline{C(z, \bar{z})}\bar{v} = 0 \tag{6.60}$$

or, in complex form,

$$\partial V/\partial z^* + C^*(z^*, z)V^*(z^*, z) = 0. \tag{6.61}$$

These singular solutions are of the form

$$\begin{aligned} X_1(z, z^*, z_0, z_0^*) &= \frac{1}{2(z-z_0)} + U_1^{(1)}(z, z^*, z_0, z_0^*) \log |z-z_0| \\ &\quad + U_2^{(1)}(z, z^*, z_0, z_0^*) \\ X_2(z, z^*, z_0, z_0^*) &= \frac{1}{2i(z-z_0)} + U_1^{(2)}(z, z^*, z_0, z_0^*) \log |z-z_0| \\ &\quad + U_2^{(2)}(z, z^*, z_0, z_0^*) \end{aligned} \tag{6.62}$$

where $U_j^{(k)}$, $j, k = 1, 2$, are entire functions of their four independent variables and

$$\log |z-z_0| = \tfrac{1}{2} \log (z-z_0)(z^*-z_0^*). \tag{6.63}$$

We shall show how to construct X_1; the construction of X_2 follows in an identical fashion. Substituting X_1 into (6.61) and suppressing the superscripts shows that X_1 will be a solution of (6.61) provided

$$\begin{aligned} &\left[\frac{\partial U_1}{\partial z^*} + C^*(z^*, z) U_1^*\right] \log |z-z_0| \\ &\quad + \left[\frac{\partial U_2}{\partial z^*} + C^*(z^*, z) U_2^*\right] + \frac{U_1 + C^*(z^*, z)}{2(z^*-z_0^*)} = 0. \end{aligned} \tag{6.64}$$

Since a logarithmic singularity cannot be cancelled by a pole-like singularity, (6.64) implies that

$$\frac{\partial U_1}{\partial z^*} + C^*(z, z^*) U_1^* = 0 \tag{6.65a}$$

$$U_1(z, z_0^*, z_0, z_0^*) + C^*(z_0^*, z) = 0 \tag{6.65b}$$

$$\frac{\partial U_2}{\partial z^*} + C^*(z, z) U_2^* = -\left(\frac{U_1 + C^*(z^*, z)}{2(z^*-z_0^*)}\right). \tag{6.65c}$$

From Theorem 6.6 we can construct U_1 and conclude that it is an entire function of its four independent complex variables. Since (6.65b) implies that the right-hand side of (6.65c) is an entire function of its four independent complex variables, we can again use Theorem 6.6 to construct a particular solution of (6.65c) that is an entire function of its independent complex variables. Hence we have established the existence of X_1, and in an identical fashion we can construct X_2. The functions X_1 and X_2 are known as *fundamental solutions* to (6.61).

We shall now use the fundamental solutions X_1 and X_2 to derive an analogue of Cauchy's integral formula for continuously differentiable solutions of (6.59) in a domain D. Let v be a solution of (6.59) in D and u a solution of the adjoint equation (6.60) in D. Then from (6.14) we have that for $\bar{D}_0$ a compact subset of D with smooth boundary ∂D_0 that

$$\frac{1}{2i}\int_{\partial D_0} v(z)u(z)\,dz = \iint_{D_0} \frac{\partial(vu)}{\partial\bar{z}}\,dx\,dy$$

$$= \iint_{D_0} \left(v\frac{\partial u}{\partial\bar{z}} + u\frac{\partial v}{\partial\bar{z}}\right) dx\,dy$$

$$= \iint_{D_0} [C(z,\bar{z})u\bar{v} - \overline{C(z,\bar{z})}\bar{u}v]\,dx\,dy \tag{6.66}$$

which implies

$$\mathrm{Re}\left[\frac{1}{2i}\int_{\partial D_0} v(z)u(z)\,dz\right] = 0. \tag{6.67}$$

Now let x and y be real (i.e. $z^* = \bar{z}$), $z_0 \in D_0$, and define

$$\begin{aligned} S_1(z, z_0) &= X_1(z, z_0) + iX_2(z, z_0) \\ S_2(z, z_0) &= X_1(z, z_0) - iX_2(z, z_0) \end{aligned} \tag{6.68}$$

(where we have suppressed the dependence on $z^* = \bar{z}$, $z_0^* = \bar{z}_0$). Let $D_\varepsilon = \{z : |z - z_0| < \varepsilon\}$ where ε is a small positive number, and in (6.67) let D_0 be $D_0 \backslash D_\varepsilon$ and $u = X_k$. Then from (6.67) we have

$$\int_{\partial D_0} [v(z)X_k(z, z_0)\,dz - \overline{v(z)}\,\overline{X_k(z, z_0)}\,d\bar{z}]$$

$$= \int_{\partial D_\varepsilon} [v(z)X_k(z, z_0)\,dz - \overline{v(z)}\,\overline{X_k(z, z_0)}\,d\bar{z}], \qquad k = 1, 2. \tag{6.69}$$

Multiplying the equation corresponding to $k = 2$ by i and adding this to the equation with $k = 1$ now gives

$$\int_{\partial D_0} [v(z)S_1(z, z_0)\,dz - \overline{v(z)}\,\overline{S_2(z, z_0)}\,d\bar{z}]$$

$$= \int_{\partial D_\varepsilon} [v(z)S_1(z, z_0)\,dz - \overline{v(z)}\,\overline{S_2(z, z_0)}\,d\bar{z}]. \tag{6.70}$$

From (6.68), (6.62), we have

$$S_1(z, z_0) - 1/(z - z_0) = 0(\log|z - z_0|) \quad \text{and} \quad S_2(z, z_0) = 0(\log|z - z_0|) \tag{6.71}$$

and hence, letting ε tend to zero on (6.70), gives

$$v(z_0)=\frac{1}{2\pi i}\int_{\partial D_0}[v(z)S_1(z,z_0)\,dz-\overline{v(z)}\,\overline{S_2(z,z_0)}\,d\bar{z}]. \tag{6.72}$$

Formula (6.72) is known as the *generalized Cauchy's integral formula.* Since D_0 is an arbitrary compact subset of D and for z on ∂D_0 $S_1(z, z_0)$ and $\overline{S_2(z, z_0)}$ can be continued to an analytic function of z_0 and z_0^*, respectively, for z_0 in D, z_0^* in D^*, we now have the following result.

Theorem 6.7 *Let v be a continuously differentiable solution of (6.59) in a domain D where the coefficient C is an entire function of z and $\bar{z}$. Then v is the restriction to D of an analytic function V of the two complex variables z and z^* defined in $D\times D^*$, i.e. $v(z)=V(z,\bar{z})$ for z in D where $V(z, z^*)$ is analytic in $D\times D^*$.*

Corollary 6.4 *Let w be a continuously differentiable solution of (6.43) in a bounded simply connected domain D where the coefficients A and B are entire functions of the independent complex variables z and $\bar{z}$. Then for $(z_0, \bar{z}_0)\in D\times D^*$, $D^*=\{z: \bar{z}\in D\}$, there exists an analytic function defined in D such that, for z in D, w can be represented in the form*

$$w(z)=\left\{\phi(z)+\int_{z_0}^{z}\Gamma_1(z,\bar{z},\zeta,\bar{z}_0)\phi(\zeta)\,d\zeta\right.$$
$$\left.+\int_{\bar{z}_0}^{\bar{z}}\Gamma_2(z,\bar{z},z_0,\zeta^*)\phi^*(\zeta^*)\,d\zeta^*\right\}\exp\left[\int_{\bar{z}_0}^{\bar{z}}A(z,\zeta^*)\,d\zeta^*\right] \tag{6.73}$$

where $\phi^*(z)=\overline{\phi(\bar{z})}$ and Γ_1 and Γ_2 are defined in (6.58).

We are now almost ready to formulate and prove a reflection principle for solutions of (6.43). The only step that needs to be taken before doing this is to extend the representation (6.73) so that it is valid on a portion of the boundary of ∂D. In particular, we assume that D is contained in the lower half plane $y<0$ and that ∂D contains a segment σ of the x-axis with the origin as an interior point (Fig. 6.2). The following theorem is due to Yu [184], [185].

Theorem 6.8 *Let w be a continuously differentiable solution of (6.43) in D such that w is continuous in $D\cup\sigma$. Then for z in $D\cup\sigma$ w can be represented in the form (6.73) where ϕ is analytic in D and continuous in $D\cup\sigma$.*

Proof Let D_0 be an arbitrary open half disk in D where ∂D_0 contains a

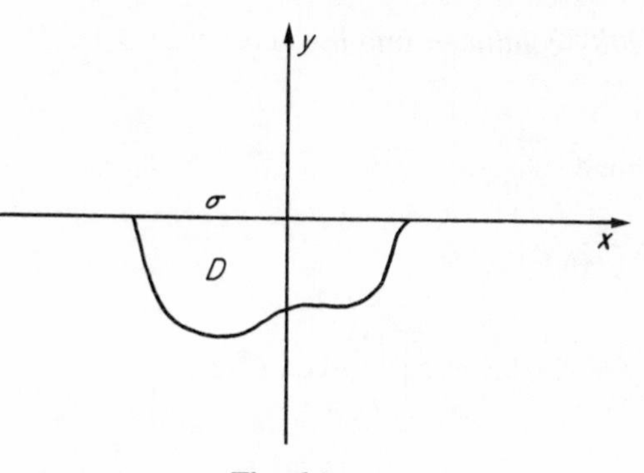

Fig. 6.2

segment σ_0 of σ and $D_0 \subset D \cup \sigma$. It suffices to prove that ϕ, as given in Corollary 6.4, is continuous in $\bar{D}_0$. To this end let

$$k(z) = w(z) \exp\left[-\int_{\bar{z}_0}^{\bar{z}} A(z, \zeta^*)\, d\zeta^*\right]. \tag{6.74}$$

Thus (6.73) becomes

$$k(z) = \phi(z) + \int_{z_0}^{z} \Gamma_1(z, \bar{z}, \zeta, \bar{z}_0)\phi(\zeta)\, d\zeta + \int_{\bar{z}_0}^{\bar{z}} \Gamma_2(z, \bar{z}, z_0, \zeta^*)\phi^*(\zeta^*)\, d\zeta^* \tag{6.75}$$

and, by the method of successive approximations, we have

$$\phi(z) = k(z) + \int_{z_0}^{z} H_1(z, \zeta)k(\zeta)\, d\zeta + \int_{\bar{z}_0}^{\bar{z}} H_2(z, \zeta^*)k^*(\zeta^*)\, d\zeta^* \tag{6.76}$$

where k^* is the analytic continuation of $\bar{k}$ into the complex domain and the resolvent kernels H_1 and H_2 are entire functions of their two independent complex variables. Since k is uniformly continuous on $\bar{D}_0$, k^* is uniformly continuous on $\overline{D_0^*}$, and hence from (6.76) we can conclude that ϕ is continuous on $\bar{D}_0$.

We can now state and prove the following reflection principle due to Yu [184].

Theorem 6.9 (*Reflection principle*) *Let $w \in C^1(D)$ be a solution of* (6.43) *in D, continuous in $D \cup \sigma$, such that for x on σ*

$$\operatorname{Re}[\overline{\lambda(x)}\, w] = \rho(x),$$

where $\lambda(x) = \alpha(x) + i\beta(x)$ and ρ, α and β are analytic in $D \cup \sigma \cup D^*$. Assume $\alpha(z) - i\beta(z) \neq 0$ for z in $D^* \cup \sigma$ and $\alpha(z) + i\beta(z) \neq 0$ for z in

$D\cup\sigma$. Then w can be analytically continued into the domain $D\cup\sigma\cup D^*$ as a solution of (6.43).

Proof By Theorem 6.8 we have

$$w(z)=\left\{\phi(z)+\int_{z_0}^{z}\Gamma_1(z,\bar z,\zeta,\bar z_0)\phi(\zeta)\,d\zeta\right.$$
$$\left.+\int_{\bar z_0}^{\bar z}\Gamma_2(z,\bar z,z_0,\zeta^*)\phi^*(\zeta)\,d\zeta\right\}\exp\left[\int_{\bar z_0}^{\bar z}A(z,\zeta^*)\,d\zeta^*\right] \tag{6.77}$$

$$\overline{w(z)}=\left\{\phi^*(\bar z)+\int_{\bar z_0}\Gamma_1^*(\bar z,z,\zeta^*,z_0)\phi^*(\zeta^*)\,d\zeta^*\right.$$
$$\left.+\int_{z_0}^{z}\Gamma_2^*(\bar z,z,\bar z_0,\zeta)\phi(\zeta)\,d\zeta\right\}\exp\left[\int_{z_0}^{z}A^*(\bar z,\zeta)\,d\zeta\right].$$

The boundary condition on σ implies that for x on σ we have

$$2\rho(x)=(\alpha(x)-i\beta(x))w(x)+(\alpha(x)+i\beta(x))\overline{w(x)}. \tag{6.78}$$

Now set

$$G_1(z)=(\alpha(z)-i\beta(z))\exp\left[\int_{\bar z_0}^{z}A(z,\zeta^*)\,d\zeta^*\right]$$
$$G_2(z)=(\alpha(z)+i\beta(z))\exp\left[\int_{z_0}^{z}A^*(z,\zeta)\,d\zeta\right] \tag{6.79}$$

and note that G_1 and G_2 are analytic in $D\cup\sigma\cup D^*$ and $G_1(z)\neq 0$ for z in $D^*\cup\sigma$, $G_2(z)\neq 0$ for z in $D\cup\sigma$. Substituting (6.77) into (6.78) shows that for z on σ (i.e. $z=\bar z$) we have

$$2\rho(z)=G_1(z)\phi(z)+G_2(z)\phi^*(z)$$
$$+\int_{z_0}^{z}\{\Gamma_1(z,z,\zeta,\bar z_0)G_1(z)+\Gamma_2^*(z,z,\bar z_0,\zeta)G_2(z)\}\phi(\zeta)\,d\zeta$$
$$+\int_{\bar z_0}^{z}\{\Gamma_2(z,z,z_0,\zeta^*)G_1(z)+\Gamma_1^*(z,z,\zeta^*,z_0)G_2(z)\}\phi^*(\zeta^*)\,d\zeta. \tag{6.80}$$

Recall now that ϕ is defined in $D\cup\sigma$ and ϕ^* in $D^*\cup\sigma$. Hence from (6.80) we have that for z in $D^*\cup\sigma$

$$f(z)=G_1(z)\phi(z)+\int_0^{z}\{\Gamma_1(z,z,\zeta,\bar z_0)G_1(z)+\Gamma_2^*(z,z,\bar z_0,\zeta)G_2(z)\}\phi(\zeta)\,d\zeta \tag{6.81}$$

where

$$f(z)=2\rho(z)-G_2(z)\phi^*(z)$$
$$-\int_{z_0}^{0}\{\Gamma_1(z,z,\zeta,\bar{z}_0)G_1(z)+\Gamma_2^*(z,z,\bar{z}_0,\zeta)G_2(z)\}\phi(\zeta)\,d\zeta$$
$$-\int_{\bar{z}_0}^{z}\{\Gamma_2(z,z,z_0,\zeta^*)G_1(z)+\Gamma_1^*(z,z,\zeta^*,z_0)G_2(z)\}\phi^*(\zeta^*)\,d\zeta^*. \tag{6.82}$$

is an integral equation for the (unknown) function ϕ in $D^*\cup\sigma$. Noting that f is analytic for z in D^*, continuous in $D^*\cup\sigma$, we can conclude by the method of successive approximations (since the kernel of (6.81) is analytic in $D\cup\sigma\cup D^*$) that there exists a unique solution ϕ of (6.81) that is analytic in D^* and continuous in $D^*\cup\sigma$. But ϕ is already known to be analytic in D and continuous in $D\cup\sigma$ and by (6.80) agrees with the solution of (6.81) on σ. Hence the solution of (6.81) provides the analytic continuation of ϕ into $D\cup\sigma\cup D^*$. In a similar manner we can continue ϕ^* analytically into $D\cup\sigma\cup D^*$ and the first equation of (6.77) now provides the analytic continuation of w as a solution of (6.43) into all of $D\cup\sigma\cup D^*$.

Exercises

(1) Let ϕ be a solution of

$$\frac{\partial^2\phi}{\partial x^2}+\frac{\partial^2\phi}{\partial y^2}+a(x,y)\frac{\partial\phi}{\partial x}+b(x,y)\frac{\partial\phi}{\partial y}+c(x,y)\phi=0 \tag{1}$$

in a domain D where there exists a positive solution ϕ_0 of (1) in D. Show that (1) can be reduced to a system of the form (6.1).

(2) Let f be an analytic function defined in a domain D with twice continuously differentiable boundary ∂D and let s be the function defined in Theorem 6.4. Show that s may be chosen so as to be real on ∂D and to vanish at a given point z_0 on ∂D. (*Hint:* Use *Privaloff's theorem* which states that if $g(z)=u(x,y)+iv(x,y)$ is analytic in the unit disk and v is Hölder continuous on the unit circle, then g is Hölder continuous on the closed unit disk.)

(3) Use the results of Exercises 1 and 2 to show the existence of a Green's function for the equation

$$\frac{\partial^2\phi}{\partial x^2}+\frac{\partial^2\phi}{\partial y^2}+a(x,y)\frac{\partial\phi}{\partial x}+b(x,y)\frac{\partial\phi}{\partial y}=0$$

defined in a bounded simply connected domain D with twice continuously differentiable boundary ∂D and where the coefficients a and b are Hölder continuous in $\bar{D}$.

(4) Let $\{w_n\}$ be a sequence of pseudoanalytic functions defined in a bounded simply connected domain D with smooth boundary ∂D. Show that if w_n converges uniformly on ∂D then w_n converges uniformly in D to a pseudoanalytic function w.

(5) Let $\tilde{S}_1(z, z_0)$ and $\tilde{S}_2(z, z_0)$ be defined by

$$\tilde{S}_1(z, z_0) = \tilde{X}_1(z, z_0) + \mathrm{i}\tilde{X}_2(z, z_0)$$
$$\tilde{S}_2(z, z_0) = \tilde{X}_1(z, z_0) - \mathrm{i}\tilde{X}_2(z, z_0)$$

where $\tilde{X}_1$ and $\tilde{X}_2$ are fundamental solutions to (6.59). Show that Cauchy's integral formula for pseudoanalytic functions (6.72) may be written in the form

$$v(z_0) = -\frac{1}{2\pi\mathrm{i}}\int_{\partial D_0} [v(z)\tilde{S}_1(z_0, z)\,\mathrm{d}z - \overline{v(z)}\,\tilde{S}_2(z_0, z)\,\mathrm{d}\bar{z}]$$

(6) Show that (6.56) is valid.

(7) Show by the method of successive approximations that there exists a unique solution of (6.81) that is analytic in D^* and continuous in $D^* \cup \sigma$.

7

The backwards heat equation

In this and the following chapter we shall show how the methods developed in the previous chapters can be used to investigate two of the classical inverse problems in applied mathematics: the backwards heat equation and the inverse scattering problem. Other inverse problems can also be studied by the use of function theoretic methods and we refer the reader to Section 4 of Chapter 4, and to Bauer, Garabedian and Korn [11], Hill [84], Garabedian [62], [65] and Garabedian and Lieberstein [67] for further examples. An interesting characteristic of inverse problems such as the ones we are about to consider is that they are in general improperly posed in the real domain, i.e. the solution either does not exist, is not unique, or does not depend continuously on the initial data. We have already met one such improperly posed problem in Chapter 3 where we showed that Cauchy's problem for Laplace's equation was improperly posed. A similar problem was also encountered in Chapter 4. Due to what now appears to be an inexplicable bias, the area of improperly posed problems in applied mathematics was left almost untouched by mathematicians until the late 1950s, when the beginnings of a theory began to be developed. Thus, in 1961 Courant wrote (Courant and Hilbert [43]): 'The stipulations about existence, uniqueness and stability of solutions dominate classical mathematical physics. They are deeply inherent in the ideal of a unique, complete and stable determination of physical events by appropriate conditions at the boundaries, at infinity, at time $t=0$, or in the past.' Laplace's vision of the possibility of calculating the whole future of the physical world from complete data of the present state is an extreme expression of this attitude. However, this rational ideal of causal–mathematical determination was gradually eroded by confrontation with physical reality. Nonlinear phenomena, quantum theory and the advent of powerful numerical methods have shown that 'properly posed' problems are by far not the only ones which appropriately reflect real phenomena. So far, unfortunately, little mathematical progress has been made in the important task of solving or even identifying and formulating such problems which are not 'properly posed' but still are important and motivated by realistic situations. Since the time of these observations by Courant, there has been a virtual explosion of interest in improperly posed problems, as evidenced by the long lists of references in the monographs by Lattes and Lions [109], Lavrentiev [110], Lavrentiev, Romanov and Vasiliev [111], Payne [142] and Tikhonov and Arsenin [165], as well as the proceedings of the conferences organized by Anger [8], Carasso and Stone [22], Knops [104] and Nashed [133]. However, as the interest has risen, the problems have multiplied, and many of the most important improperly posed problems arising in applications still require a solution that is suitable for practitioners.

In order to investigate a problem that is improperly posed, we must answer two basic questions: (1) What do we mean by a solution? and (2) How do we construct this solution? The answers to these questions are by no means trivial. For example, as initially posed a solution may not even exist in the classical sense, or if it does exist, may not be defined in a large enough domain to be of practical use. In this context it is worthwhile recalling the remark of Lanczos: 'A lack of information cannot be remedied by any mathematical trickery'. Hence, in order to determine what we mean by a solution it is often necessary to introduce 'nonstandard' information gained from an intimate knowledge of the physical situation that one is trying to model. Even after we have resolved the question of what we mean by a solution, there remains the problem of actually constructing such a solution, and this is often complicated by the fact that the above-mentioned 'nonstandard' information has been incorporated into the mathematical model, thus leading to 'nonstandard' problems in analysis.

In this chapter we shall consider one of the classical improperly posed problems in applied mathematics: the backwards heat equation. For an excellent survey of the numerous methods that have been used to investigate this problem we refer the reader to the lecture notes by Payne [142]. In this chapter we shall only discuss two of these methods: logarithmic convexity and the method of quasireversibility. Physically, the problem we are about to consider is to determine the temperature of a solid in the past from a knowledge of its temperature in the present and the temperature on the boundary of the solid in the past. Mathematically, we can formulate this problem in the following manner (assuming zero boundary data and a homogeneous medium): to find a solution u of

$$\Delta_n u = u_t \quad \text{in} \quad D\times[0, T] \tag{7.1a}$$

$$u = 0 \quad \text{on} \quad \partial D\times[0, T] \tag{7.1b}$$

$$u(\mathbf{x}, T) = f(\mathbf{x}) \quad \text{for} \quad \mathbf{x}\in D \tag{7.1c}$$

for a prescribed function f, where D denotes the given solid. We note that from Theorem 3.9 it follows that no solution exists to this problem unless f is analytic. Furthermore, even if f is analytic, the solution, if it exists, does not depend continuously on the data f. To see this let ϕ_k be a (normalized) eigenfunction corresponding to an eigenvalue λ_k of

$$\Delta_n\phi + \lambda\phi = 0 \quad \text{in} \quad D \tag{7.2a}$$

$$\phi = 0 \quad \text{on} \quad \partial D. \tag{7.2b}$$

Then $u_k(\mathbf{x}, t) = (1/\lambda_k)\phi_k(\mathbf{x}) \exp(-\lambda_k(t-T))$ is a solution of (7.1a)–(7.1c) for $f(\mathbf{x}) = (1/\lambda_k)\phi_k(\mathbf{x})$, i.e. if ϕ_k is normalized such that $\|\phi_k\| = 1$ then

$$\|f\| = 1/\lambda_k. \tag{7.3}$$

But for $0 \leq t < T$

$$\|u_k(\mathbf{x}, t)\| = (1/\lambda_k) \exp(\lambda_k(T-t)) \tag{7.4}$$

and since $\lim_{k\to\infty} \lambda_k = \infty$ (cf. Garabedian [61]), we have $\|f\|$ tends to zero as k tends to infinity whereas $\|u_k(\mathbf{x}, t)\|$ tends to infinity. Thus the solution of (7.1a)–(7.1c) does not depend continuously on the data f.

In this chapter we shall show how the above problems can be avoided if we look for a solution of (7.1a)–(7.1c) in the class of solutions to (7.1a), (7.1b) that satisfy an *a priori* bound, and define a 'solution' to be a function that is either in this class and best approximates the data (7.1c), or else satisfies a well-posed perturbed equation where the perturbation parameter depends on the *a priori* bound. This leads us to a discussion of logarithmic convexity methods (Agmon [2]; Payne [142]) and the method of quasireversibility (Lattes and Lions [109]). The reader should be warned that these are only two of many methods for treating improperly posed problems such as (7.1a)–(7.1c) and for a discussion of some of the other approaches the reader should consult Cannon [21], John [96], Nashed [134], Payne [142], and the references contained therein.

We note that having resolved the problems associated with (7.1a)–7.1c), we can then easily handle the nonhomogeneous problem

$$\Delta_n u = u_t \qquad \text{in} \quad D\times[0, T] \tag{7.5a}$$

$$u = g(\mathbf{x}, t) \quad \text{on} \quad \partial D\times[0, T] \tag{7.5b}$$

$$u(\mathbf{x}, T) = f(\mathbf{x}) \qquad \text{for} \quad \mathbf{x}\in D. \tag{7.5c}$$

To see this we observe that if we first solve the well-posed 'forward' problem

$$\Delta_n u^{(1)} = u_t^{(1)} \qquad \text{in} \quad D\times[0, T] \tag{7.6a}$$

$$u^{(1)} = g(\mathbf{x}, t) \quad \text{on} \quad \partial D\times[0, T] \tag{7.6b}$$

$$u^{(1)}(\mathbf{x}, 0) = h(\mathbf{x}) \qquad \text{for} \quad \mathbf{x}\in D \tag{7.6c}$$

where $h(\mathbf{x})$ is continuous in $\bar{D}$ such that $h(\mathbf{x})=g(\mathbf{x},0)$ for $\mathbf{x}\in\partial D$, and then 'solve' (7.1a)–(7.1c) with $f(\mathbf{x})$ replaced by $f(\mathbf{x})-u^{(1)}(\mathbf{x},T)$ to obtain $u^{(2)}(\mathbf{x},t)$, then the solution of (7.5a)–(7.5c) is given by $u=u^{(1)}+u^{(2)}$. It is perhaps worthwhile to point out that although existence and continuous dependence fails for (7.1a)–(7.1c), a solution, if it exists, is unique (cf. Theorem 7.1).

7.1 The method of logarithmic convexity

Our aim in this section is to show how continuous dependence of u on f, i.e. *stability*, can be restored to problem (7.1a)–(7.1c) if we require the solution to lie in the set M defined by

$$\mathsf{M}=\{\phi\colon \phi\in C^2(D\times(0,T))\cap C^{\circ}(\bar{D}\times[0,T]) \quad\text{and}\quad \|\phi(x,0)\|^2\leqslant M\}$$

where $\|\cdot\|$ denotes the L_2 norm over D and M is a prescribed positive constant. Physically this means that we have an *a priori* bound on the temperature of the solid at time $t=0$ (and hence for $0\leqslant t\leqslant T$). We begin by defining the functional $F(t)=F(t;u)$ by

$$F(t)=\iint_D [u(\mathbf{x},t)]^2\,\mathrm{d}\mathbf{x}=\|u(t)\|^2 \tag{7.7}$$

where u satisfies (7.1a), (7.1b). We assume that $F(0)\neq 0$ since if this were not true it follows from the maximum principle that $u(\mathbf{x},t)=0$ for all $(\mathbf{x},t)\in D\times(0,T)$. Under this assumption it is easy to verify that $F(t)\neq 0$ for any $t\in[0,T]$. Now assume further that D is a bounded simply connected domain with smooth boundary so that we can apply Green's theorem. Then from (7.7) we have

$$F'(t)=2\iint_D u(\partial u/\partial t)\,\mathrm{d}\mathbf{x}=2(u,\partial u/\partial t) \tag{7.8}$$

and by Green's theorem

$$F'(t)=2\iint_D u\Delta_n u\,\mathrm{d}\mathbf{x}=-2\iint_D |\mathrm{grad}\, u|^2\,\mathrm{d}\mathbf{x}. \tag{7.9}$$

Hence, we have from (7.9) that

$$\begin{aligned}F''(t)&=-4\iint_D \mathrm{grad}\, u\cdot\mathrm{grad}\,(\partial u/\partial t)\,\mathrm{d}\mathbf{x}\\ &=4\iint_D (\partial u/\partial t)\Delta_n u\,\mathrm{d}\mathbf{x}\\ &=4\iint_D (\partial u/\partial t)^2\,\mathrm{d}\mathbf{x}\\ &=4\|\partial u/\partial t\|^2.\end{aligned} \tag{7.10}$$

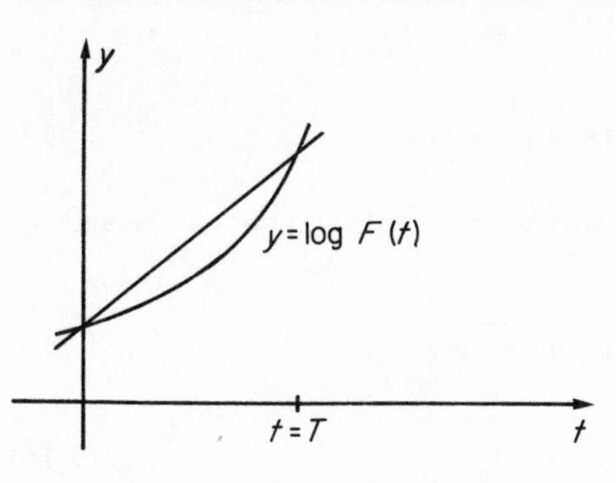

Fig. 7.1

From (7.8), (7.10) we now have

$$FF'' - (F')^2 = 4[\|u\|^2 \, \|\partial u/\partial t\|^2 - (u, \partial u/\partial t)^2] \tag{7.11}$$

and by Schwartz's inequality

$$FF'' - (F')^2 \geqslant 0. \tag{7.12}$$

(7.12) can be rewritten as

$$(\log F)'' \geqslant 0 \tag{7.13}$$

which implies that $\log F$ is a convex function of t. Hence (see Fig. 7.1)

$$\log F(t) \leqslant \log F(0)((T-t)/T) + \log F(T)(t/T) \tag{7.14}$$

or

$$F(t) \leqslant [F(0)]^{(T-t)/T} [F(T)]^{t-T}. \tag{7.15}$$

Now suppose $u \in \mathsf{M}$. Then from (7.15) we have the following theorem.

Theorem 7.1 *Let* $u \in C^2(D \times (0, T)) \cap C^0(\bar{D} \times [0, T])$ *be a solution of* (7.1a)–(7.1c) *belonging to* M. *Then*

$$\|u(t)\| \leqslant M^{(1-t/T)} \|f\|^{t/T}$$

for $0 \leqslant t \leqslant T$. *Hence from Theorem* 7.1 *we can conclude that for* $t > 0$ *the solution of* (7.1a)–(7.1c) *depends Hölder continuously on the data f provided we assume a priori that* $u \in \mathsf{M}$.

To obtain an approximate solution to (7.1a)–(7.1c) in the class M we can now use the method of least squares (Miller [129]). Let $\{u_n\}$ be a complete family of solutions to (7.1a) in $D \times [0, T]$, for example the heat polynomials constructed in Chapter 5. Then for fixed N we want to minimize the functional

$$\left\| \sum_{n=0}^{N} a_n u_n \right\|^2_{\partial D \times (0,T)} + \left\| \sum_{n=0}^{N} a_n u_n(\mathbf{x}, T) - f(\mathbf{x}) \right\|^2_D \tag{7.16}$$

subject to

$$\left\|\sum_{n=0}^{N} a_n u_n\right\|^2_{\partial D\times(0,T)} + \left\|\sum_{n=0}^{N} a_n u_n(\mathbf{x},0)\right\|^2_D \leq M. \tag{7.17}$$

(7.16) and (7.17) can be combined if we fix $\varepsilon > 0$ and choose N and a_n, $n = 0, 1, \ldots, N$, such that

$$\left\|\sum_{n=0}^{N} a_n u_n\right\|^2_{\partial D\times(0,T)} + \left\|\sum_{n=0}^{N} a_n u_n(\mathbf{x},T) - f(\mathbf{x})\right\|^2_D + \frac{\varepsilon}{M}\left\|\sum_{n=0}^{N} a_n u_n(\mathbf{x},0)\right\|^2_D \leq \varepsilon. \tag{7.18}$$

For computational purposes we would fix N and ε and minimize the left-hand side of (7.18). This can be accomplished by standard methods (Miller [129]). From Theorem 7.1 and the discussion following equation (7.5) we can now conclude that

$$u^N(\mathbf{x},t) = \sum_{n=0}^{N} a_n u_n(\mathbf{x},t) \tag{7.19}$$

approximates the solution $u \in \mathsf{M}$ of (7.1a)–(7.1c) in an L_2 sense. We note that the minimization problem (7.18) is essentially Tikhonov's regularization method (cf. Tikhonov and Arsenin [165]).

7.2 Quasireversibility and pseudoparabolic equations

We now turn to an alternate method for constructing solutions to (7.1a)–(7.1c), the method of quasireversibility. This approach was initially developed by Lattes and Lions [109] and is based on the idea of perturbing the heat equation in order to stabilize the problem and then investigating the limiting behavior as the perturbation vanishes. For the sake of simplicity we shall consider only the case of one-space dimension. However, we shall consider the case when the heated rod is nonhomogeneous. In this case, by a change of dependent and independent variables (cf. Colton [35]), we can reduce our problem to that of finding a solution u of

$$u_{xx} + q(x)u = u_t; \qquad -1 < x < 1, \quad 0 < t < T \tag{7.20a}$$

$$u(-1,t) = u(1,t) = 0; \qquad 0 \leq t \leq T \tag{7.20b}$$

$$u(x,T) = f(x); \qquad -1 \leq x \leq 1 \tag{7.20c}$$

where q is assumed to be negative and continuous for $-1 \leq x \leq 1$. We first outline the original approach of Lattes and Lions and point out some of its limitations.

Instead of considering the improperly posed problem (7.20a)–(7.20c) we consider the properly posed problem (Friedman [58])

$$\varepsilon\left(\frac{\partial^2}{\partial x^2}+q(x)\right)^2 u+\frac{\partial^2 u}{\partial x^2}+q(x)u=\frac{\partial u}{\partial t};\qquad -1<x<1,\ 0<t<T \tag{7.21a}$$

$$u(-1,t)=u(1,t)=0;\qquad 0\leq t\leq T \tag{7.21b}$$

$$\left(\frac{\partial^2 u}{\partial x^2}\right)(-1,t)=\left(\frac{\partial^2 u}{\partial x^2}\right)(1,t)=0;\qquad 0\leq t\leq T \tag{7.21c}$$

$$u(x,T)=f(x);\qquad -1\leq x\leq 1 \tag{7.21d}$$

where ε is a small positive parameter. If ϕ_n and λ_n are the eigenfunctions and eigenvalues, respectively, of

$$u_{xx}+[\lambda_n+q(x)]u=0;\qquad -1<x<1 \tag{7.22a}$$

$$u(-1)=u(1)=0 \tag{7.22b}$$

and f is sufficiently smooth then we can represent the solution of (7.21a)–(7.21d) in the form

$$u(x,t)=\sum_{n=1}^{\infty} f_n\phi_n(x)\exp\left[\lambda_n(\varepsilon\lambda_n-1)(t-T)\right] \tag{7.23}$$

where f_n are the Fourier coefficients of f. Note that for sufficiently small ε the first few terms of (7.23) are essentially the same as the corresponding formal series expansion of the solution to (7.20a)–(7.20c). At $t=0$ we have

$$u(x,0)=\sum_{n=1}^{\infty} f_n\phi_n(x)\exp\left[-\lambda_n(\varepsilon\lambda_n-1)T\right]. \tag{7.24}$$

Our aim now is to find a new problem whose solution approximates the solution of (7.20a)–(7.20c). To this end let U satisfy

$$U_{xx}+q(x)U=U_t;\qquad -1<x<1,\qquad 0<t<T \tag{7.25a}$$

$$U(-1,t)=U(1,t)=0;\qquad 0\leq t\leq T \tag{7.25b}$$

$$U(x,0)=\sum_{n=1}^{\infty} f_n\phi_n(x)\exp\left[-\lambda_n(\varepsilon\lambda_n-1)T\right]. \tag{7.25c}$$

Then this problem is well posed and its solution is given by

$$U(x,t)=\sum_{n=1}^{\infty} f_n\phi_n(x)\exp\left[-\varepsilon\lambda_n^2T-\lambda_n(t-T)\right]. \tag{7.26}$$

We therefore have

$$\|U(x,T)-f(x)\|^2=\int_{-1}^{1}[U(x,T)-f(x)]^2\,\mathrm{d}x=\sum_{n=1}^{\infty} f_n^2[1-\exp(-\varepsilon\lambda_n^2T)^2], \tag{7.27}$$

and as ε tends to zero $U(x,T)$ tends to $f(x)$ in the L_2 norm.

The perturbation (7.21a) is far from unique, and an alternate equation has been suggested by Gajewski and Zacharias [60] and Showalter [160]. In their work the improperly posed problem (7.20a)–(7.20c) is replaced by

$$\varepsilon\frac{\partial}{\partial t}\left(\frac{\partial^2 u}{\partial x^2}+q(x)u\right)+\frac{\partial^2 u}{\partial x^2}+q(x)u=\frac{\partial u}{\partial t}; \qquad -1<x<1, \qquad 0<t<T \tag{7.28a}$$

$$u(-1,t)=u(1,t)=0; \qquad 0\leqslant t\leqslant T \tag{7.28b}$$

$$u(x,T)=f(x) \tag{7.28c}$$

where $\varepsilon>0$. Equation (7.28a) is known as an equation of pseudoparabolic or Sobolev type and has been extensively studied in recent years by a large number of mathematicians (see the references in Carroll and Showalter [23]). We shall show shortly that (7.28a)–(7.28c) is well posed. The solution of (7.28a)–(7.28c) is given by

$$u(x,t)=\sum_{n=1}^{\infty} f_n\phi_n(x)\exp\left[-\frac{\lambda_n(t-T)}{1+\varepsilon\lambda_n}\right] \tag{7.29}$$

where f_n and ϕ_n are defined as above. We now define V to be the solution of

$$V_{xx}+q(x)V=V_t; \qquad -1<x<1, \qquad 0<t<T \tag{7.30a}$$

$$V(-1,t)=V(1,t)=0; \qquad 0\leqslant t\leqslant T \tag{7.30b}$$

$$V(x,0)=\sum_{n=1}^{\infty} f_n\phi_n(x)\exp\left[\frac{\lambda_n T}{1+\varepsilon\lambda_n}\right] \tag{7.30c}$$

and easily verify that

$$V(x,t)=\sum_{n=1}^{\infty} f_n\phi_n(x)\exp\left[\frac{\lambda_n T}{1+\varepsilon\lambda_n}-\lambda_n t\right]. \tag{7.31}$$

From (7.31) we have

$$\|V(x,T)-f(x)\|=\sum_{n=1}^{\infty} f_n^2\left[1-\exp\left(-\frac{\varepsilon\lambda_n^2 T}{1+\varepsilon\lambda_n}\right)\right]^2 \tag{7.32}$$

and hence as ε tends to zero $V(x,T)$ tends to $f(x)$ in the L_2 norm. Indeed, for $\varepsilon>0$, $V(x,T)$ is a better approximation to $f(x)$ than is $U(x,T)$. Other perturbations besides (7.28a) and (7.21a) have also been suggested, and a particularly attractive approach has recently been developed by Buzbee and Carasso [20].

Although both of the above methods of perturbing the heat equation lead to stable methods for constructing a solution to (7.20a)–(7.20c) that approximate f at time $t=T$, there are serious disadvantages to both approaches as they are thus far presented. To begin with, although both

U and V tend to f as ε tends to zero at time $t=T$, for preceding times the solutions in general fail to exist as ε tends to zero. Since this is precisely where we wish to determine the solution u of (7.20a)–(7.20c), in a sense we are back to the same improperly posed problem we began with! A second problem is computational in the sense that for ε small the convergence of the series for U and V is very slow and hence it is necessary to compute a large number of eigenfunctions and eigenvalues to the Sturm–Liouville problem (7.22a), (7.22b). For q not equal to a constant this is by no means an easy task! Alternatively, if finite difference methods are used numerically to solve (7.21a)–(7.21d) or (7.28a)–(7.28c) instabilities occur for time intervals not of the same order as ε. The first of the above problems has been studied by Ewing [54] and Miller [130], and an approach for dealing with the second problem has been suggested by Colton [27]. In what follows we shall describe the methods of Ewing and Colton, which are based on the pseudoparabolic equation (7.28a). However, before entering into this subject we shall establish that the initial-boundary value problem (7.28a)–(7.28c) is well posed. Our approach to this question is based on Colton [27], [30]; for alternate methods the reader is referred to Carroll and Showalter [23], Lagnese [108], Rundell [151], Showalter [159] and Ting [166].

Instead of considering the first initial-boundary value problem for (7.28a) we shall consider the corresponding problem for the slightly more general pseudoparabolic equation

$$L[u]\equiv\frac{\partial^3 u}{\partial x^2\,\partial t}+d(x,t)\frac{\partial u}{\partial t}+\eta\frac{\partial^2 u}{\partial x^2}+a(x)\frac{\partial u}{\partial x}+b(x)u=p(x,t) \tag{7.33}$$

defined in the rectangle $R=\{(x,t): -1<x<1,\ 0<t<T\}$ and make the assumption that $d\in C^1(\bar{R})$, $p\in C(\bar{R})$, $a\in C^1[-1,1]$, $b\in C[-1,1]$. In (7.33), η is a constant. Following Colton [27], [30] we shall introduce a special solution of (7.33) analogous to the complex Riemann function for elliptic equations introduced in Chapter 4. This function can be constructed by iteration and we shall refer to it as the Riemann function for (7.33). We shall then use this Riemann function to reduce the solution of the first initial-boundary value problem for (7.33) to that of solving a one-dimensional Volterra integral equation. This is made possible by the fact that both the lines $t=$ constant and $x=$ constant are characteristics of (7.33). Our approach leads in a direct and natural manner to sufficient conditions for the existence, uniqueness, and continuous dependence on the initial-boundary data of the solution to the first initial-boundary value problem for (7.33).

We begin by defining the adjoint equation to $L[u]=0$ to be

$$M[v]\equiv\frac{\partial^3 v}{\partial x^2\,\partial t}+d(x,t)\frac{\partial v}{\partial t}-\eta\frac{\partial^2 v}{\partial x^2}+\frac{\partial}{\partial x}(a(x)v)-b(x)v=0. \tag{7.34}$$

Now let $(\xi, \tau) \in R$ and integrate the identity

$$v_t L[u] - u_t M[v] = (\partial/\partial x)[u_{xt}v_t - u_t v_{xt} - a u_t v + \eta u_x v_t + \eta u_t v_x] + (\partial/\partial t)[a u_x v + b u v - \eta u_x v_x] \tag{7.35}$$

over the rectangle $R_{\xi\tau}$ which is bounded by the lines $x = -1$, $x = \xi$, $t = 0$ and $t = \tau$. An application of Green's formula gives

$$\int_0^\tau \int_{-1}^\xi (v_t L[u] - u_t M[v])\, dx\, dt = \int_{\partial R_{\xi\tau}} (u_{xt}v_t - u_t v_{xt} - a u_t v + \eta u_x v_t + \eta u_t v_x)\, dt - (a u_x v + b u v - \eta u_x v_x)\, dx. \tag{7.36}$$

Suppose there exists a function $v(x, t; \xi, \tau)$ satisfying $M[v] = 0$ in $R_{\xi\tau}$ and the initial conditions

$$v_x(\xi, t; \xi, \tau) = (1/\eta)[1 - e^{\eta(t-\tau)}] \tag{7.37a}$$

$$v(\xi, t; \xi, \tau) = 0, \tag{7.37b}$$

$$v(x, \tau; \xi, \tau) = 0, \tag{7.37c}$$

where if $\eta = 0$ the initial condition (7.37a) is to be interpreted in its limiting form as η tends to zero. Then if there exists a solution u of $L[u] = p$ in R satisfying

$$u(-1, t) = f(t) \tag{7.38a}$$

$$u_x(-1, t) = g(t) \tag{7.38b}$$

$$u(x, 0) = h(x) \tag{7.38c}$$

where $f, g \in C^1[0, T]$, $h \in C^2[-1, 1]$, we have from (7.36) that

$$\begin{aligned} u(\xi, \tau) = h(\xi) + \int_{-1}^{\xi} & [a(x)h'(x)v(x, 0; \xi, \tau) - \eta h'(x)v_x(x, 0; \xi, \tau) \\ & + b(x)h(x)v(x, 0; \xi, \tau)]\, dx + \int_0^\tau [g'(t)v_t(-1, t; \xi, \tau) \\ & - f'(t)v_{xt}(-1, t; \xi, \tau) - a(-1)f'(t)v(-1, t; \xi, \tau) \\ & + \eta g(t)v_t(-1, t; \xi, \tau) + \eta f'(t)v_x(-1, t; \xi, \tau)]\, dt \\ & + \int_0^\tau \int_{-1}^\xi p(x, t)v_t(x, t; \xi, \tau)\, dx\, dt. \end{aligned} \tag{7.39}$$

(7.39) gives the solution of the Goursat problem (7.33), (7.38) in terms of the *Riemann function* v. In particular, if we can show that v exists and is sufficiently smooth, then we can use (7.39) to verify directly the existence of a function u satisfying $L[u] = p$ and the initial data (7.38). We now turn our attention to this construction, basing our approach on the methods first introduced in Chapter 3.

We first rewrite (7.34) in the form

$$v_{xxt} = F(x, t, v, v_t, v_x, v_{xx}) \tag{7.40}$$

where

$$F(x, t, v_t, v_x, v_{xx}) = \eta v_{xx} - d(x, t)v_t - (a(x)v)_x + b(x)v. \tag{7.41}$$

Let

$$s(x, t) = v_{xxt}(x, t) \tag{7.42}$$

and define the operators $\mathbf{B}_1, \mathbf{B}_2, \mathbf{B}_3, \mathbf{B}_4$ by

$$\begin{aligned}
\mathbf{B}_1[s] &= v(x, t) = \int_\xi^x \int_\tau^t (x - x_1)s(x_1, t_1)\,dx_1\,dt_1 + (1/\eta)(x-\xi)(1 - e^{\eta(t-\tau)}) \\
\mathbf{B}_2[s] &= v_t(x, t) = \int_\xi^x (x - x_1)s(x_1, t)\,dx_1 - (x-\xi)e^{\eta(t-\tau)} \\
\mathbf{B}_3[s] &= v_x(x, t) = \int_\xi^x \int_\tau^t s(x_1, t_1)\,dx_1\,dt_1 + (1/\eta)(1 - e^{\eta(t-\tau)}) \\
\mathbf{B}_4[s] &= v_{xx}(x, t) = \int_\tau^t s(x, t_1)\,dt_1.
\end{aligned} \tag{7.43}$$

Let $C(R_{\xi\tau})$ be the Banach space of continuous functions defined and continuous in the closed rectangle $R_{\xi\tau}$ with norm

$$\|s\|_\lambda = \max_{(x,t)\in R_{\xi\tau}} \{e^{-\lambda[(\xi-x)+(\tau-t)]}|s(x, t)|\} \tag{7.44}$$

where $\lambda > 0$ is fixed, $s \in C(R_{\xi\tau})$. The existence of a Riemann function v is now reduced to finding a fixed point of the operator $\mathbf{T}$ in $C(R_{\xi\tau})$ where

$$\mathbf{T}[s] = F(x, t, \mathbf{B}_1[s], \mathbf{B}_2[s], \mathbf{B}_3[s], \mathbf{B}_4[s]). \tag{7.45}$$

Due to the linearity and continuity of the coefficients of $M[v] = 0$ we have that for $(x, t) \in R_{\xi\tau}$ there exists a constant C such that

$$\begin{aligned}
\|\mathbf{T}[s_1] - \mathbf{T}[s_2]\|_\lambda \leq C\{&\|\mathbf{B}_1[s_1] - \mathbf{B}_1[s_2]\|_\lambda + \|\mathbf{B}_2[s_1] + \mathbf{B}_2[s_2]\|_\lambda \\
&+ \|\mathbf{B}_3[s_1] - \mathbf{B}_3[s_2]\|_\lambda + \|\mathbf{B}_4[s_1] - \mathbf{B}_4[s_2]\|_\lambda\}
\end{aligned} \tag{7.46}$$

From estimates of the form

$$\begin{aligned}
\left|\int_\xi^x s(x_1, t)\,dx_1\right| &\leq \int_x^\xi \|s\|_\lambda e^{\lambda[(\xi-x_1)+(\tau-t)]}\,dx_1 \\
&\leq (1/\lambda)\,\|s\|_\lambda\,[e^{\lambda[(\xi-x)+(\tau-t)]} - e^{\lambda(\tau-t)}]
\end{aligned} \tag{7.47}$$

i.e.

$$\begin{aligned}
\left\|\int_\xi^x s(x_1, t)\,dx_1\right\|_\lambda &\leq (1/\lambda)\,\|s\|_\lambda\,[1 - e^{-\lambda(\xi-x)}] \\
&\leq (1/\lambda)\,\|s\|_\lambda,
\end{aligned} \tag{7.48}$$

we have

$$\|\mathbf{B}_i[s_1]-\mathbf{B}_i[s_2]\| \leqslant (C_i/\lambda)\,\|s_1-s_2\|_\lambda, \qquad i=1,2,3,4 \tag{7.49}$$

where the C_i are positive constants independent of λ. From (7.46) this implies that

$$\|\mathbf{T}[s_1]-\mathbf{T}[s_2]\|_\lambda \leqslant (M/\lambda)\,\|s_1-s_2\|_\lambda \tag{7.50}$$

where M is a positive constant independent of λ. (7.50) now implies that for λ sufficiently large $\mathbf{T}$ takes $C(R_{\xi\tau})$ into itself and is a contraction mapping. Thus, by the Banach contraction mapping principle, there exists an $s \in C(R_{\xi\tau})$ such that $s=\mathbf{T}[s]$ and we have established the existence of a Riemann function v. Note that by construction we have shown that with respect to its first two variables v is a (classical) solution of $M[v]=0$ and is furthermore continuous with respect to its four independent variables for $(x, t)\in R_{\xi\tau}$, $x \leqslant \xi \leqslant 1$, $t<\tau\leqslant T$.

We now wish to establish some further regularity properties of v. Let $(\alpha, \beta)\in R_{\xi\tau}$ and let $R_{\alpha\beta}$ be the rectangle bounded by the lines $x=\alpha$, $x=\xi$, $t=\beta$ and $t=\tau$. By the same method we used to construct v we can construct a solution $w(x, t; \alpha, \beta)$ of $L[u]=0$ in $R_{\alpha\beta}$ which satisfies the initial conditions

$$w_x(\alpha, t; \alpha, \beta)=(1/\eta)[e^{-\eta(t-\beta)}-1] \tag{7.51a}$$

$$w(\alpha, t; \alpha, \beta)=0 \tag{7.51b}$$

$$w(x, \beta; \alpha, \beta)=0. \tag{7.51c}$$

Integrating the identity (7.35) over $R_{\alpha\beta}$ (setting $u=w$) and applying Green's formula gives

$$w(\xi, \tau; \alpha, \beta)=v(\alpha, \beta; \xi, \tau), \tag{7.52}$$

i.e. as a function of its last two variables $v(x, t; \xi, \tau)$ is a solution of $L[u]=0$. Following the proof of Lemma 3.1, we can furthermore easily show that if $s=w_{\xi\xi\tau}$ then s_α, s_β and $s_{\alpha\beta}$ are continuous for $\alpha\leqslant\xi\leqslant 1$, $\beta\leqslant\tau\leqslant T$. It is now possible to show directly that (7.39) is the unique solution of $L[u]=p$ satisfying the Goursat data (7.38) and that u depends continuously on the Goursat data f, g and h and their derivatives.

Before proceeding to the solution of the first initial-boundary value problem for $L[u]=p$, we first pause to observe that (7.39) implies that the analytic behavior of solutions to the pseudoparabolic equation (7.33) differs dramatically from that of parabolic equations. Recall from Holmgren's theorem we have that solutions of parabolic equations with analytic coefficients satisfy the unique continuation property with respect to the space variables. In particular, it is not possible for a solution of a parabolic equation with analytic coefficients to vanish for $x<0$ without the solution vanishing identically. However, from (7.39) we see that this is possible for pseudoparabolic equations (set $f=g=p=0$ and let $h=0$ for $x<0$, but be nonzero for $x>0$). On the other hand, the following result of Rundell and Stecher [152] shows that it is not possible for solutions of pseudoparabolic equations to have compact support in x.

Theorem 7.2 *Let u be a classical solution of*

$$u_{xxt}-u_t+u_{xx}=0$$

for $-\infty<x<\infty$, $t\geq 0$, such that $u(x,t)=0$ for $x<0$ and $x>\pi$. Then u vanishes identically.

Proof Expanding u in a Fourier series and using the differential equation to determine the functional dependence of the Fourier coefficients on t shows that

$$u(x,t)=\sum_{n=1}^{\infty} a_n \exp(-\lambda_n t)\sin nx; \qquad \lambda_n=n^2/(1+n^2) \tag{7.53}$$

and hence

$$u(x,0)=\sum_{n=1}^{\infty} a_n \sin nx. \tag{7.54}$$

From the regularity assumption on u we have (see the Exercises) that

$$\sum_{n=1}^{\infty} n\,|a_n|<\infty. \tag{7.55}$$

Hence

$$0=u_x(0,t)=\sum_{n=1}^{\infty} na_n \exp(-\lambda_n t) \tag{7.56}$$

and thus

$$0=\lim_{t\to\infty}\sum_{n=1}^{\infty} na_n \exp((\lambda_1-\lambda_n)t)=a_1. \tag{7.57}$$

Proceeding recursively we have $a_n=0$, $n=1,2,3,\ldots$ and hence $u\equiv 0$, thus proving the theorem.

For further results in this direction the reader is referred to Rundell and Stecher [152].

We now turn to the first initial-boundary value problem for $L[u]=p$, i.e. to find a solution of

$$L[u]=p(x,t) \quad \text{in} \quad R \tag{7.58a}$$

$$u(-1,t)=f(t), \qquad 0\leq t\leq T \tag{7.58b}$$

$$u(1,t)=\phi(t) \qquad 0\leq t\leq T \tag{7.58c}$$

$$u(x,0)=h(x), \qquad -1\leq x\leq 1. \tag{7.58d}$$

Theorem 7.3 *Let $d \leqslant 0$ in $\bar{R}$, $d \in C^1(\bar{R})$, $p \in C(\bar{R})$, $a \in C^1[-1, 1]$, $b \in C[-1, 1]$ and f, $\phi \in C^1[0, T]$, $h \in C^2[-1, 1]$. Then there exists a unique solution to $L[u]=p$ in R satisfying the initial-boundary data* (7.58b)–(7.58d) *and u depends continuously on f, ϕ, h and h'* (*with respect to the maximum norm*).

Proof We set $\xi = 1$ in (7.39). After an integration by parts in the second integral on the right-hand side we arrive at

$$\gamma(\tau) = g(\tau)v_t(-1, \tau; 1, \tau) + \int_0^\tau [\eta v_t(-1, t; 1, \tau) - v_{tt}(-1, t; 1, \tau)]g(t)\, dt \tag{7.59}$$

where

$$\begin{aligned}\gamma(\tau) = {} & \phi(\tau) - h(1) + h'(-1)v_t(-1, 0; 1, \tau) \\ & - \int_{-1}^1 [h'(x)(a(x)v(x, 0; 1, \tau) - \eta v_x(x, 0; 1, \tau)) \\ & \qquad + h(x)b(x)v(x, 0; 1, \tau)]\, dx \\ & + \int_0^\tau f'(t)[v_{xt}(-1, t; 1, \tau) - a(-1)v(-1, t; 1, \tau) \\ & + \eta v_x(-1, t; 1, \tau)]\, dt - \int_0^\tau \int_{-1}^1 p(x, t)v_t(x, t; 1, \tau)\, dx\, dt.\end{aligned} \tag{7.60}$$

Note that due to the assumption that f, $\phi \in C^1[0, T]$, $d \in C^1(\bar{R})$, we can conclude that $\gamma(\tau)$ and the kernel of the integral equation (7.59) are continuously differentiable with respect to τ for $0 \leqslant \tau \leqslant T$ and hence, if a solution g of the integral equation exists, then $g \in C^1[0, T]$. The continuous dependence of u on the initial-boundary data now follows from an integration by parts in (7.60) and (7.39). To show the existence of a unique solution to the integral equation (7.59) on the interval $0 \leqslant \tau \leqslant T$ it is sufficient to show that $v_t(-1, \tau; 1, \tau)$ is never equal to zero on this interval, for in this case we can divide by $v_t(-1, \tau; 1, \tau)$ and solve (7.59) by successive approximations. To show that $v_t(-1, \tau; 1, \tau)$ is never zero we consider the function

$$\mu(x) = v_t(x, \tau; 1, \tau) \tag{7.61}$$

for an arbitrary (but fixed) τ in the interval $0 \leqslant \tau \leqslant T$. Then, from the differential equation (7.58a) and the boundary condition (7.37c), we have

$$\mu_{xx} + d(x, \tau)\mu = 0. \tag{7.62}$$

Hence since $d \leqslant 0$ in $\bar{R}$ we can conclude that if $\mu(-1) = 0$ then $\mu(x) = 0$ for $-1 \leqslant x \leqslant 1$ since $\mu(1) = 0$ by (7.37b). But this then implies that

$\mu_x(1)=v_{xt}(1,\tau;1,\tau)=0$, contradicting (7.37a). Therefore $\mu(-1)=v_t(1,\tau;1,\tau)$ does not equal zero for any τ on the interval $0\leq\tau\leq T$, and hence the integral equation (7.59) is invertible. Solving (7.59) for g and substituting into (7.39) now gives the (unique) solution of (7.58a)–(7.58d).

Note that if we decompose the solution u of (7.58a)–(7.58d) into the sum of two solutions, one of which vanishes on $x=\pm 1$ and the other on $t=0$, then from Theorem 7.3 and the method of eigenfunction expansions we can conclude that u depends continuously on f, ϕ and h in the L_2 norm.

It is easy to show that in general the assumption that $d\leq 0$ in $\bar{R}$ is necessary (cf. the Exercises and Coleman, Duffin and Mizel [26]).

We now return to the initial-boundary value problem (7.20a)–(7.20c) and make the assumption that

$$\int_{-1}^{1} |u(x,0)|^2\,dx=\|u(0)\|^2\leq M \tag{7.63}$$

where M is a positive constant. Under the assumption (7.63) we have exactly as in Theorem 7.1 (see the Exercises) that

$$\|u(t)\|\leq M^{(1-t/T)}\,\|f\|^{t/T} \tag{7.64}$$

for $0\leq t\leq T$. The following theorem of Ewing [54] now shows that for ε sufficiently small the solution of the pseudoparabolic initial-boundary value problem (7.28a)–(7.28c) approximates any approximate solution to (7.20a)–(7.20c) satisfying (7.63).

Theorem 7.4 *Let u be any solution of* (7.20a), (7.20b), (7.63) *such that $\|u(T)-f\|<\delta$. Let $\varepsilon=T[\log M/\delta]^{-1}$ and v be the solution of* (7.28a)–(7.28c). *Then for $0<t\leq T$*

$$\|u(t)-v(t)\|=0([\log M/\delta]^{-1}).$$

Proof We can write

$$u(x,t)=\sum_{n=1}^{\infty}\alpha_n\phi_n(x)\exp(-\lambda_n t) \tag{7.65}$$

where ϕ_n, λ_n are defined as in (7.22a), (7.22b) and the α_n are the Fourier coefficients of $u(x,0)$. By hypothesis we have

$$\sum_{n=1}^{\infty}[\alpha_n\exp(-\lambda_n T)-f_n]^2<\delta^2 \tag{7.66}$$

where the f_n are the Fourier coefficients of f. We write

$$\begin{aligned} u(x,t)-v(x,t) &= \sum_{n=1}^{\infty} \alpha_n\phi_n(x)\exp(-\lambda_n t) \\ &\quad -\sum_{n=1}^{\infty} f_n\phi_n(x)\exp\left[-\frac{\lambda_n(t-T)}{1+\varepsilon\lambda_n}\right] \\ &= \sum_{n=1}^{\infty} \alpha_n\phi_n(x)\left[\exp(-\lambda_n t)-\exp\left(-\lambda_n T-\frac{\lambda_n(t-T)}{1+\varepsilon\lambda_n}\right)\right] \\ &\quad +\sum_{n=1}^{\infty} (\alpha_n\exp(-\lambda_n T)-f_n)\phi_n(x)\exp\left(-\frac{\lambda_n(t-T)}{1+\varepsilon\lambda_n}\right), \end{aligned} \tag{7.67}$$

and hence

$$\|u(t)-v(t)\|\leqslant B(t)\,\|u(0)\|+C(t)\,\|u(T)-f\| \tag{7.68}$$

where

$$\begin{aligned} B(t) &= \max_{n\geqslant 1}|B_n(t)| = \max_{n\geqslant 1}\left|\exp(-\lambda_n t)-\exp\left(-\lambda_n T-\frac{\lambda_n(t-T)}{1+\varepsilon\lambda_n}\right)\right| \\ C(t) &= \max_{n\geqslant 1}\exp\left[-\frac{\lambda_n(t-T)}{1+\varepsilon\lambda_n}\right]. \end{aligned} \tag{7.69}$$

Note that since the coefficient q in (7.20a) is strictly negative, the eigenvalues λ_n are all positive. We now proceed to estimate $B(t)$ and $C(t)$. Since $x/(1+\varepsilon x)$ is an increasing function of x, its maximum occurs in the limit as x tends to infinity and hence

$$C(t)\leqslant\exp((T-t)/\varepsilon). \tag{7.70}$$

We next consider $B_n(t)=B_n(t,\varepsilon)$ as a function of ε. Since $B_n(t,0)=0$ we have by the mean value theorem

$$B_n(t,\varepsilon)=\varepsilon\frac{\lambda_n^2(t-T)}{(1+\xi\lambda_n)^2}\exp\left(-\lambda_n T-\frac{\lambda_n(t-T)}{1+\xi\lambda_n}\right) \tag{7.71}$$

for some ξ, $0<\xi<\varepsilon$. Hence if $t>0$

$$|B_n(t,\varepsilon)|\leqslant\varepsilon T\lambda_n^2\exp(-\lambda_n t)\leqslant C\varepsilon \tag{7.72}$$

where C is a positive constant independent of n and ε. Since by hypothesis $\|u(0)\|\leqslant M$, $\|u(T)-f\|<\delta$, we have from (7.68), (7.70), (7.72) that for $t>0$

$$\begin{aligned} \|u(t)-v(t)\| &\leqslant CM\varepsilon+\delta\exp((T-t)/\varepsilon) \\ &= CMT[\log M/\delta]^{-1}+M^{1-t/T}\delta^{t/T} \\ &= 0([\log M/\delta]^{-1}). \end{aligned} \tag{7.73}$$

In view of Theorems 7.4 and 7.3, we can now obtain an approximation to (7.20a)–(7.20c), (7.63) by obtaining an approximate solution to (7.28a)–(7.28c) for ε chosen as in Theorem 7.4 and we shall now turn our attention to this problem. There is clearly a variety of possible approaches to this problem. In particular, the method of eigenfunction expansions is essentially the approach of constructing a complete family of solutions for the class of solutions to pseudoparabolic equations that vanish on $x=\pm 1$. We shall continue with this approach, but in order to avoid the problem of constructing the eigenfunctions of a Sturm–Liouville problem with variable coefficients we shall, in the spirit of Chapter 5, construct a complete family of solutions to (7.28a) without having any knowledge of the eigenvalues and eigenfunctions of the above-mentioned Sturm–Liouville problem.

We first consider the problem of constructing a set of solutions to (7.28a) that approximates given *initial* data at $t=0$. Since we are not using the method of partial eigenfunction expansions to construct our approximation, members of our set do not in general vanish on the sides $x=\pm 1$ of the rectangle $-1<x<1$, $0<t<T$. We shall therefore also construct a set of solutions to (7.28a) that vanish at $t=0$ and approximate given boundary data on $x=\pm 1$. The union of the above two sets provides us with a complete family of solutions to (7.28a) in the rectangle $-1<x<1$, $0<t<T$. We note that from Theorem 7.3 the first initial-boundary value problem for (7.28a) is well posed in both the ‘backward’ and ‘forward’ directions in time. Our method of constructing a complete family of solutions is based on the ‘forward’ initial-boundary value problem since, as we shall see, it allows us to approximate members of this family easily for small values of the parameter ε.

To construct a set of solutions $\{u_n\}$ such that $\{u_n(x, 0)\}$ is complete in $L_2(-1, 1)$ we first separate variables in (7.28a) to obtain solutions of the form

$$u_n(x, t)=y_n(x)\exp\left\{-\frac{\lambda_n^2 t}{1+\varepsilon\lambda_n^2}\right\} \tag{7.74}$$

where y_n is a solution of

$$\mathrm{d}^2y/\mathrm{d}x^2+(\lambda_n^2+q(x))y=0 \tag{7.75}$$

and the λ_n are as of yet arbitrary real parameters. We now observe that from Chapter 4 any solution of (7.75) can be represented in the form

$$y_n(x)=(\mathbf{I}+\mathbf{T})[h_n]=h_n(x)+\int_{-x}^{x} E(s, x)h_n(s)\,\mathrm{d}s \tag{7.76}$$

where E is the solution of the hyperbolic characteristic initial-value problem

$$E_{xx} - E_{ss} + q(x)E = 0 \tag{7.77a}$$

$$E(x, x) = -\frac{1}{2}\int_0^x q(s)\,\mathrm{d}s \tag{7.77b}$$

$$E(-x, x) = 0, \tag{7.77c}$$

and h_n is a solution of

$$\mathrm{d}^2h_n/\mathrm{d}x^2 + \lambda_n^2 h_n = 0. \tag{7.78}$$

We note that numerical approximations to E can be obtained by using the Cauchy–Euler polygon method (see Chapter 2). We now make use of the following theorem due to Levinson (see Levin [115], p. 415).

Theorem 7.5 *Let $\{\lambda_n\}$ be a sequence of distinct real numbers such that $0 < \lambda_n \leq \pi(n + \frac{1}{4})$, $n = 1, 2, \ldots$. Then the set $\{\exp(\pm i\lambda_n x)\}$ is complete in $L_2(-1, 1)$.*

From (7.76) and the invertibility of the operator $\mathbf{I}+\mathbf{T}$ we can now conclude that if $\{\lambda_n\}$ satisfy the hypothesis of Theorem 7.5 then the set

$$y_n^{\pm}(x) = (\mathbf{I}+\mathbf{T})[\exp(\pm i\lambda_n x)] \tag{7.79}$$

is also complete in $L_2(-1, 1)$, and hence the set of solutions $\{u_n\}$ of (7.28a) defined by

$$u_n(x, t) = y_n^{\pm}(x)\exp\left\{-\frac{\lambda_n^2 t}{1+\varepsilon\lambda_n^2}\right\} \tag{7.80}$$

can be used to construct a solution of (7.28a) that approximates $u(x, 0)$ arbitrarily closely in $L_2(-1, 1)$. Note, however, that in general we do not have $u_n(-1, t) = u_n(1, t) = 0$, and hence we now turn our attention to constructing a set of solutions $\{v_n\}$ to (7.28a) that vanish at $t = 0$ and are such that the set $\{v_n(-1, t)\} \times \{v_n(1, t)\}$ is complete in $L_2(0, T) \times L_2(0, T)$.

In order to construct the set $\{v_n\}$ we shall first develop a potential theory for solutions of (7.28a) that vanish at $t = 0$. To this end we first define two solutions of (7.75) for $\lambda_n = i\omega$ by the formulae

$$f_1(x; \omega) = e^{-(1+\delta_1)\omega}(\mathbf{I}+\mathbf{T})[e^{-\omega x}] \tag{7.81a}$$

$$f_2(x; \omega) = e^{-(1+\delta_2)\omega}(\mathbf{I}+\mathbf{T})[e^{\omega x}] \tag{7.81b}$$

where $\delta_j > 0$, $j = 1, 2$, are arbitrary constants. These constants are introduced in order to ensure that the functions f_1 and f_2 decay exponentially to zero as ω tends to infinity and this fact will be used subsequently to obtain asymptotic estimates on the functions v_n. We now define fundamental solutions for (7.28a) by the formulae

$$K^{(j)}(x, t) = -\frac{1}{\pi i} \oint_{|\omega - \sqrt{1/\varepsilon}| = \delta} f_j(x; \omega) \exp \frac{\omega^2 t}{1 - \varepsilon\omega^2} \, d\omega \tag{7.82}$$

where the path of integration is a circle of radius $\delta > 0$ traversed counterclockwise around the point $\omega = \sqrt{1/\varepsilon}$. An elementary residue computation shows that

$$K^{(j)}(x, 0) = 0 \tag{7.83}$$

and

$$K_t^{(j)}(x, 0) = \varepsilon^{-3/2} f_j(x, 1/\sqrt{\varepsilon}). \tag{7.84}$$

Now let u be a solution of (7.28a) such that $u(x, 0) = 0$, $u(-1, t) = g(t)$, $u(1, t) = h(t)$, where g and h are continuously differentiable functions prescribed on $[0, T]$. We shall show that for ε sufficiently small there exist continuously differentiable densities μ_1 and μ_2 defined on $[0, T]$, $\mu_1(0) = \mu_2(0) = 0$, such that u can be represented in the form

$$u(x, t) = \int_0^t K_t^{(1)}(x, t - \tau)\mu_1'(\tau) \, d\tau + \int_0^t K_t^{(2)}(x, t - \tau)\mu_2'(\tau) \, d\tau. \tag{7.85}$$

Indeed, if we integrate (7.85) by parts we see that μ_1 and μ_2 satisfy the system of Volterra integral equations

$$\begin{aligned} g(t) &= K_t^{(1)}(-1, 0)\mu_1(t) + K_t^{(2)}(-1, 0)\mu_2(t) + \int_0^t K_{tt}^{(1)}(-1, t - \tau)\mu_1(\tau) \, d\tau \\ &\quad + \int_0^t K_{tt}^{(2)}(-1, t - \tau)\mu_2(\tau) \, d\tau \\ h(t) &= K_t^{(1)}(1, 0)\mu_1(t) + K_t^{(2)}(1, 0)\mu_2(t) + \int_0^t K_{tt}^{(1)}(1, t - \tau)\mu_1(\tau) \, d\tau \\ &\quad + \int_0^t K_{tt}^{(2)}(1, t - \tau)\mu_2(\tau) \, d\tau. \end{aligned} \tag{7.86}$$

The system (7.86) will have a unique set of solutions μ_1, μ_2 provided

$$\begin{aligned} 0 \neq \det \begin{vmatrix} K_t^{(1)}(-1, 0) & K_t^{(2)}(-1, 0) \\ K_t^{(1)}(1, 0) & K_t^{(2)}(1, 0) \end{vmatrix} \\ = \varepsilon^{-3}[f_1(-1, 1/\sqrt{\varepsilon})f_2(1, 1/\sqrt{\varepsilon}) - f_1(1, 1/\sqrt{\varepsilon})f_2(-1, 1/\sqrt{\varepsilon})]. \end{aligned} \tag{7.87}$$

However, from (7.81a), (7.81b) we have, integrating $\mathbf{T}[e^{-\omega x}]$ and $\mathbf{T}[e^{\omega x}]$

by parts,

$$\begin{aligned} f_1(-1, \omega) &= e^{-\delta_1\omega}[1+0(1/\omega)] \\ f_2(1; \omega) &= 0\left(\frac{1}{\omega} e^{-\delta_1\omega}\right) \\ f_2(-1; \omega) &= 0\left(\frac{1}{\omega} e^{-\delta_2\omega}\right) \\ f_2(1; \omega) &= e^{-\delta_2\omega}[1+0(1/\omega)]. \end{aligned} \tag{7.88}$$

From (7.87), (7.88) we can conclude that

$$\det \begin{vmatrix} K_t^{(1)}(-1, 0) & K_t^{(2)}(-1, 0) \\ K_t^{(1)}(1, 0) & K_t^{(2)}(1, 0) \end{vmatrix} = \varepsilon^{-3} \exp\left(-\frac{\delta_1+\delta_2}{\sqrt{\varepsilon}}\right)[1+0(\sqrt{\varepsilon})] \tag{7.89}$$

which is nonzero for ε sufficiently small. Hence for ε sufficiently small there exist densities μ_1 and μ_2 such that u can be represented in the form (7.85).

From the point of view of constructing a complete family of solutions there is no loss of generality in assuming that the boundary data satisfied by u are in fact twice continuously differentiable, since by Theorem 7.3 the solution of initial-boundary value problems for pseudoparabolic equations depend continuously on their initial-boundary data. In this case we can integrate (7.85) by parts and use numerical quadrature to conclude that on the rectangle $-1 \leq x \leq 1$, $0 \leq t \leq T$, u can be approximated uniformly on this rectangle by a linear combination of functions of the form

$$\begin{aligned} v_{2n}(x, t) &= \begin{cases} K^{(1)}(x, t-\tau_n); & \tau_n \leq t \\ 0 \quad ; & t < \tau_n \end{cases} \\ v_{2n+1}(x, t) &= \begin{cases} K^{(2)}(x, t-\tau_n); & \tau_n \leq t \\ 0 \quad ; & t < \tau_n \end{cases} \end{aligned} \tag{7.90}$$

where $\{\tau_n\}$ is a dense equidistributed set of points on $[0, T]$ (cf. Davis [44], p. 355). We note that by (7.83) the functions v_n are continuous. We have now proved the following theorem (Colton [35]).

Theorem 7.6 *Let $\{\lambda_n\}$ be a sequence of distinct real numbers such that $0 < \lambda_n \leq \pi(n+\frac{1}{4})$ and let $\{\tau_n\}$ be a dense equidistributed set of points on $[0, T]$. Then for ε sufficiently small the set $\{u_n\} \cup \{v_n\}$, where u_n and v_n are defined by (7.80) and (7.90), respectively, is a complete set of solutions to (7.28a) with respect to the L_2 norm defined over the base or top and the vertical sides of the rectangle $-1 \leq x \leq 1$, $0 \leq t \leq T$.*

Now let $\{\psi_n\}$ denote the complete set of functions defined in Theorem 6.6. In order to approximate the solution of (7.28a)–(7.28c) we minimize the functional

$$\left\|\sum_{n=1}^{N} c_n\psi_n(x, T)-f(x)\right\| + \left\|\sum_{n=1}^{N} c_n\psi_n(-1, t)\right\| + \left\|\sum_{n=1}^{N} c_n\psi_n(1, t)\right\| \tag{7.91}$$

and define the approximate solution to be $\sum_{n=1}^{N} c_n\psi_n(x, t)$. Due to the fact that (7.28a)–(7.28c) is well posed, the minimization procedure indicated above is stable. In this minimization process one should allow not only for the variation of the coefficients c_n but also for the basis functions ψ_n themselves, i.e. by considering τ_n, $n=1,\ldots,N$, and λ_n, $n=1,\ldots,N$, to be variables as well. We are thus led to the problem of finding the minimum of a function of the $4N$ variables $\operatorname{Re} c_n$, $\operatorname{Im} c_n$, λ_n and τ_n, where $1\leq n\leq N$, and this can be accomplished by a variety of known methods, for example the conjugate gradient method.

In order to implement the above-described approximation scheme we need actually to construct the basis functions u_n and v_n. The construction of the functions u_n depends on constructing an approximation to the kernel E of the transformation operator $\mathbf{I}+\mathbf{T}$ which, as has already been mentioned, can be accomplished by appealing to the Cauchy–Euler polygon method. However, the construction of the functions v_n is a different type of problem in the sense that we must obtain approximations to the fundamental solutions $K^{(j)}$, $j=1, 2$ for small values of the parameter ε, and to complete our discussion we now address ourselves to this problem. We shall consider only the function $K^{(1)}$ since the asymptotic analysis of $K^{(2)}$ proceeds in exactly the same manner. We note that in practice it is only necessary to consider the case where $t-\tau_n \gg \varepsilon$ since for $t=0(\varepsilon)$, $K^{(1)}(x, t)$ is exponentially small (see the Exercises). From (7.81a), (7.81b) and (7.82) it is seen that the behavior of $K^{(1)}$ for $t\gg\varepsilon$ can be accomplished by evaluating

$$\Gamma(\alpha, t)=-\frac{1}{\pi i}\int_{|\omega-\sqrt{1/\varepsilon}|=\delta} \exp\left[-\alpha\omega+\frac{\omega^2 t}{1-\varepsilon\omega^2}\right] d\omega \tag{7.92}$$

for small ε, uniformly for α on compact subsets of $(0, \infty)$, and hence we now turn our attention to this problem (Colton and Wimp [42]).

We first deform the path of integration in (7.92) onto the contour pictured in Fig. 7.2. We then have, setting $\omega=z/\sqrt{t}$, $N=t/\varepsilon$, $b=\alpha/\sqrt{t}$, that

$$\Gamma(\alpha, t)=\frac{1}{\pi i\sqrt{t}}\int_C \exp[-bz+z^2]g_N(z)\, dz \tag{7.93}$$

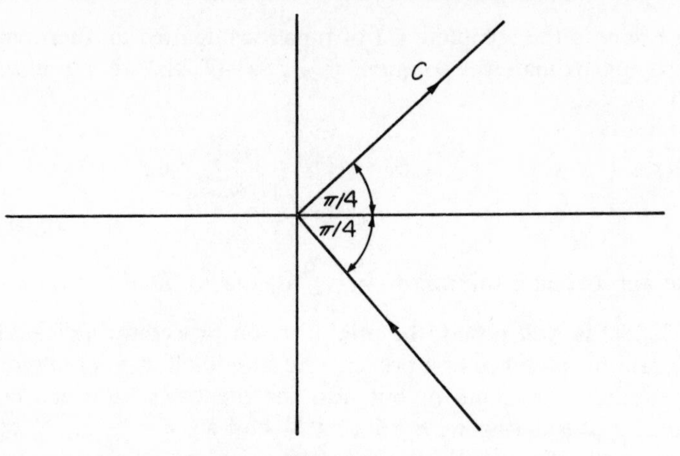

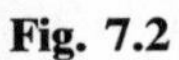

Fig. 7.2

where

$$g_N(z) = \exp\left[z^4/(N - z^2)\right] \tag{7.94}$$

We now want to examine the behavior of (7.93) for N tending to infinity. For $|z| < \sqrt{N}$ we have

$$\begin{aligned} g_N(z) &= \exp\left[N \sum_{m=2}^{\infty} \frac{z^{2m}}{N^m}\right] \\ &= 1 + \left(\frac{z^4}{N} + \frac{z^6}{N^2} + \cdots\right) + \frac{1}{2}\left(\frac{z^4}{N} + \frac{z^6}{N^2} + \cdots\right)^2 + \cdots \\ &= 1 + c_4 z^4 + c_6 z^6 + \cdots \end{aligned} \tag{7.95}$$

where the c_{2j}, $j = 2, 3, \ldots$, are easily computable constants. We now write

$$g_N(z) = S_N(z) + T_N(z) \tag{7.96}$$

where

$$\begin{aligned} S_N(z) &= 1 + c_4 z^4 + \cdots + c_{2m} z^{2m}; \qquad m \geq 2 \\ T_N(z) &= \sum_{n=m+1}^{\infty} c_{2n} z^{2n}. \end{aligned} \tag{7.97}$$

Since for $|z| < \sqrt{N}$,

$$|g_N(z)| \leq \exp\left[\frac{|z|^4}{N - |z|^2}\right] \tag{7.98}$$

we have by Cauchy's inequality that for $|z| < a < \sqrt{N}$

$$|T_N(z)| < \exp\left[\frac{a^4}{N - a^2}\right] \sum_{n=m+1}^{\infty} \left|\frac{z}{a}\right|^{2n} = \frac{|z/a|^{2m+2}}{1 - |z/a|^2} \exp\left[\frac{a^4}{N - a^2}\right]. \tag{7.99}$$

We now write $\Gamma(\alpha, t)$ in the form

$$\Gamma(\alpha, t) = \frac{1}{\pi i \sqrt{t}}[I_1 + I_2 + I_3 + I_4] \tag{7.100}$$

where

$$\begin{aligned} I_1 &= \int_C \exp[-bz + z^2] S_N(z)\, dz \\ I_2 &= \int_{C \cap |z| < A} \exp[-bz + z^2] T_N(z)\, dz \\ I_3 &= -\int_{C \cap |z| > A} \exp[-bz + z^2] S_N(z)\, dz \\ I_4 &= \int_{C \cap |z| > A} \exp[-bz + z^2] g_N(z)\, dz \end{aligned} \tag{7.101}$$

where A is a constant depending on N which will be chosen shortly. First, let $A < a$ and consider the integral I_2. Then, since on C

$$|\exp[-bz + z^2]| \leq 1, \tag{7.102}$$

we have

$$|I_2| \leq 2 \int_0^A \left|\frac{z}{a}\right|^{2m+2} \left[1 - \frac{A^2}{a^2}\right]^{-1} \exp\left[\frac{a^4}{N - a^2}\right] |dz| \tag{7.103}$$

and choosing $a = N^{1/4}$, $A = N^{1/(2(2m+3))}$, we have

$$|I_2| = 0(N^{-m/2}). \tag{7.104}$$

Now consider I_3. For $N > N_0$ we can find a constant M, independent of N, such that

$$|S_N(z)| \leq M |z|^{2m}. \tag{7.105}$$

Hence

$$|I_3| \leq 2M \int_A^\infty r^{2m} \exp[-br/\sqrt{2}]\, dr, \tag{7.106}$$

and choosing A as before it is easily seen (see the Exercises) that there exist positive constants ρ and δ, independent of N, such that

$$|I_3| = 0(\exp[-\rho N^\delta]). \tag{7.107}$$

Since on C

$$|e^{z^2} g_N(z)| \leq 1 \tag{7.108}$$

a similar result is also seen to hold for the integral I_4. Hence from

(7.100)–(7.108) we have

$$\Gamma(\alpha, t)=\frac{1}{\pi i\sqrt{t}}\int_C \exp[-bz+z^2]S_N(z)\,dz+0(N^{-m/2})$$

$$=a_0+a_4c_4+\cdots+a_{2m}c_{2m}+0(N^{-m/2}) \qquad (7.109)$$

where

$$a_{2j}=\frac{1}{\pi i\sqrt{t}}\int_C \exp[-bz+z^2]z^{2j}\,dz \qquad (7.110)$$

and the c_{2j} are defined in (7.95). The coefficients may be expressed in terms of Hermite polynomials. To see this we note that the contour C may be deformed onto the imaginary axis and hence, from the well-known integral representation of Hermite polynomials (cf. Erdélyi *et al.* [53]), we have

$$a_{2j}=\frac{1}{\pi i\sqrt{t}}\int_{-i\infty}^{i\infty} \exp[-bz+z^2]z^{2j}\,dz$$

$$=\frac{(-1)^j}{\pi i\sqrt{t}}\int_{-\infty}^{\infty} \exp[-ibt-t^2]t^{2j}\,dt$$

$$=\frac{1}{4^j\sqrt{\pi t}}\exp(-b^2/4)H_{2j}(b/2) \qquad (7.111)$$

where H_{2j} denotes Hermite's polynomial of order $2j$. Hence

$$\Gamma(\alpha, t)=\frac{1}{\sqrt{\pi t}}\exp(-b^2/4)\left[1+\frac{c_4H_4(b/2)}{4^2}+\frac{c_6H_6(b/2)}{4^3}+\cdots\right.$$

$$\left.+\frac{c_{2m}H_{2m}(b/2)}{4^m}+0(N^{-m/2})\right]. \qquad (7.112)$$

Now let n be a positive integer and choose m such that $m\geq 2n+2$. Putting all terms with powers of $1/N$ greater than $n+1$ into the error term shows that the series (7.112) may be rearranged as a valid asymptotic series in $1/N$. Hence, returning to our original variables, we have our final result that for $\alpha>0$, $t>0$,

$$\Gamma(\alpha, t)=\frac{1}{\sqrt{\pi t}}\exp\left(-\frac{\alpha^2}{4t}\right)$$

$$\times\left[1+d_1\left(\frac{\varepsilon}{t}\right)+d_2\left(\frac{\varepsilon}{t}\right)^2+\cdots+d_n\left(\frac{\varepsilon}{t}\right)^n+0\left(\left(\frac{\varepsilon}{t}\right)^{n+1}\right)\right]$$

(7.113)

where

$$d_1 = \frac{1}{4^2} H_4\left(\frac{\alpha}{2\sqrt{t}}\right)$$

$$d_2 = \frac{1}{2\cdot 4^4} H_8\left(\frac{\alpha}{2\sqrt{t}}\right) + \frac{1}{4^3} H_6\left(\frac{\alpha}{2\sqrt{t}}\right) \tag{7.114}$$

$$\vdots$$

From the above analysis it is seen that the expansion (7.113) is uniformly valid for α on compact subsets of $(0, \infty)$.

Exercises

(1) By using the Green's function for (7.1a) show how the final value problem (7.1a)–(7.1c) can be reformulated as a Fredholm integral equation of the first kind. Construct an example to show that the solution of such an integral equation does not depend continuously on the given data.

(2) Use logarithmic convexity arguments to show that if the solution u of (7.1a)–(7.1c) exists for $-\infty < t < T$ then $\|u(t)\|$ must grow exponentially as t lends to $-\infty$.

(3) Let u be a (classical) solution of

$$u_{xxt} - u_t + u_{xx} = 0; \qquad -1 < x < 1, \qquad 0 < t < T$$
$$u(-1, t) = u(1, t) = 0; \qquad 0 \leqslant t \leqslant T$$
$$u(x, 0) = \phi(x); \qquad -1 \leqslant x \leqslant 1$$

that is continuous for $-1 \leqslant x \leqslant 1$, $0 \leqslant t \leqslant T$. Show that if ϕ is positive for $-1 < x < 1$, then $u(x, t) > 0$ for $-1 < x < 1$, $0 < t < T$.

(4) Give an example to show that in Theorem 7.3 the assumption $d \leqslant 0$ in $\bar{R}$ is in general a necessary condition for the theorem to be valid.

(5) Show that for $t = 0(\varepsilon)$ we have $K^{(j)}(x, t) = 0(\exp(-\gamma_j/\sqrt{\varepsilon}))$ for positive constants γ_j, $j = 1, 2$.

(6) Show that equation (7.107) is valid.

(7) Verify equation (7.55).

(8) Establish the inequality (7.64).

(9) Solve the initial-boundary value problem (7.58a)–(7.58d) in the special case when $L[u] = u_{xxt}$.

8

The inverse scattering problem

We have already encountered a variety of problems associated with the scattering of acoustic waves. In particular, in Chapter 2 we studied the scattering of acoustic waves by a nonhomogeneous medium of compact support, in Chapter 4 a reflection principle for solutions of the Helmholtz equation defined outside a sphere, and in Chapter 5 a Runge approximation theorem for solutions of the Helmholtz equation defined in an exterior domain. In this chapter we shall consider one of the classical inverse problems of mathematical physics: the inverse scattering problem. Since the literature in this area is enormous, we shall set ourselves modest goals and consider only the use of function theoretic methods to study the low-frequency inverse scattering problem for acoustic waves. In particular, we shall assume that an incoming acoustic plane wave is scattered by a bounded obstacle, or nonhomogeneous medium of compact support, and from a knowledge of the far field pattern (see Chapter 2) we want to determine the shape or composition of the scattering object. It is with some reluctance that we have decided not to consider the many fascinating problems arising in the inverse scattering problem in quantum mechanics, and we refer the reader to Chadan and Sabatier [24] and De Alfaro and Regge [45] for details in this area.

The plan of this chapter is as follows. We first consider the problem of determining the location of the sources that generate a given far field pattern under the assumption that the far field pattern is known exactly for a single frequency and that the field is axially symmetric. We next consider the Born approximation as derived in Chapter 2 and show how, by using the Backus–Gilbert method, an approximation can be found for the speed of sound in a nonhomogeneous medium even if only incomplete far field data are known. Finally, we use potential theoretic methods to investigate the exterior Dirichlet and Neumann problem for the Helmholtz equation defined in the plane, and use these results to investigate the inverse scattering problem for a cylinder.

8.1 Equivalent sources and their location

We shall consider axially symmetric solutions of the three-dimensional Helmholtz equation which are regular in the exterior of a bounded domain D. Since we are concerned only with a single fixed frequency, we shall without loss of generality assume that the wave number is equal to one. Then in cylindrical coordinates (r, z, ϕ) such solutions satisfy the equation

$$\frac{\partial^2 u}{\partial z^2}+\frac{\partial^2 u}{\partial r^2}+\frac{1}{r}\frac{\partial u}{\partial r}+u=0 \tag{8.1}$$

where we have assumed the axis of symmetry to be $r=0$. If we further assume that u satisfies the Sommerfeld radiation condition (Chapter 2)

$$\lim_{R\to\infty} R(\partial u/\partial R-\mathrm{i}u)=0; \qquad R=(r^2+z^2)^{1/2} \tag{8.2}$$

then u may be regarded as being generated by a set of equivalent sources all of which are contained in $\bar{D}$. The equivalent sources are not in general unique, although they are confined to some finite region of space, and one of the fundamental problems in scattering theory is to determine the extent and location of this region (cf. Bates and Wall [10]; Colton [27], [29]; Hartman and Wilcox [81]; Millar [126], [128]; Müller [132]; Sleeman [161]; Weston, Bowman and Ar [178]).

A solution u of (8.1) and (8.2) behaves asymptotically like (cf. Theorem 2.9 and the Exercises)

$$u\sim\frac{\mathrm{e}^{\mathrm{i}R}}{R}f(\cos\theta); \qquad R\to\infty, \tag{8.3}$$

where (R, θ) are polar coordinates defined by $z=R\cos\theta$, $r=R\sin\theta$. Since regular solutions of (8.1) are even functions of θ, we shall assume that u and the far field pattern f are defined for $\theta\in[0, 2\pi]$. Given f, our problem is to determine the location of the sources which generate u. Our first problem is to characterize the class of functions which can be far field patterns. This class is neither the class of continuous functions nor the class of analytic functions, but can best be characterized by introducing a certain class of harmonic functions (Müller [132]). Before presenting our theorem, recall (cf. Levin [115]) that if f is an entire function of a complex variable and

$$M_f(r)=\max_{|z|=r}|f(z)|$$

then the *order* ρ of f is defined by

$$\rho=\overline{\lim_{r\to\infty}}\frac{\ln\ln M_f(r)}{\ln r}$$

and the *type* σ is given by

$$\sigma = \overline{\lim_{r\to\infty}} \frac{\ln M_f(r)}{r^\rho}.$$

Theorem 8.1 *A necessary and sufficient condition for a function f to be a far field pattern is that there exists an axially symmetric harmonic function* $h(z, r) = \tilde{h}(R, \theta)$ *which is regular in the entire space and is such that* $\tilde{h}(1, \theta) = f(\cos\theta)$ *and further has the property that* $\int_0^\pi |\tilde{h}(R, \theta)|^2 \sin\theta \, d\theta$ *is an entire function of R of order one and finite type C. When this condition holds there exists a unique function* $u(z, r) = \tilde{u}(R, \theta)$ *which satisfies the Sommerfeld radiation condition* (8.2) *and is a regular solution of the axially symmetric Helmholtz equation* (8.1) *for* $R > C$ *such that*

$$\tilde{u}(R, \theta) = \frac{e^{iR}}{R} f(\cos\theta) + 0(1/R^2); \qquad R \to \infty.$$

Proof Suppose that f is a far field pattern and expand f in a Legendre series

$$f(\cos\theta) = \sum_{n=0}^{\infty} a_n P_n(\cos\theta). \tag{8.4}$$

Then from (8.3) and the fact that u is a solution of (8.1) in the exterior of a bounded domain D, we can write

$$\tilde{u}(R, \theta) = \sum_{n=0}^{\infty} a_n i^{n+1} h_n^{(1)}(R) P_n(\cos\theta) \tag{8.5}$$

for $R \geq C$, where C is some positive constant, and conclude that the series (8.5) is absolutely and uniformly convergent in this region. (In (8.5) $h_n^{(1)}$ denotes a spherical Hankel function.) From the relations (cf. Erdélyi *et al.* [53])

$$\lim_{n\to\infty} \frac{R^{n+1} h_n^{(1)}(R)}{2^n \Gamma(n+\frac{1}{2})} = -i$$

$$\max_{\theta\in[0,\pi]} |P_n(\cos\theta)| = 1$$

we can now conclude that $|a_n| \Gamma(n+\frac{1}{2}) 2^n C^{-n}$ is bounded, i.e. (using Stirling's formula)

$$\overline{\lim_{n\to\infty}} n |a_n|^{1/n} = \tfrac{1}{2} e C \tag{8.6}$$

Since

$$\tilde{h}(R, \theta) = \sum_{n=0}^{\infty} a_n R^n P_n(\cos\theta)$$

we have from the orthogonality properties of the Legendre polynomials that

$$\int_0^{\pi} |\tilde{h}(R,\theta)|^2 \sin\theta \, d\theta = \sum_{n=0}^{\infty} \frac{2}{2n+1} |a_n|^2 R^{2n}. \tag{8.7}$$

We now recall a basic theorem from the theory of entire functions, that is if an entire function has Taylor coefficients $c_0, c_1, c_2, \ldots$, then the order and type of the function are given by (Levin [115])

$$\begin{aligned} \rho &= \overline{\lim_{n\to\infty}} \frac{n \ln n}{\ln 1/|c_n|} \\ (\sigma e \rho)^{1/\rho} &= \overline{\lim_{n\to\infty}} (n^{1/\rho} |c_n|^{1/n}). \end{aligned} \tag{8.8}$$

Hence from (8.6), (8.8) we can conclude that (8.7) defines an entire function of order one and type C.

Now suppose that (8.7) is an entire function of order one and type C. Then from (8.6) the series (8.5) converges for all $R > C$ and from Theorem 2.9 defines a solution of (8.1) satisfying the radiation condition (8.2) and having (8.4) as its far field pattern. This completes the proof of the theorem.

Suppose now that f is a far field pattern corresponding to the solution u of (8.1) defined by (8.5) for $R \geq C$. We want to determine the maximal domain of regularity of u, i.e. we want to continue u across the circle $R = C$. We shall determine such a region in terms of the entire harmonic function h defined in Theorem 8.1. In particular, recall from Chapter 4 that since h is a solution of

$$h_{zz} + h_{rr} + (1/r)h_r = 0, \tag{8.9}$$

h is uniquely determined by the function

$$h(z,0) = \sum_{n=0}^{\infty} a_n z^n, \tag{8.10}$$

which, from Theorem 8.1, can be viewed as an entire function of the complex variable z of order one and type $C/2$. Let B be the Borel transform of $h(2iz, 0)$ defined by (cf. Levin [115])

$$B(z) = \sum_{n=0}^{\infty} a_n 2^n i^n n! \, z^{-n-1}. \tag{8.11}$$

Then from (8.10) we have that B is analytic for $|z| > C$. However, much more can be said about the domain of regularity of B. In particular, let

$$K(\theta) = \overline{\lim_{R\to\infty}} R^{-1} \log |h(2iRe^{i\theta}, 0)| \tag{8.12}$$

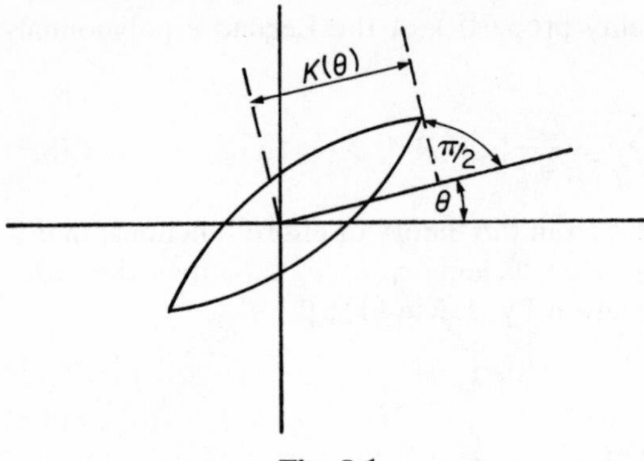

Fig. 8.1

Then it can be shown (Levin [115]) that K is the supporting function of a convex set, called the indicator diagram of $h(2iz, 0)$ (see Fig. 8.1), and a result of Polya says that $B(z)$ is analytic in the exterior of the conjugate of the indicator diagram of $h(2iz, 0)$. Note that since h is of order one and type $C/2$, the indicator diagram of $h(2iz, 0)$ is contained in the disk $\Omega = \{z: |z| \leq C\}$. Considering the complex z plane as superimposed over the Euclidean plane, we shall show that if G is the indicator diagram of $h(2iz, 0)$ and G^* its conjugate, then u can be continued as a solution of (8.1) into the exterior of $G \cup G^*$.

We first define the analytic function

$$g(z) = \sum_{n=0}^{\infty} a_n i^{n+1} h_n^{(1)}(C)(z/C)^{-n-1}. \tag{8.13}$$

Lemma 8.1 *$g(z)$ is analytic in the exterior of G^*.*

Proof From Polya's theorem and the Hadamard multiplication of singularities theorem (Chapter 4) it suffices to show that the singularities of

$$G(z) = \sum_{n=0}^{\infty} \frac{h_n^{(1)}(C)}{n!\,2^n} \left(\frac{z}{C}\right)^{-n} \tag{8.14}$$

lie on the closed interval [0, 1]. But from Erdélyi *et al.* [53] we can actually sum (8.14) to give

$$G(z) = -(i/C)(1 - 1/z)^{-1/2} \exp[iC(1 - 1/z)^{1/2}], \tag{8.15}$$

i.e. the only singularities of G are branch points at $z = 0$ and $z = +1$.

We now construct an axially symmetric harmonic function v such that $v(z, 0) = g(z)$:

$$v(z, r) = \sum_{n=0}^{\infty} a_n i^{n+1} h_n^{(1)}(C)(R/C)^{-n-1} P_n(\cos\theta). \tag{8.16}$$

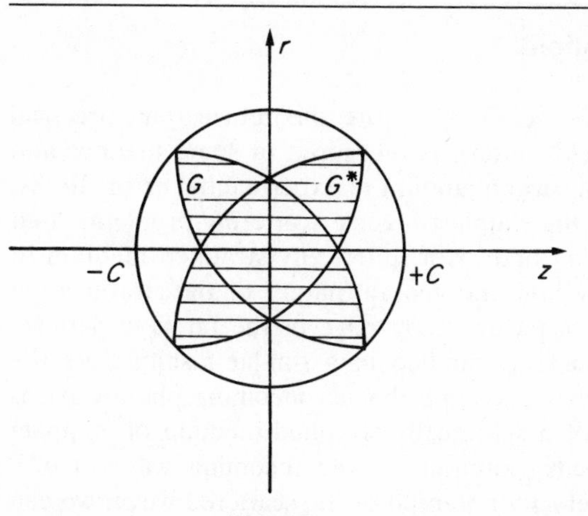

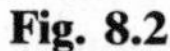
Fig. 8.2

Note that $(1/R)\tilde{v}(C^2/R, \theta)$ (where $\tilde{v}(R, \theta) = v(z, r)$) is an axially symmetric harmonic function in a neighborhood of the origin, and hence from Lemma 8.1 and Theorem 4.12 we can conclude that v is regular in the exterior of $G \cup G^*$ (Fig. 8.2). Note that by construction we have $\tilde{v}(c, \theta) = \tilde{u}(c, \theta)$.

Theorem 8.2 *u is regular in the exterior of $G \cup G^*$.*

Proof Define the axially symmetric harmonic function h by

$$h(R, \theta) = \tfrac{1}{2}[\tilde{v}(R, \theta) + (C/R)\tilde{v}(C^2/R, \theta)]. \tag{8.17}$$

Then h is regular in the domain exterior to $G \cup G^*$ and interior to the inversion of $G \cup G^*$ across the circle $R = C$. Furthermore, we have

$$h_R(C, \theta) + (1/2C)h(C, \theta) = 0. \tag{8.18}$$

Hence from (4.49) of Chapter 4 we have that

$$u_1(R, \theta) = h(R, \theta) + \int_C^R \tilde{K}(R, s; 1)h(s, \theta)\, ds \tag{8.19}$$

is a solution of (8.1) in the same region in which h is harmonic and $u_1(C, \theta) = h(C, \theta)$. Hence $w = u_1 - u$ is a solution of (8.1) in the region bounded by the circle $R = C$ and the inversion of $\partial G \cup \partial G^*$ across this circle. Furthermore, $w(C, \theta) = 0$. Hence by Theorem 4.5 we can continue w into the complement of $G \cup G^*$ with respect to the disk $R \leqslant C$. Since u_1 is already known to be regular in this region we can conclude that u must be regular there also, and since we already know that u is regular for $R \geqslant C$, the theorem is proved.

8.2 The Born approximation

The results of the previous section are more theoretical than practical since in practice the far field pattern is measured in some manner and thus only known to within a certain amount of experimental error. In this section we consider one of the simplest inverse scattering problems, that when we know the scattering body is a spherically stratified medium of compact support, and show how the general nature of the stratification can be obtained from only a partial knowledge of the far field pattern. Certain related problems can be handled in a similar fashion (see the Exercises). To be more precise, assume that an incoming plane wave is scattered by the presence of a spherically stratified medium of compact support. Then if the velocity potential of the incoming wave is e^{ikz}, $\mathbf{x}=(x, y, z)$, and u_s is the velocity potential of the scattered wave, we can write the following equations for the determination of the total field u (see Chapter 2):

$$u(\mathbf{x}) = e^{ikz} + u_s(\mathbf{x}) \tag{8.20a}$$

$$\Delta_3 u + k^2(1-m(r))u = 0 \tag{8.20b}$$

$$\lim_{r\to\infty} r(\partial u_s/\partial r - iku_s) = 0 \tag{8.20c}$$

where $m(r)=0$ for $r \geq r_0$, $r=|\mathbf{x}|$, and reflects the fact that the medium is spherically stratified for $r \leq r_0$. If we define the far field pattern $F(\theta; k) = f(\cos\theta; k)$ by

$$f(\cos\theta; k) = \lim_{r\to\infty} re^{-ikr}u_s(\mathbf{x}) = \sum_{n=0}^{\infty} A_n(k)P_n(\cos\theta), \tag{8.21}$$

then from Theorem 2.8 of Chapter 2 we have for

$$\mu_n = -\lim_{k\to 0} \frac{(2n+1)A_n(k)}{k^{2n+2}}\left(\frac{(2n)!}{2^n n!}\right)^2 \tag{8.22}$$

that the *Born approximation*

$$\mu_n = \int_0^{r_0} m(r)r^{2n+2}\,dr \tag{8.23}$$

is valid. Our problem now is to determine m from the moment problem (8.23). Since in practice only a finite number of the μ_n can be measured from the far field pattern f, the real problem to be considered is to determine m from the relation (8.23) where only a finite number of moments μ_n, $n=0, 1, 2, \ldots, N$, are given. (We could also consider the case where these μ_n are only known to within a certain experimental error; however, for simplicity of presentation we shall assume that these

$N+1$ moments are known exactly.) The problem as thus formulated is obviously improperly posed, since in general m lies in some infinite-dimensional function space and cannot be uniquely determined from knowing only a finite number of its moments. Nevertheless, by assuming an *a priori* bound on the derivative of m is known, we shall show that an approximation to m can be found provided N is sufficiently large (cf. Backus and Gilbert [9]).

Our starting point is to construct a function A_N of the form

$$A_N(r, s) = \sum_{n=0}^{N} r^{2n+2} a_n(s) \tag{8.24}$$

where the a_n are chosen such that A_N is a delta sequence, i.e. A_N is an approximation in some sense to the delta function (cf. Korevaar [105]). More precisely, we require that for $s \in (0, r_0)$

$$\int_0^{r_0} A_N(r, s)\, dr = 1 \tag{8.25}$$

and

$$\int_0^{r_0} (r-s)^2 [A_N(r, s)]^2\, dr < \varepsilon(N, s) \tag{8.26}$$

where $\varepsilon(N, s)$ is as 'small' as possible. Suppose now that the a_n have been chosen such that (8.25) and (8.26) are valid and that there exists a positive constant M such that

$$\max_{[0, r_0]} |m'(r)| \leqslant M. \tag{8.27}$$

We shall show that an approximation to m is given by

$$m_N(r) = \sum_{n=0}^{N} \mu_n a_n(r). \tag{8.28}$$

Indeed, from (8.23)–(8.28) we have that for $s \in (0, r_0)$

$$\begin{aligned} m_N(s) - m(s) &= \int_0^{r_0} m(r) A_N(r, s)\, dr - m(s) \\ &= \int_0^{r_0} [m(r) - m(s)] A_N(r, s)\, dr \\ &= \int_0^{r_0} \left(\frac{m(r) - m(s)}{r - s}\right)(r-s) A_N(r, s)\, dr \\ &\leqslant \left[\int_0^{r_0} \left(\frac{m(r) - m(s)}{r - s}\right)^2 dr\right]^{1/2} \left[\int_0^{r_0} (r-s)^2 [A_N(r, s)]^2\, dr\right]^{1/2}. \end{aligned} \tag{8.29}$$

Hence from the mean value theorem

$$|m_N(s) - m(s)| \leq \sqrt{Mr_0}\sqrt{\varepsilon}. \tag{8.30}$$

Thus our problem can be considered to be solved provided we can construct the functions a_n. One approach for doing this is as follows. For fixed s we can write the left-hand side of (8.26) as

$$E(s, A_N) = \sum_{i,j=0}^{N} E_{ij}(s) a_i a_j \tag{8.31}$$

where

$$E_{ij}(s) = \int_0^{r_0} (r-s)^2 r^{2i+2j+4}\,dr$$
$$a_i = a_i(s) \tag{8.32}$$

and rewrite (8.25) as

$$\sum_{i=0}^{N} a_i \frac{r_0^{2i+3}}{2i+3} = 1. \tag{8.33}$$

Hence our problem is to minimize (8.31) subject to the constraint (8.33) or, in matrix notation,

$$\text{minimize } a^T E a, \qquad \text{subject to } u^T a = 1, \tag{8.34}$$

where the superscript denotes transpose, $E = (E_{ij})$, $a = (a_i)$, $u = (r_0^{2i+3}/(2i+3))$. Geometrically, as γ varies, $a^T E a = \gamma$ represents a family of multidimensional ellipsoids and $u^T a = 1$ a hyperplane. Thus our extremum problem is to find the smallest ellipsoid of the family $a^T E a = \gamma$ which has a nonempty intersection with the hyperplane $u^T a = 1$ (see Fig. 8.3). The solution is, of course, that point a^* on the ellipsoid which

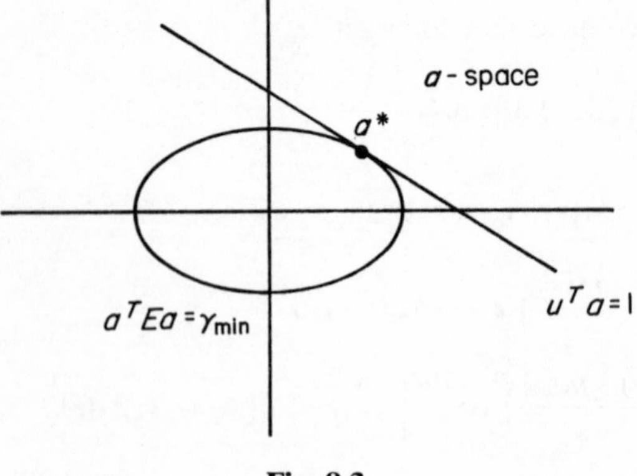

Fig. 8.3

touches the hyperplane, and analytically this can be found through the use of Lagrange multipliers.

In closing, we note that we have only touched the surface of the Backus–Gilbert method for solving improperly posed moment problems, and the reader is referred to Backus and Gilbert [9] and Sabatier [155] for more information and references.

8.3 Potential theory

We now want to develop some methods that will allow us to solve the inverse scattering problem for a cylinder having a bounded, simply connected cross-section. To this end we shall first need to develop some basic results on potential theory in the plane, in particular, the continuity properties of the *single layer potential*

$$u(\mathbf{x}) = \int_{\partial D} \phi(\mathbf{y}) \log \frac{1}{|\mathbf{x}-\mathbf{y}|} \, ds(y) \tag{8.35}$$

and the *double layer potential*

$$v(\mathbf{x}) = \int_{\partial D} \psi(\mathbf{y}) \frac{\partial}{\partial \nu(y)} \log \frac{1}{|\mathbf{x}-\mathbf{y}|} \, ds(y), \tag{8.36}$$

where D is a bounded, simply connected domain in R^2 with a twice continuously differentiable boundary ∂D, ν is the unit outward normal, and ϕ, ψ are continuous functions defined on ∂D. We shall, in addition, be interested in the spectrum of the integral operator defined by (8.36). Our presentation is based on some unpublished lecture notes by Kress [106].

One of the problems in developing a potential theory for (8.35), (8.36) is where to start. We shall assume the reader is familiar with elementary functional analysis and the concept of a compact operator. In particular, we shall simply state the following definition and theorem and refer the reader to Stakgold [162] or Taylor [164] for more details. In what follows, C denotes the complex numbers.

Definition 8.1 *A linear operator $A: X \to Y$ from a normed space X into a normed space Y is called* compact *if it maps any bounded set in X into a relatively compact set in Y.*

Let $C(\partial D)$ be the Banach space of complex-valued continuous functions defined on ∂D equipped with the maximum norm

$$\|\phi\| = \sup_{\mathbf{x} \in \partial D} |\phi(\mathbf{x})| \tag{8.37}$$

and consider the integral operator $\mathbf{A}: C(\partial D) \to C(\partial D)$ defined by

$$(\mathbf{A}\phi)(\mathbf{x}) = \int_{\partial D} K(\mathbf{x}, \mathbf{y})\phi(\mathbf{y})\, ds(y) \tag{8.38}$$

where K is *weakly singular*, i.e. K is defined and continuous for all $\mathbf{x}, \mathbf{y} \in \partial D$, $\mathbf{x} \neq \mathbf{y}$, and there exist positive constants M and α such that for $\mathbf{x}, \mathbf{y} \in \partial D$, $\mathbf{x} \neq \mathbf{y}$, we have

$$|K(\mathbf{x}, \mathbf{y})| \leq M\, |\mathbf{x} - \mathbf{y}|^{\alpha-1}. \tag{8.39}$$

Theorem 8.3 **A** *is a compact operator on* $C(\partial D)$.

Since we want to study operators of the form (8.38) on the Banach space $C(\partial D)$, it is convenient to introduce the concept of a dual system (cf. Jörgens [99]; Kress [106]).

Definition 8.2 *Let X and Y be normed spaces, and $\langle \cdot, \cdot \rangle: X \times Y \to C$ be a bounded, nondegenerate bilinear form, i.e.*

(1) *there exists a positive real number γ such that $|\langle u, v\rangle| \leq \gamma\, \|u\|\, \|v\|$;*

(2) $\sup_{\|v\|=1} |\langle u, v\rangle| > 0$ *for* $u \neq 0$, *and* $\sup_{\|u\|=1} |\langle u, v\rangle| > 0$ *for* $v \neq 0$;

(3) *for constants α_1, α_2, β_1, β_2 we have*

$$\langle \alpha_1 u_1 + \alpha_2 u_2, v\rangle = \alpha_1 \langle u_1, v\rangle + \alpha_2 \langle u_2, v\rangle$$
$$\langle u, \beta_1 v_1 + \beta_2 v_2\rangle = \beta_1 \langle u, v_1\rangle + \beta_2 \langle u, v_2\rangle.$$

Two normed spaces equipped with a bounded nondegenerate bilinear form is called a *dual system* and is denoted by $\langle X, Y\rangle$.

Theorem 8.4 *$\langle C(\partial D), C(\partial D)\rangle$ is a dual system with the bilinear form*

$$\langle \phi, \psi\rangle = \int_{\partial D} \phi(\mathbf{y})\psi(\mathbf{y})\, ds(y).$$

Proof This follows immediately from Definition 8.2.

Definition 8.3 *Let $\langle X, Y\rangle$ be a dual system. Then two operators $\mathbf{A}: X \to X$, $\mathbf{B}: Y \to Y$ are called* adjoint *if for every $u \in X$, $v \in Y$,*

$$\langle \mathbf{A}u, v\rangle = \langle u, \mathbf{B}v\rangle.$$

Theorem 8.5 *Let $\langle X, Y\rangle$ be a dual system. If an operator $\mathbf{A}: X \to X$ has an adjoint $\mathbf{B}: Y \to Y$ then B is uniquely determined and **A** and **B** are linear.*

Proof Suppose there existed two adjoints to $\mathbf{A}$ and denote these by $\mathbf{B}_1$ and $\mathbf{B}_2$. Let $\mathbf{B}=\mathbf{B}_1-\mathbf{B}_2$. Then for every $v \in Y$ we have $\langle u, \mathbf{B}v\rangle = \langle u, \mathbf{B}_1 v - \mathbf{B}_2 v\rangle = \langle \mathbf{A}u, v\rangle - \langle \mathbf{A}u, v\rangle = 0$ for all $u \in X$. Hence, since $\langle \cdot, \cdot \rangle$ is nondegenerate we have $\mathbf{B}v=0$ for every $v \in Y$, i.e. $\mathbf{B}_1=\mathbf{B}_2$. To show that $\mathbf{B}$ is linear we simply observe that for every $u \in X$,

$$\begin{aligned}\langle u, \alpha_1\mathbf{B}v_1+\alpha_2\mathbf{B}v_2\rangle &= \alpha_1\langle u, \mathbf{B}v_1\rangle + \alpha_2\langle u, \mathbf{B}v_2\rangle \\ &= \alpha_1\langle \mathbf{A}u, v_1\rangle + \alpha_2\langle \mathbf{A}u, v_2\rangle \\ &= \langle \mathbf{A}u, \alpha_1 v_1 + \alpha_2 v_2\rangle \\ &= \langle u, \mathbf{B}(\alpha_1 v_1 + \alpha_2 v_2)\rangle, \end{aligned} \tag{8.40}$$

i.e. $\alpha_1\mathbf{B}v_1+\alpha_2\mathbf{B}v_2 = \mathbf{B}(\alpha_1 v_1+\alpha_2 v_2)$. In a similar manner it is seen that $\mathbf{A}$ is linear.

Theorem 8.6 *Let $K(\mathbf{x}, \mathbf{y})$ be weakly singular. Then in $\langle C(\partial D), C(\partial D)\rangle$ the (compact) operators defined by*

$$(\mathbf{A}\phi)(x) = \int_{\partial D} K(\mathbf{x}, \mathbf{y})\phi(\mathbf{y})\, ds(y)$$

$$(\mathbf{B}\psi)(\mathbf{x}) = \int_{\partial D} K(\mathbf{y}, \mathbf{x})\psi(\mathbf{y})\, ds(y)$$

are adjoint.

Proof The theorem follows from

$$\begin{aligned}\langle \mathbf{A}\phi, \psi\rangle &= \int_{\partial D} (\mathbf{A}\phi)(\mathbf{x})\psi(\mathbf{x})\, ds(x) \\ &= \int_{\partial D}\left[\int_{\partial D} K(\mathbf{x}, \mathbf{y})\phi(\mathbf{y})\, ds(y)\right]\psi(\mathbf{x})\, ds(\mathbf{x}) \\ &= \int_{\partial D}\phi(\mathbf{y})\left[\int_{\partial D} K(\mathbf{x}, \mathbf{y})\psi(\mathbf{x})\, ds(x)\right] ds(y) \\ &= \int_{\partial D}\phi(\mathbf{y})(\mathbf{B}\psi)(\mathbf{y})\, ds(y) \\ &= \langle \phi, \mathbf{B}\psi\rangle. \end{aligned} \tag{8.41}$$

We are now in a position to state the Fredholm alternative for dual systems (Jörgens [99]; Kress [106]; Wendland [176], [177]).

Theorem 8.7 (*Fredholm alternative*) *Let $\langle X, Y\rangle$ be a dual system and $\mathbf{A}: X \to X$, $\mathbf{B}: Y \to Y$, be compact adjoint operators. Then* either *the homogeneous equations*

$$u - \mathbf{A}u = 0 \quad \textit{and} \quad v - \mathbf{B}v = 0$$

have only the trivial solutions $u=0$ *and* $v=0$ *and the nonhomogeneous equations*

$$u-\mathbf{A}u=w \quad \textit{and} \quad v-\mathbf{B}v=z$$

have a unique solution for any right-hand side $w\in X$ *and* $z\in Y$, or *the homogeneous equations have the same finite number of linearly independent solutions and the nonhomogeneous equations are solvable if and only if* $\langle w, v\rangle=0$ *for all solutions* v *of* $v-\mathbf{B}v=0$ *and* $\langle u, z\rangle=0$ *for all solutions* u *of* $u-\mathbf{A}u=0$, *respectively.*

Our aim now is to formulate the exterior Dirichlet and Neumann problems for Laplace's equations as integral equations for which the Fredholm alternative is applicable. To this end we first derive the continuity properties of single- and double-layer potentials.

Theorem 8.8 *Let* $\phi\in C(\partial D)$. *Then the single-layer potential*

$$u(\mathbf{x})=\int_{\partial D}\phi(\mathbf{y})\log\frac{1}{|\mathbf{x}-\mathbf{y}|}\,ds(y)$$

exists (as an improper integral) for $\mathbf{x}$ *on* ∂D *and* u *is continuous for all* $\mathbf{x}$ *in* R^2.

Proof Let $\mathbf{x}=\mathbf{x}(s)$ be a parametric representative of ∂D where s denotes arc length. Then

$$\chi(t, s)=\left|\frac{\mathbf{x}(t)-\mathbf{x}(s)}{t-s}\right|$$

is a continuous function of t and s since ∂D is assumed to be smooth. Since $\mathbf{x}'(s)$ is the unit tangent vector, we see that there exists a positive constant γ such that $1\geqslant\chi(t, s)\geqslant\gamma>0$ for t and s restricted to a compact subset of $(0, L)$ where L is the arc length of ∂D. Hence for such values of t and s,

$$|\log 1/|\mathbf{x}-\mathbf{y}||\leqslant|\log 1/\gamma\,|t-s|| \tag{8.42}$$

and we can conclude that $\log 1/|\mathbf{x}-\mathbf{y}|$ is a weakly singular kernel. Thus, from Theorem 8.3 u exists and is continuous for $\mathbf{x}$ on ∂D. We must now show that u is continuous in a neighborhood of ∂D (u is trivially continuous for $\mathbf{x}$ outside a neighborhood of ∂D). To this end let $\mathbf{x}, \mathbf{y}\in\partial D$ and $\alpha=\alpha(\mathbf{x}, \mathbf{y})$ be the angle between the line segment joining $\mathbf{x}$ and $\mathbf{y}$ and the tangent to ∂D at $\mathbf{x}$. Let $\mathbf{x}_0$ be a point not on ∂D and $\beta=\beta(\mathbf{x}, \mathbf{y})$ the angle between the line segments joining $\mathbf{x}_0$ to $\mathbf{x}$ and $\mathbf{x}_0$ to $\mathbf{y}$ (see Fig. 8.4). Assume $\mathbf{x}$ is such that the line segment joining $\mathbf{x}_0$ to $\mathbf{x}$ is perpendicular to ∂D at $\mathbf{x}$. Since α depends continuously on $\mathbf{x}$ and $\mathbf{y}$, $\cos\alpha(\mathbf{x}, \mathbf{x})=1$, there

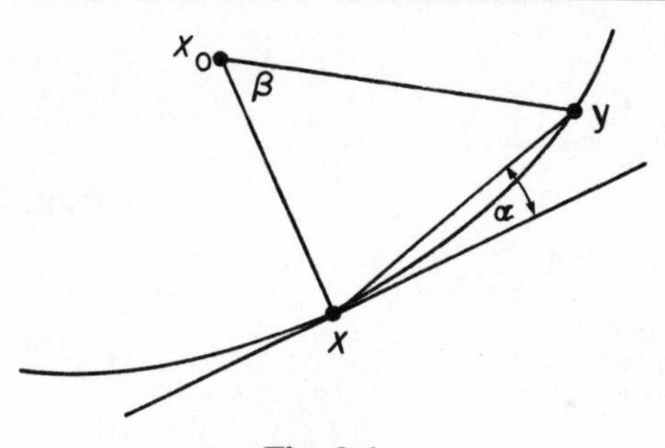

Fig. 8.4

exists a ρ, $0<\rho\leqslant\frac{1}{4}$ such that $\cos\alpha(\mathbf{x},\mathbf{y})\geqslant\frac{1}{2}$ for all $\mathbf{x},\mathbf{y}\in\partial D$ with $|\mathbf{x}-\mathbf{y}|\leqslant\rho$. Then, from the law of sines,

$$\frac{|\mathbf{x}-\mathbf{y}|}{|\mathbf{x}_0-\mathbf{y}|}=\frac{\sin\beta}{\sin(\pi/2-\alpha)}\leqslant\frac{1}{\cos\alpha}\leqslant 2. \tag{8.43}$$

Here $|\mathbf{x}-\mathbf{y}|\leqslant 2\,|\mathbf{x}_0-\mathbf{y}|$ and for $|\mathbf{x}_0-\mathbf{x}|\leqslant\rho$ we have

$$|\mathbf{x}_0-\mathbf{y}|\leqslant|\mathbf{x}-\mathbf{y}|+|\mathbf{x}_0-\mathbf{x}|\leqslant 2\rho\leqslant\tfrac{1}{2}. \tag{8.44}$$

Thus,

$$\left|\log\frac{1}{2\,|\mathbf{x}_0-\mathbf{y}|}\right|\leqslant\left|\log\frac{1}{|\mathbf{x}-\mathbf{y}|}\right|. \tag{8.45}$$

Now let $C_\rho=\{\mathbf{y}:\mathbf{y}\in\partial D,\ |\mathbf{y}-\mathbf{x}|\leqslant\rho\}$. Then, if $\|\phi\|=\max_{\mathbf{x}\in\partial D}|\phi(\mathbf{x})|$ we have, from the mean value theorem,

$$\begin{aligned}|u(\mathbf{x}_0)-u(\mathbf{x})|\leqslant{}&\|\phi\|\int_{C_\rho}\left\{\left|\log\frac{1}{|\mathbf{x}_0-\mathbf{y}|}\right|+\left|\log\frac{1}{|\mathbf{x}-\mathbf{y}|}\right|\right\}ds(y)\\&+\|\phi\|\int_{\partial D\setminus C_\rho}\left|\log\frac{1}{|\mathbf{x}_0-\mathbf{y}|}-\log\frac{1}{|\mathbf{x}-\mathbf{y}|}\right|ds(y)\\\leqslant{}&\|\phi\|\int_{C_\rho}\left(2\log\frac{1}{|\mathbf{x}-\mathbf{y}|}+\log 2\right)ds(y)\\&+\|\phi\|\int_{\partial D\setminus C_\rho}\left|\nabla_x\log\frac{1}{|\mathbf{z}-\mathbf{y}|}\cdot(\mathbf{x}-\mathbf{x}_0)\right||\mathbf{x}_0-\mathbf{x}|\,ds(y)\end{aligned} \tag{8.46}$$

where $\mathbf{z}=\mathbf{z}(y)$ is a point lying on the line segment joining $\mathbf{x}_0$ to $\mathbf{x}$. However, on $\partial D\setminus C_\rho$ we have

$$\left|\nabla_x\log\frac{1}{|\mathbf{z}-\mathbf{y}|}\cdot(\mathbf{x}-\mathbf{x}_0)\right|\leqslant\frac{1}{|\mathbf{y}-\mathbf{z}|}\leqslant\frac{2}{\rho} \tag{8.47}$$

and hence

$$|u(\mathbf{x}_0)-u(\mathbf{x})|\le\|\phi\|\int_{|\mathbf{y}|\le\rho}\left(2\log\frac{1}{|\mathbf{y}|}+2\right)ds(y)+\|\phi\|\frac{2L}{\rho}|\mathbf{x}_0-\mathbf{x}|. \tag{8.48}$$

Now, for given $\varepsilon>0$ choose ρ such that

$$\int_{|\mathbf{y}|\le\rho}\left(2\log\frac{1}{|\mathbf{y}|}+2\right)ds(y)<\frac{\varepsilon}{2\,\|\phi\|} \tag{8.49}$$

and let

$$\delta=\min\left(\frac{\rho}{2},\frac{\varepsilon\rho}{4L\,\|\phi\|}\right).$$

Then, if $\mathbf{x}_0=\mathbf{x}+h\boldsymbol{\nu}$ where $\boldsymbol{\nu}$ is the unit outward normal and $|h|<\delta$, we have that

$$|u(\mathbf{x}_0)-u(\mathbf{x})|<\varepsilon,$$

i.e. u is (uniformly) continuous along the normal direction. Since we have now shown that u is continuous both along ∂D and in the normal direction to ∂D, we can conclude, by the triangle inequality, that u is continuous in a neighborhood of ∂D and hence in all of R^2.

Theorem 8.9 *Let $\psi\in C(\partial D)$. Then the double-layer potential*

$$v(\mathbf{x})=\int_{\partial D}\psi(\mathbf{y})\frac{\partial}{\partial\nu(y)}\log\frac{1}{|\mathbf{x}-\mathbf{y}|}ds(y)$$

exists and is continuous for $\mathbf{x}$ on ∂D. If for $\mathbf{x}\in\partial D$ we define

$$v_+(\mathbf{x})=\lim_{\mathbf{y}\to\mathbf{x}}v(\mathbf{y}),\qquad \mathbf{y}\in R^2\backslash\bar{D}$$

$$v_-(\mathbf{x})=\lim_{\mathbf{y}\to\mathbf{x}}v(\mathbf{y}),\qquad \mathbf{y}\in D$$

then

$$v_\pm(\mathbf{x})=\int_{\partial D}\psi(\mathbf{y})\frac{\partial}{\partial\nu(y)}\log\frac{1}{|\mathbf{x}-\mathbf{y}|}ds(y)\pm\pi\psi(\mathbf{x}).$$

Proof First consider the case when $\psi=1$:

$$v_0(\mathbf{x})=\int_{\partial D}\frac{\partial}{\partial\nu(y)}\log\frac{1}{|\mathbf{x}-\mathbf{y}|}ds(y).$$

If $\mathbf{x}\in D$ then let K_R be a disk of radius R with center at $\mathbf{x}$ such that $K_R\subset D$, and if $\mathbf{x}\in\partial D$ let $H_R=K_R\cap D$. Then from Green's formula we have that for $\mathbf{x}\in R^2\backslash\bar{D}$

$$v_0(\mathbf{x})=\int_{\partial D}\Delta_y \log\frac{1}{|\mathbf{x}-\mathbf{y}|}\,ds(y)=0,$$

for $\mathbf{x}\in D$,

$$v_0(\mathbf{x})=\int_{\partial K_R}\frac{\partial}{\partial\nu(y)}\log\frac{1}{|\mathbf{x}-\mathbf{y}|}\,ds(y)=-\int_{\partial K_R}\frac{1}{R}\,ds=-2\pi$$

and for $\mathbf{x}\in\partial D$,

$$v_0(\mathbf{x})=\lim_{R\to 0}\int_{\partial H_R}\frac{\partial}{\partial\nu(y)}\log\frac{1}{|\mathbf{x}-\mathbf{y}|}\,ds(y)=-\lim_{R\to 0}\int_{\partial H_R}\frac{1}{R}\,ds=-\pi.$$

Hence

$$\int_{\partial D}\frac{\partial}{\partial\nu(y)}\log\frac{1}{|\mathbf{x}-\mathbf{y}|}\,ds(y)=\begin{cases}0, & \mathbf{x}\in R^2\backslash\bar{D}\\ -\pi, & \mathbf{x}\in\partial D\\ -2\pi, & \mathbf{x}\in D.\end{cases}\tag{8.50}$$

Now consider $v(\mathbf{x})$ for $\mathbf{x}$ on ∂D. Then since for $\mathbf{x}\neq\mathbf{y}$

$$\left|\frac{\partial}{\partial\nu(y)}\log\frac{1}{|\mathbf{x}-\mathbf{y}|}\right|=\frac{|\nu(y)\cdot(\mathbf{y}-\mathbf{x})|}{|\mathbf{y}-\mathbf{x}|^2}=\frac{|\cos(\nu,\mathbf{y}-\mathbf{x})|}{|\mathbf{y}-\mathbf{x}|}\leqslant K\tag{8.51}$$

where K is a positive constant, we have by Theorem 8.3 that v exists and is continuous for $\mathbf{x}$ on ∂D. We now establish the behavior of v as $\mathbf{x}$ tends to ∂D along a normal direction. Let $\mathbf{x}_0\notin\partial D$ and $\mathbf{x}\in\partial D$ such that the line segment joining $\mathbf{x}_0$ to $\mathbf{x}$ is normal to ∂D at $\mathbf{x}$. Then

$$\begin{aligned}v(\mathbf{x}_0)=\psi(\mathbf{x})&\int_{\partial D}\frac{\partial}{\partial\nu(y)}\log\frac{1}{|\mathbf{x}_0-\mathbf{y}|}\,ds(y)\\&+\int_{\partial D}[\psi(\mathbf{y})-\psi(\mathbf{x})]\frac{\partial}{\partial\nu(y)}\log\frac{1}{|\mathbf{x}_0-\mathbf{y}|}\,ds(y).\end{aligned}\tag{8.52}$$

and

$$\left|\frac{\partial}{\partial\nu(y)}\log\frac{1}{|\mathbf{x}_0-\mathbf{y}|}\right|=\frac{|\nu(y)\cdot(\mathbf{y}-\mathbf{x}_0)|}{|\mathbf{y}-\mathbf{x}_0|^2}\leqslant\frac{|\nu(y)\cdot(\mathbf{y}-\mathbf{x})|}{|\mathbf{y}-\mathbf{x}_0|^2}+\frac{|\mathbf{x}-\mathbf{x}_0|}{|\mathbf{y}-\mathbf{x}_0|^2}.\tag{8.53}$$

Now assume $|\mathbf{y}-\mathbf{x}|\leqslant\rho<1/2K$ and $|\mathbf{x}-\mathbf{x}_0|\leqslant\rho/2$. Then

$$\begin{aligned}|\mathbf{y}-\mathbf{x}_0|^2&=|\mathbf{y}-\mathbf{x}|^2+2(\mathbf{y}-\mathbf{x})\cdot(\mathbf{x}-\mathbf{x}_0)+|\mathbf{x}-\mathbf{x}_0|^2\\&\geqslant|\mathbf{y}-\mathbf{x}|^2-K\,|\mathbf{y}-\mathbf{x}|^2\,\rho+|\mathbf{x}-\mathbf{x}_0|^2\\&\geqslant\tfrac{1}{2}\,|\mathbf{y}-\mathbf{x}|^2+|\mathbf{x}-\mathbf{x}_0|^2\\&\geqslant\tfrac{1}{2}\,|\mathbf{y}-\mathbf{x}|^2.\end{aligned}\tag{8.54}$$

Hence from (8.53) we have

$$\left|\frac{\partial}{\partial\nu(y)}\log\frac{1}{|\mathbf{x}_0-\mathbf{y}|}\right| \leq 2K+\frac{|\mathbf{x}-\mathbf{x}_0|}{|\mathbf{y}-\mathbf{x}_0|^2}. \tag{8.55}$$

Since from Theorem 8.8 we know that there exists a $\gamma>0$ such that $|\mathbf{y}-\mathbf{x}|\geq\gamma|t-s|$ for $|\mathbf{y}-\mathbf{x}|\leq\rho$ we have from (8.54), (8.55) that for $C_\rho=\{\mathbf{y}:\mathbf{y}\in\partial D, |\mathbf{y}-\mathbf{x}|\leq\rho\}$

$$\begin{aligned}\int_{C_\rho}\left|\frac{\partial}{\partial\nu(y)}\log\frac{1}{|\mathbf{x}_0-\mathbf{y}|}\right| ds(y) &\leq 2KL+\int_{-\infty}^{\infty}\frac{|\mathbf{x}-\mathbf{x}_0|\,dt}{(\gamma^2/2)t^2+|\mathbf{x}-\mathbf{x}_0|^2}\\ &= 2KL+\frac{2}{\gamma^2}\tan^{-1}\left(\frac{t}{|\mathbf{x}-\mathbf{x}_0|}\frac{\gamma}{\sqrt{2}}\right)\Bigg|_{-\infty}^{\infty}\\ &= M\end{aligned} \tag{8.56}$$

where M is a positive constant. From the mean value theorem we have

$$\left|\frac{\partial}{\partial\nu(y)}\log\frac{1}{|\mathbf{x}_0-\mathbf{y}|}-\frac{\partial}{\partial\nu(y)}\log\frac{1}{|\mathbf{x}-\mathbf{y}|}\right| \leq \frac{N}{\rho^2}|\mathbf{x}-\mathbf{x}_0| \tag{8.57}$$

for $|\mathbf{y}-\mathbf{x}|\geq\rho$, $|\mathbf{x}-\mathbf{x}_0|\leq\rho/2$, where N is a positive constant. Hence, if we define w by

$$w(\mathbf{x}_0)=\int_{\partial D}[\psi(\mathbf{y})-\psi(\mathbf{x})]\frac{\partial}{\partial\nu(y)}\log\frac{1}{|\mathbf{x}_0-\mathbf{y}|}\,ds(y)$$

we have

$$|w(\mathbf{x}_0)-w(\mathbf{x})|\leq 2M\sup_{\mathbf{y}\in C_\rho}|\psi(\mathbf{y})-\psi(\mathbf{x})|+\frac{2N}{\rho^2}\|\psi\|\,|\mathbf{x}-\mathbf{x}_0|, \tag{8.58}$$

and since ψ is uniformly continuous for every $\varepsilon>0$ there exists a δ such that if $\mathbf{x}=\mathbf{x}+\mathbf{v}h$, $|h|<\delta$, then

$$|w(\mathbf{x}_0)-w(\mathbf{x})|<\varepsilon.$$

Hence from (8.50), (8.52) we have

$$|v(\mathbf{x}\pm\mathbf{v}h)-(v(\mathbf{x})\pm\pi\psi(\mathbf{x}))|=|w(\mathbf{x}+\mathbf{v}h)-w(\mathbf{x})|<\varepsilon \tag{8.59}$$

for $0<h<\delta$, uniformly for $\mathbf{x}\in\partial D$. Finally, since $v(\mathbf{x})$ is continuous for $\mathbf{x}$ on ∂D, the conclusion of the theorem follows from the triangle inequality.

The following theorems can be proved in a similar way as Theorems 8.8 and 8.9, and for this reason we leave their proofs to the Exercises.

Theorem 8.10 *Let $\phi\in C(\partial D)$. Then, with $\pm$ having the same meaning* *as in Theorem 8.9, the single-layer potential*

$$u(\mathbf{x}) = \int_{\partial D} \phi(\mathbf{y}) \log \frac{1}{|\mathbf{x}-\mathbf{y}|} \, ds(y)$$

satisfies

$$\frac{\partial u}{\partial \nu_\pm}(\mathbf{x}) = \int_{\partial D} \phi(\mathbf{y}) \frac{\partial}{\partial \nu(x)} \log \frac{1}{|\mathbf{x}-\mathbf{y}|} \, ds(y) \mp \pi\phi(\mathbf{x})$$

for $\mathbf{x}$ *on* ∂D.

Theorem 8.11 *Let* $\psi \in C(\partial D)$. *Then for the double-layer potential*

$$v(\mathbf{x}) = \int_{\partial D} \psi(\mathbf{y}) \frac{\partial}{\partial \nu(y)} \log \frac{1}{|\mathbf{x}-\mathbf{y}|} \, ds(y)$$

we have that

$$\lim_{h\to 0} \left[\frac{\partial v}{\partial \nu}(\mathbf{x}+h\mathbf{v}) - \frac{\partial v}{\partial \nu}(\mathbf{x}-h\mathbf{v}) \right] = 0$$

uniformly for $\mathbf{x}$ *on* ∂D.

We now want to use the potential theory we have just developed to show the existence of solutions to the exterior Dirichlet and Neumann problems for Laplace's equation. The corresponding interior problems can be treated in the same way, and we leave these problems to the Exercises. We first consider the exterior Neumann problem, i.e. to determine a function $u \in C^2(R^2 \backslash \bar{D}) \cap C^1(R^2 \backslash \bar{D})$ such that for given $g \in C(\partial D)$ we have

$$\Delta_2 u = 0 \quad \text{in} \quad R^2 \backslash \bar{D} \tag{8.60a}$$

$$\partial u / \partial \nu = g \quad \text{on} \quad \partial D \tag{8.60b}$$

$$u \text{ is regular at infinity} \tag{8.60c}$$

where u is said to be regular at infinity if

$$\lim_{r\to\infty} |ru(\mathbf{x})| < \infty$$
$$\lim_{r\to\infty} |r^2(\partial/\partial r)u(\mathbf{x})| < \infty \tag{8.61}$$

uniformly for $0 \leqslant \phi \leqslant 2\pi$ where (r, ϕ) are polar coordinates. We note that from Green's formula we have that if a solution to (8.60a)–(8.60c) exists then

$$\int_{\partial D} g \, ds = \int_{\partial D} (\partial u / \partial \nu) \, ds = \int_{\partial \Omega_R} (\partial u / \partial \nu) \, ds = 0(1/R) \tag{8.62}$$

where $\Omega_R = \{\mathbf{x}: |\mathbf{x}| \leq R\}$. Hence, letting R tend to infinity, we see that a necessary condition for a solution of the exterior Neumann problem to exist is that

$$\int_{\partial D} g \, ds = 0. \tag{8.63}$$

We now look for a solution of (8.60a)–(8.60c) in the form

$$u(\mathbf{x}) = \int_{\partial D} \phi(\mathbf{y}) \log \frac{1}{|\mathbf{x}-\mathbf{y}|} \, ds(y) \tag{8.64}$$

where $\phi \in C(\partial D)$ is an unknown density to be determined such that

$$\int_{\partial D} \phi \, ds = 0. \tag{8.65}$$

Note that (8.64), (8.65) implies that u is regular at infinity. From Theorem 8.10 we have that ϕ must satisfy the integral equation

$$\phi(\mathbf{x}) - \frac{1}{\pi} \int_{\partial D} \phi(\mathbf{y}) \frac{\partial}{\partial \nu(x)} \log \frac{1}{|\mathbf{x}-\mathbf{y}|} \, ds(y) = -\frac{1}{\pi} g(\mathbf{x}) \tag{8.66}$$

for $\mathbf{x}$ on ∂D. By the Fredholm alternative (Theorem 8.7) the existence of a solution to (8.66) will be assured, provided the homogeneous equation

$$\phi(\mathbf{x}) - \frac{1}{\pi} \int_{\partial D} \phi(\mathbf{y}) \frac{\partial}{\partial \nu(x)} \log \frac{1}{|\mathbf{x}-\mathbf{y}|} \, ds(y) = 0 \tag{8.67}$$

has only the trivial solution $\phi \equiv 0$. We first note that from (8.67) we have

$$\begin{aligned} 0 &= \int_{\partial D} \phi(\mathbf{x}) \, ds(x) - \frac{1}{\pi} \int_{\partial D} \phi(\mathbf{y}) \int_{\partial D} \frac{\partial}{\partial \nu(x)} \log \frac{1}{|\mathbf{x}-\mathbf{y}|} \, ds(x) \, ds(y) \\ &= 2 \int_{\partial D} \phi(\mathbf{x}) \, ds(\mathbf{x}). \end{aligned} \tag{8.68}$$

Hence, if ϕ is a solution of (8.67) we have that

$$u(\mathbf{x}) = \int_{\partial D} \phi(\mathbf{y}) \log \frac{1}{|\mathbf{x}-\mathbf{y}|} \, ds(y) \tag{8.69}$$

is a harmonic function in $R^2 \backslash \bar{D}$ that is regular at infinity. From (8.67) we see that $\partial u/\partial \nu_+ = 0$ on ∂D. Hence from Green's formula

$$\int_{\Omega_R \backslash D} |\text{grad } u|^2 \, dx = \int_{\partial \Omega_R} \bar{u} (\partial u/\partial \nu) \, ds = 0(1/R^2) \tag{8.70}$$

where again $\Omega_R = \{\mathbf{x}: |\mathbf{x}| \leq R\}$. Hence, letting R tend to infinity, we see that $u(\mathbf{x}) \equiv$ constant for $\mathbf{x}$ in $R^2 \backslash D$. But since u is regular at infinity this constant must be zero, i.e. $u \equiv 0$ in $R^2 \backslash D$. From Theorem 8.8 we can now conclude that u, as defined by (8.69), is a solution of Laplace's equation in D and vanishes on ∂D and hence, by the maximum principle, $u \equiv 0$ in D. But now Theorem 8.10 implies that

$$0 = \partial u/\partial \nu_- - \partial u/\partial \nu_+ = 2\pi\phi(\mathbf{x}) \tag{8.71}$$

for $\mathbf{x}$ on ∂D and hence the only solution of (8.67) is $\phi \equiv 0$. Thus, by the Fredholm alternative, there exists a unique $\phi \in C(\partial D)$ to (8.66). From (8.63), (8.66) we have

$$\int_{\partial D} \phi(\mathbf{x})\, ds(x) - \frac{1}{\pi}\int_{\partial D} \phi(\mathbf{y}) \int_{\partial D} \frac{\partial}{\partial \nu(x)} \log \frac{1}{|\mathbf{x}-\mathbf{y}|}\, ds(x)\, ds(y)$$

$$= \frac{1}{\pi}\int_{\partial D} g(\mathbf{x})\, ds(x) = 0 \tag{8.72}$$

and hence (8.65) is satisfied. We have now established the existence (and from (8.70), uniqueness) of the solution to the exterior Neumann problem.

We now consider the exterior Dirichlet problem, i.e. to determine a function $u \in C^2(R^2 \backslash \bar{D}) \cap C(R^2 \backslash D)$ such that for given $f \in C(\partial D)$ we have

$$\Delta_2 u = 0 \quad \text{in} \quad R^2 \backslash \bar{D} \tag{8.73a}$$

$$u = f \quad \text{on} \quad \partial D \tag{8.73b}$$

$$u \text{ is bounded at infinity.} \tag{8.73c}$$

We note that since $v(r, \phi) = u(1/r, \phi)$ is the solution of an interior Dirichlet problem, we can use the maximum principle to conclude that the solution of (8.73a)–(8.73c), if it exists, is unique. We now look for a solution of (8.73a)–(8.73c) in the form

$$u(\mathbf{x}) = \int_{\partial D} \psi(\mathbf{y}) \frac{\partial}{\partial \nu(y)} \log \frac{1}{|\mathbf{x}-\mathbf{y}|}\, ds(y) \tag{8.74}$$

where $\psi \in C(\partial D)$ is an unknown density to be determined. From Theorem 8.9 we have that ψ must satisfy the integral equation

$$\psi(\mathbf{x}) + \frac{1}{\pi}\int_{\partial D} \psi(\mathbf{y}) \frac{\partial}{\partial \nu(y)} \log \frac{1}{|\mathbf{x}-\mathbf{y}|}\, ds(y) = \frac{1}{\pi} f(\mathbf{x}) \tag{8.75}$$

for $\mathbf{x}$ on ∂D. But from (8.50) we see that $\psi =$ constant is a solution of the homogeneous equation

$$\psi(\mathbf{x}) + \frac{1}{\pi}\int_{\partial D} \psi(\mathbf{y}) \frac{\partial}{\partial \nu(y)} \log \frac{1}{|\mathbf{x}-\mathbf{y}|}\, ds(y) = 0 \tag{8.76}$$

and hence, by the Fredholm alternative, a solution of (8.75) will not exist unless

$$\int_{\partial D} f\phi \, ds = 0 \tag{8.77}$$

where ϕ is a solution of the homogeneous adjoint equation. Since in general we cannot expect that f satisfies (8.77) we must modify our approach. One possibility is as follows (Kress [106]). Assume, without loss of generality, that D contains the origin and let $\Omega_R = \{\mathbf{x}: |\mathbf{x}| \leq R\} \subset D$. Then look for a solution of (8.73a)–(8.73c) in the form

$$u(\mathbf{x}) = \int_{\partial D} \psi(\mathbf{y}) \frac{\partial}{\partial \nu(y)} \left[\log \frac{1}{|\mathbf{x}-\mathbf{y}|} + \log \left(\frac{|\mathbf{y}|}{R} \left| \mathbf{x} - \frac{R^2 \mathbf{y}}{|\mathbf{y}|^2} \right| \right) \right] ds(y) \tag{8.78}$$

for $\mathbf{x} \in R^2 \backslash \bar{D}$. Then u is harmonic in $R^2 \backslash \bar{D}$, bounded at infinity, and satisfies the boundary condition (8.73b) provided

$$\psi(\mathbf{x}) + \frac{1}{\pi} \int_{\partial D} \psi(\mathbf{y}) \frac{\partial}{\partial \nu(y)} \left[\log \frac{1}{|\mathbf{x}-\mathbf{y}|} + \log \left(\frac{|\mathbf{y}|}{R} \left| \mathbf{x} - \frac{R^2 \mathbf{y}}{|\mathbf{y}|^2} \right| \right) \right] ds(y) = \frac{1}{\pi} f(\mathbf{x}) \tag{8.79}$$

for $\mathbf{x}$ on ∂D. By the Fredholm alternative a solution of (8.79) will exist provided the homogeneous equation has only the trivial solution. Let ψ be a solution of the homogeneous integral equation and u the corresponding potential. Then $u = 0$ on ∂D and by the uniqueness of the solution to the exterior Dirichlet problem $u \equiv 0$ in $R^2 \backslash D$. Hence we have $\partial u / \partial \nu_+ = 0$ on ∂D, and from Theorem 8.11 we can conclude that $\partial u / \partial \nu_- = 0$ on ∂D. But for $|\mathbf{x}| = R$ and $|\mathbf{y}| > R$ we have

$$\left(|\mathbf{y}| \left| \mathbf{x} - \frac{R^2 \mathbf{y}}{|\mathbf{y}|^2} \right| \right)^2 = |\mathbf{y}|^2 R^2 - 2R^2 \mathbf{x} \cdot \mathbf{y} + R^4 = (R\, |\mathbf{x}-\mathbf{y}|)^2 \tag{8.80}$$

and hence

$$\frac{\partial}{\partial \nu(y)} \left[\log \frac{1}{|\mathbf{x}-\mathbf{y}|} + \log \left(\frac{|\mathbf{y}|}{R} \left| \mathbf{x} - \frac{R^2 \mathbf{y}}{|\mathbf{y}|^2} \right| \right) \right] = 0 \tag{8.81}$$

for $|\mathbf{x}| = R$, $\mathbf{y}$ on ∂D. Therefore, $u = 0$ on $\partial \Omega_R$. But then, from Green's formula, we have

$$\int_{D \backslash \Omega_R} |\text{grad}\, u|^2 \, dx = \int_{\partial D} u (\partial \bar{u} / \partial \nu) \, ds - \int_{\partial \Omega_R} u (\partial \bar{u} / \partial \nu) \, ds = 0 \tag{8.82}$$

and hence grad $u = 0$ in $D \backslash \Omega_R$, i.e. $u =$ constant in $D \backslash \Omega_R$. However, since $u = 0$ on $\partial \Omega_R$ we have $u = 0$ in $D \backslash \Omega_R$ and thus, from Theorem 8.9,

$$0 = u_+(\mathbf{x}) - u_-(\mathbf{x}) = 2\pi \psi(\mathbf{x}) \tag{8.83}$$

for $\mathbf{x}$ on ∂D, i.e. $\psi = 0$. From the Fredholm alternative we can now conclude that there exists a unique solution $\psi \in C(\partial D)$ of (8.79) and this establishes the existence of a solution to the exterior Dirichlet problem.

We now want to determine the spectrum of the compact integral operators $\mathbf{A}: C(\partial D) \to C(\partial D)$ and $\mathbf{B}: C(\partial D) \to C(\partial D)$ defined by

$$(\mathbf{A}\psi)(\mathbf{x}) = \frac{1}{\pi} \int_{\partial D} \psi(\mathbf{y}) \frac{\partial}{\partial \nu(y)} \log \frac{1}{|\mathbf{x}-\mathbf{y}|} \, ds(y)$$

$$(\mathbf{B}\phi)(\mathbf{x}) = \frac{1}{\pi} \int_{\partial D} \phi(\mathbf{y}) \frac{\partial}{\partial \nu(x)} \log \frac{1}{|\mathbf{x}-\mathbf{y}|} \, ds(y) \tag{8.84}$$

which belong to the Dirichlet and Neumann problems, respectively. We first recall the following definitions and theorems (cf. Stakgold [162]; Taylor [164]).

Definition 8.4 *Let $\mathbf{A}: X \to X$ be a bounded linear operator mapping a Banach space X into itself. Then a complex number $\lambda \in C$ is called an* eigenvalue *of $\mathbf{A}$ if there exists an element $x \in X$, $x \neq 0$, such that $Ax = \lambda x$. A complex number $\lambda \in C$ is called a* regular value *if $(\lambda \mathbf{I} - \mathbf{A})^{-1}$ exists and is bounded. The set of all regular values of $\mathbf{A}$ is called the* resolvent set *$\rho(\mathbf{A})$ and the complement of $\rho(\mathbf{A})$ is called the* spectrum *$\sigma(\mathbf{A})$.*

Theorem 8.12 *Let X be an infinite-dimensional Banach space and let $\mathbf{A}: X \to X$ be a compact linear operator. Then $\lambda = 0$ belongs to the spectrum $\sigma(\mathbf{A})$ and $\sigma(\mathbf{A}) \backslash \{0\}$ consists of at most a countable set of eigenvalues with no point of accumulation except, possibly, $\lambda = 0$.*

Definition 8.5 *Let $\mathbf{A}: X \to X$ be a bounded linear operator mapping the Banach space X into itself. Then*

$$r(\mathbf{A}) = \sup_{\lambda \in \sigma(\mathbf{A})} |\lambda|$$

is called the spectral radius *of $\mathbf{A}$.*

Theorem 8.13 *Let $\mathbf{A}: X \to X$ be a bounded linear operator mapping the Banach space X into itself. Then $r(\mathbf{A}) = \lim \|\mathbf{A}^n\|^{1/n}$ and the Neumann series*

$$\sum_{n=0}^{\infty} \lambda^{-n-1} \mathbf{A}^n$$

converges for all $\lambda > r(\mathbf{A})$.

Note that in particular if $r(\mathbf{A})<1$ then

$$(\mathbf{I}-\mathbf{A})^{-1}=\sum_{n=0}^{\infty}\mathbf{A}^n$$

and the equation

$$y=x-\mathbf{A}x, \qquad x, y\in X$$

can be solved by successive approximations. As a corollary to Theorem 8.13 we have the following result (Taylor [164]).

Corollary 8.1 *Let $\mathbf{A}: X\to X$, $\mathbf{B}: X\to X$ be bounded linear operators mapping the Banach space X into itself. Suppose $\mathbf{A}^{-1}: X\to X$ exists, is a bounded linear operator on X, and $\|\mathbf{A}-\mathbf{B}\|<1/\|\mathbf{A}^{-1}\|$. Then $\mathbf{B}^{-1}: X\to X$ exists, is a bounded linear operator on X, and*

$$\|\mathbf{B}^{-1}\|\leq\frac{\|\mathbf{A}^{-1}\|}{1-\|\mathbf{A}^{-1}\|\,\|\mathbf{A}-\mathbf{B}\|}$$

$$\|\mathbf{B}^{-1}-\mathbf{A}^{-1}\|\leq\frac{\|\mathbf{A}^{-1}\|^2\,\|\mathbf{A}-\mathbf{B}\|}{1-\|\mathbf{A}^{-1}\|\,\|\mathbf{A}-\mathbf{B}\|}.$$

Proof We have

$$\mathbf{B}=\mathbf{A}-(\mathbf{A}-\mathbf{B})=\mathbf{A}[\mathbf{I}-\mathbf{A}^{-1}(\mathbf{A}-\mathbf{B})].$$

Since $\|\mathbf{A}^{-1}(\mathbf{A}-\mathbf{B})\|\leq\|\mathbf{A}^{-1}\|\,\|\mathbf{A}-\mathbf{B}\|<1$, it follows that $[\mathbf{I}-\mathbf{A}^{-1}(\mathbf{A}-\mathbf{B})]^{-1}: X\to X$ exists and is given by

$$[\mathbf{I}-\mathbf{A}^{-1}(\mathbf{A}-\mathbf{B})]^{-1}=\sum_{n=0}^{\infty}[\mathbf{A}^{-1}(\mathbf{A}-\mathbf{B})]^n. \tag{8.85}$$

Hence $\mathbf{B}^{-1}: X\to X$ exists and is given by

$$\mathbf{B}^{-1}=[\mathbf{I}-\mathbf{A}^{-1}(\mathbf{A}-\mathbf{B})]^{-1}\mathbf{A}^{-1}. \tag{8.86}$$

(8.85) and the hypothesis of the theorem now imply that $\mathbf{B}^{-1}$ is bounded and from (8.85), (8.86) we easily deduce the inequalities of the theorem.

We now apply the above results to the integral operators of potential theory.

Theorem 8.14 *Let* **A** *and* **B** *be as in* (8.84). *Then* $\sigma(\mathbf{A})=\sigma(\mathbf{B})\subset[-1, 1)$ *and* -1 *is an eigenvalue.*

Proof The fact that $\sigma(\mathbf{A})=\sigma(\mathbf{B})$ follows from the Fredholm alternative. From Theorem 8.12 we have that $\sigma(\mathbf{A})\backslash\{0\}$ consists only of eigenvalues. Now let $\lambda\neq 0$ be an eigenvalue of **A**, $\mathbf{A}\psi=\lambda\phi$, and define

$$u(\mathbf{x})=\frac{1}{\pi}\int_{\partial D}\psi(\mathbf{y})\frac{\partial}{\partial\nu(y)}\log\frac{1}{|\mathbf{x}-\mathbf{y}|}\,ds(y)$$

for $\mathbf{x} \in R^2 \backslash \partial D$. Then by Theorem 8.9

$$u_{\pm} = \mathbf{A}\psi \pm \psi = (\lambda \pm 1)\psi \tag{8.87}$$

for $\mathbf{x}$ on ∂D. Hence

$$(\lambda - 1)u_+ = (\lambda + 1)u_- \tag{8.88}$$

for $\mathbf{x}$ on ∂D. Now multiply both sides of (8.88) by $\partial \bar{u}/\partial \nu_+ = \partial \bar{u}/\partial \nu_-$ and integrate over ∂D to obtain

$$(\lambda - 1)\int_{\partial D} u_+(\partial \bar{u}/\partial \nu_+)\,ds = (\lambda + 1)\int_{\partial D} u_-(\partial \bar{u}/\partial \nu_-)\,ds.$$

Noting that u is regular at infinity and applying Green's theorem we obtain

$$(1 - \lambda)\hat{I} = (1 + \lambda)I \tag{8.89}$$

where

$$I = \int_D |\text{grad } u|^2\,dx, \qquad \hat{I} = \int_{R^2 \backslash D} |\text{grad } u|^2\,dx. \tag{8.90}$$

If $I + \hat{I} > 0$ then it follows that

$$\lambda = (\hat{I} - I)/(\hat{I} + I)$$

and hence λ is real and $|\lambda| \leq 1$. If $I + \hat{I} = 0$ then $u = 0$ in $R^2 \backslash \bar{D}$ (noting that u is regular at infinity) and hence if $\lambda \neq -1$ we have from (8.88) that $u_- = 0$ on ∂D. Then $2\psi = u_+ - u_- = 0$ and ψ cannot be an eigenfunction. From (8.76) we see that -1 is an eigenvalue of $\mathbf{A}$ whereas from (8.67)–(8.71) we see that 1 is not an eigenvalue of $\mathbf{B}$ and hence, by the Fredholm alternative, not of $\mathbf{A}$. This completes the proof of the theorem.

We now want to extend the results developed thus far for Laplace's equation to the case of the Helmholtz equation. More precisely, we shall establish the existence and uniqueness of the solutions to the following boundary value problems for k real and sufficiently small.

Neumann problem

Let $g \in C(\partial D)$ be prescribed on ∂D. Then we want to determine a function $u \in C^2(R^2 \backslash \bar{D}) \cap C^1(R^2 \backslash D)$ such that

$$\Delta_2 u + k^2 u = 0 \quad \text{in} \quad R^2 \backslash \bar{D} \tag{8.91a}$$

$$\partial u/\partial \nu = g \quad \text{on} \quad \partial D \tag{8.91b}$$

$$\lim_{r \to \infty} \sqrt{r}(\partial u/\partial r - iku) = 0 \tag{8.91c}$$

where the radiation condition holds uniformly in ϕ as r tends to infinity.

Dirichlet problem

Let $f \in C(\partial D)$ be prescribed on ∂D. Then we want to determine a function $u \in C^2(R^2 \backslash \bar{D}) \cap C(R^2 \backslash D)$ such that

$$\Delta_2 u + k^2 u = 0 \quad \text{in} \quad R^2 \backslash \bar{D} \tag{8.92a}$$

$$u = f \quad \text{on} \quad \partial D \tag{8.92b}$$

$$\lim_{r\to\infty} \sqrt{r}(\partial u/\partial r - iku) = 0. \tag{8.92c}$$

We first make several observations. Firstly, the restriction to small real values of k can be removed (cf. Kleinman and Roach [102]; Ursell [170]; Vekua [172]). However, for our purposes this restriction is unimportant and hence we do not prove this more general result. Secondly, the uniqueness of the solution to both the Dirichlet and Neumann problems can be proved by modifying the proof of Theorem 5.6 and we leave this to the Exercises. In this connection we note that since ∂D is twice continuously differentiable if $f \equiv 0$ in (8.92b) then we can conclude that the solution of (8.92a)–(8.92c) (which we want to show is identically zero) is continuously differentiable up to the boundary and hence we can apply Green's formulae to the problem (cf. Vekua [172]). Hence the only problem remaining is to establish the existence of a solution to the above boundary value problems.

We first consider the Neumann problem and look for a solution in the form of a single-layer metaharmonic potential defined by

$$u(\mathbf{x}) = \frac{\pi i}{2} \int_{\partial D} \phi(\mathbf{y}) H_0^{(1)}(kR) \, ds(y) \tag{8.93}$$

where $R = |\mathbf{x} - \mathbf{y}|$, $\phi \in C(\partial D)$ is a density to be determined, and $H_0^{(1)}$ denotes a Hankel function of the first kind defined by

$$H_0^{(1)}(kR) = J_0(kR) + i\left\{\frac{2}{\pi}\left(\gamma + \log\frac{kR}{2}\right) J_0(kR) - \frac{4}{\pi}\sum_{n=1}^{\infty}\frac{(-1)^n}{n} J_{2n}(kr)\right\} \tag{8.94}$$

with J_n being a Bessel function and γ Euler's constant. Note that $H_0^{(1)}(kR)$ satisfies the Helmholtz equation (8.91a) for $R \neq 0$ and the radiation condition (8.91c). Note also that $(\pi i/2)H_0^{(1)}(kR) + \log kR$ remains bounded as kR tends to zero and hence the discontinuity properties of the single-layer metaharmonic potential (8.93) are the same as the single-layer potential for Laplace's equation. Hence, in order for (8.93) to satisfy the boundary condition (8.91b), we must have that ϕ satisfies the integral equation

$$\phi(\mathbf{x})-\frac{1}{\pi}\int_{\partial D}\phi(\mathbf{y})\frac{\partial}{\partial\nu(x)}H_0^{(1)}(kR)\,\mathrm{d}s(y)=-\frac{1}{\pi}g(\mathbf{x}) \tag{8.95}$$

for $\mathbf{x}$ on ∂D. By the Fredholm alternative a unique solution to (8.95) will exist, provided the homogeneous equation has only the trivial solution $\phi\equiv 0$. Suppose, on the contrary, that (8.95) with $g\equiv 0$ has a nontrivial solution ϕ and define u by (8.93). Then u is a solution of the Helmholtz equation in $R^2\backslash\bar{D}$ satisfying the radiation condition (8.91c) and such that $\partial u/\partial\nu_+=0$ on ∂D. Hence, by the uniqueness of the solution to the exterior Neumann problem, $u\equiv 0$ in $R^2\backslash\bar{D}$. By Theorem 8.9 $u_-=0$ on ∂D, and since for $\mathbf{x}\in D$ u is also a solution of the Helmholtz equation, $u\equiv 0$ in D, provided k^2 is not an eigenvalue of the interior Dirichlet problem. This will certainly be the case provided k is sufficiently small (see the last paragraph of Chapter 1). But now Theorem 8.10 implies that for k sufficiently small

$$0=\partial u/\partial\nu_- - \partial u/\partial\nu_+ = 2\pi\phi(\mathbf{x}) \tag{8.96}$$

for $\mathbf{x}$ on ∂D and hence $\phi\equiv 0$, a contradiction. Thus, the homogeneous equation (8.95) with $g\equiv 0$ has only the trivial solution and hence, by the Fredholm alternative, a unique solution exists to (8.95) and (8.93) provides us with a solution to the Neumann problem.

We now consider the Dirichlet problem and look for a solution in the form of a double-layer metaharmonic potential defined by

$$u(\mathbf{x})=\frac{\pi\mathrm{i}}{2}\int_{\partial D}\psi(\mathbf{y})\frac{\partial}{\partial\nu(y)}H_0^{(1)}(kR)\,\mathrm{d}s(y) \tag{8.97}$$

where $\psi\in C(\partial D)$ is a density to be determined. Then u, as defined by (8.97), is a solution of the Helmholtz equation in $R^2\backslash\bar{D}$ and satisfies the radiation condition (8.92c). In order for (8.97) to satisfy the boundary condition (8.92b) we use Theorem 8.9 to deduce that ψ must satisfy the integral equation

$$\psi(\mathbf{x})+\frac{1}{\pi}\int_{\partial D}\psi(\mathbf{y})\frac{\partial}{\partial\nu(y)}H_0^{(1)}(kR)\,\mathrm{d}s(y)=\frac{1}{\pi}f(x) \tag{8.98}$$

for $\mathbf{x}$ on ∂D. By the Fredholm alternative a unique solution to (8.98) will exist provided the homogeneous equation has only the trivial solution $\psi\equiv 0$. Suppose then, on the contrary, that (8.98) with $f\equiv 0$ has a nontrivial solution ψ and define u by (8.97). Then u is a solution of the Helmholtz equation in $R^2\backslash\bar{D}$ satisfying the radiation condition (8.92c) and such that $u_+=0$ on ∂D. Hence, by the uniqueness of the solution to the exterior Dirichlet problem, $u\equiv 0$ in $R^2\backslash D$ and hence $\partial u/\partial\nu_+=0$ on

∂D. From Theorem 8.11 we can now conclude that $\partial u/\partial \nu_- = 0$ on ∂D. But, for $\mathbf{x} \in D$, u is also a solution of the Helmholtz equation and hence $u \equiv 0$ in D unless k^2 is an eigenvalue of the interior Neumann problem. This will again not be the case if k is sufficiently small and hence, from Theorem 8.9, we have for k sufficiently small

$$0 = v_+ - v_- = 2\pi\psi(\mathbf{x}) \tag{8.99}$$

for $\mathbf{x}$ on ∂D and hence $\psi \equiv 0$, a contradiction. Thus, the homogeneous equation (8.98) with $f \equiv 0$ has only the trivial solution and hence by the Fredholm alternative a unique solution exists to (8.98) and (8.97) provides us with a solution to the Dirichlet problem.

We shall now proceed to use the results derived in this section to discuss the inverse scattering problem for a cylinder.

8.4 The inverse scattering problem for a cylinder

We now turn our attention to the problem of determining the shape of a cylindrical object from a knowledge of the far field data of a scattered acoustic wave. We shall assume that the incoming wave is a plane wave, the cylinder has a bounded, simply connected cross-section, and that the far field data are known for all angles and low values of the frequency. Our approach is based on the fact that since the scattering object is a cylinder, the problem is basically two-dimensional, and hence is amenable to function theoretic methods and the use of conformal mappings. Our presentation is based on the work of Colton [36] and Colton and Kleinman [38]; for other examples of the use of conformal mapping techniques in scattering theory the reader is referred to Garabedian [63], Hill, Kleinman and Pfaff [86] and Hong and Goodrich [88].

We first consider the case of a 'hard' obstacle. Assume D is a bounded, simply connected domain with C^2 boundary ∂D and let ν denote the outward unit normal to ∂D. Let a plane acoustic wave moving in the x_1 direction (for $\mathbf{x} = (x_1, x_2) \in R^2$) be perturbed by the presence of the 'hard' obstacle D. Then, if we denote the velocity potential of the total field by u and let u_s denote the velocity potential of the scattered wave, we can mathematically formulate the 'direct' problem as follows: given D to determine $u \in C^2(R^2 \backslash \bar{D}) \cap C^1(R^2 \backslash D)$ such that

$$u(\mathbf{x}) = e^{ikx_1} + u_s(\mathbf{x}) \tag{8.100a}$$

$$\Delta_2 u + k^2 u = 0 \quad \text{in} \quad R^2 \backslash \bar{D} \tag{8.100b}$$

$$\partial u/\partial \nu = 0 \quad \text{on} \quad \partial D \tag{8.100c}$$

$$\lim_{r \to \infty} \sqrt{r}(\partial u_s/\partial r - iku_s) = 0 \tag{8.100d}$$

where k is the wave number, $r=(x_1^2+x_2^2)^{1/2}$, and the limit in the 'radiation condition' (8.100d) is assumed to hold uniformly in all directions. We shall show shortly that if (r, θ) are polar coordinates,

$$u_s(\mathbf{x})=\tfrac{1}{4}e^{i(kr+\pi/4)}\left(\frac{2}{\pi kr}\right)^{1/2}F(\theta;k)+0\left(\frac{1}{r^{3/2}}\right), \tag{8.101}$$

where F denotes the far field pattern. Then the 'inverse' problem is, given F, to determine the shape of D. To reformulate the inverse problem we first consider the direct problem. Note that the existence of a solution to (8.100a)–(8.100d) for 'small' k follows from the discussion at the end of Section 8.3. Our aim is to obtain this solution by iteration. To this end we have, from Green's formula, that (cf. Ahner [4]; Ahner and Hsiao [6]; Ahner and Kleinman [7]):

$$\frac{i}{4}\int_{\partial D}\left[H_0^{(1)}(kR)\frac{\partial u_s}{\partial\nu(y)}-u_s\frac{\partial}{\partial\nu(y)}H_0^{(1)}(kR)\right]ds(y)=\begin{cases}-u_s(\mathbf{x}); & \mathbf{x}\in R^2\backslash\bar{D}\\ -\tfrac{1}{2}u_s(\mathbf{x}); & \mathbf{x}\in\partial D\end{cases}$$

$$\frac{i}{4}\int_{\partial D}\left[H_0^{(1)}(kR)\frac{\partial e^{iky_1}}{\partial\nu(y)}-e^{iky_1}\frac{\partial}{\partial\nu(y)}H_0^{(1)}(kR)\right]ds(y)=\begin{cases}0; & \mathbf{x}\in R^2\backslash\bar{D}\\ \tfrac{1}{2}e^{ikx_1}; & \mathbf{x}\in\partial D\end{cases} \tag{8.102}$$

where $R=|\mathbf{x}-\mathbf{y}|$. Hence from (8.100), (8.102) we have

$$e^{ikx_1}+\frac{i}{4}\int_{\partial D}u\frac{\partial}{\partial\nu(y)}H_0^{(1)}(kR)\,ds(y)=\begin{cases}u(\mathbf{x}); & \mathbf{x}\in R^2\backslash\bar{D}\\ \tfrac{1}{2}u(\mathbf{x}); & \mathbf{x}\in\partial D.\end{cases} \tag{8.103}$$

But from (8.103) and (8.50) we can now write

$$e^{ikx_1}-\frac{1}{2\pi}\int_{\partial D}[u(\mathbf{y})-u(\mathbf{x})]\frac{\partial}{\partial\nu(y)}\log R\,ds(y)$$
$$-\frac{1}{2\pi}\int_{\partial D}u(\mathbf{y})\frac{\partial}{\partial\nu(y)}\left\{-\frac{\pi i}{2}H_0^{(1)}(kR)-\log R\right\}ds(y)=u(\mathbf{x}) \tag{8.104}$$

for $\mathbf{x}$ in $R^2\backslash D$, in particular for $\mathbf{x}$ on ∂D. For convenience we rewrite (8.104) as the operator equation

$$e^{ikx_1}+\mathbf{L}_k[u]=u;\qquad \mathbf{x}\text{ on }\partial D \tag{8.105}$$

and note that if

$$(\mathbf{L}_0u)(\mathbf{x})=-\frac{1}{2\pi}\int_{\partial D}[u(\mathbf{y})-u(\mathbf{x})]\frac{\partial}{\partial\nu(y)}\log R\,ds(y) \tag{8.106}$$

then from (8.94) we have that for k positive and sufficiently small

$$\|\mathbf{L}_k - \mathbf{L}_0\| \leq M\,|k^2 \log k| \tag{8.107}$$

where $\|\cdot\|$ denotes the maximum operator norm and M is a positive constant independent of k. We observe that if $\mathbf{A}$ is the integral operator of classical potential theory as defined in (8.84), then $\mathbf{A} = 2\mathbf{L}_0 - \mathbf{I}$. Hence λ is an eigenvalue of $\mathbf{A}$ if and only if $\frac{1}{2}(\lambda+1)$ is an eigenvalue of $\mathbf{L}_0$. But by Theorem 8.14 the eigenvalues of $\mathbf{A}$ are all contained in the interval $[-1, 1)$ and thus the eigenvalues of the compact operator $\mathbf{L}_0$ are all contained in $[0, 1)$. This implies that the spectral radius of $\mathbf{L}_0$ is less than one, i.e. the Neumann series for $(\mathbf{I}-\mathbf{L}_0)^{-1}$ is convergent.

Theorem 8.15 *There exists a positive number k_0 such that for $|k| \leq k_0$ the spectral radius of $\mathbf{L}_k$ is less than one. Thus* (8.105) *can be solved by successive approximations.*

Proof Our proof is based on Kleinman and Wendland [103]. From (8.107) we have

$$\|(\lambda\mathbf{I}+\mathbf{L}_k)-(\lambda\mathbf{I}+\mathbf{L}_0)\| = \|\mathbf{L}_k - \mathbf{L}_0\| = 0(|k^2 \log k|). \tag{8.108}$$

But $(\lambda\mathbf{I}+\mathbf{L}_0)^{-1}$ exists for $|\lambda| \geq \lambda_0$ where $\lambda_0 < 1$. Let

$$M = \max_{|\lambda| \geq \lambda_0} \|(\lambda\mathbf{I}+\mathbf{L}_0)^{-1}\|. \tag{8.109}$$

Then from (8.108), (8.109) we see that there exists a positive constant k_0 such that for $|k| < k_0$

$$\|(\lambda\mathbf{I}+\mathbf{L}_k)-(\lambda\mathbf{I}+\mathbf{L}_0)\| \leq M^{-1} \leq \|(\lambda\mathbf{I}+\mathbf{L}_0)^{-1}\|^{-1} \tag{8.110}$$

for all $|\lambda| \geq \lambda_0$. Hence, from Corollary 8.1, we have that $(\lambda\mathbf{I}+\mathbf{L}_k)^{-1}$ exists for all $|\lambda| \geq \lambda_0$, $|k| \leq k_0$, i.e. for $|k| \leq k_0$ the spectral radius of $\mathbf{L}_k$ is less than one and this implies (8.105) can be solved by successive approximations.

From Theorem 8.15 we can now conclude that the solution of (8.105) for k sufficiently small is given by

$$\begin{aligned} u(\mathbf{x}) &= \sum_{n=0}^{\infty} \mathbf{L}_k^n[e^{ikx_1}] \\ &= \sum_{n=0}^{\infty} \mathbf{L}_k^n[1+ikx_1] + 0(|k|^2) \\ &= \sum_{n=0}^{\infty} \mathbf{L}_0^n[1+ikx_1] + 0(|k^2 \log k|) \\ &= 1 + iku_0(\mathbf{x}) + 0(|k^2 \log k|) \end{aligned} \tag{8.111}$$

for $\mathbf{x}$ on ∂D where u_0 is a harmonic function defined in the exterior of D and assuming the Dirichlet data

$$u_0(\mathbf{x}) = \sum_{n=0}^{\infty} L_0^n[x_1]; \qquad \mathbf{x} \text{ on } \partial D. \tag{8.112}$$

Hence from (8.105)–(8.107) we see that u_0 is the solution of

$$u_0(\mathbf{x}) = x_1 + u_s(\mathbf{x}) \tag{8.113a}$$

$$\Delta_2 u_0 = 0 \quad \text{in} \quad R^2 \backslash \bar{D} \tag{8.113b}$$

$$\partial u_0 / \partial \nu = 0 \quad \text{on} \quad \partial D \tag{8.113c}$$

$$u_s \text{ is regular at infinity,} \tag{8.113d}$$

i.e. u_0 is the velocity potential for an incompressible irrotational fluid flow past D having constant velocity in the x_1 direction at infinity (cf. Bergman and Schiffer [15]). If

$$w = f(z) = az + b + c/z + d/z^2 + \cdots; \qquad a > 0, \tag{8.114}$$

$z = x_1 + ix_2$, maps the exterior of D conformally onto the exterior of the unit disk Ω then it is easily verified that module is a constant

$$u_0(\mathbf{x}) = a^{-1} \operatorname{Re}[f(z) + 1/f(z)] \tag{8.115}$$

where a^{-1} is known as the *mapping radius* of D. Note that $f(z) + 1/f(z)$ is real on ∂D and hence, from the Cauchy–Riemann equations, u_0 as defined by (8.115) satisfies the boundary condition (8.113c). As will be seen in what follows, the fact that the representation (8.115) is only valid up to a constant will not affect us in the determination of D when we consider the inverse problem.

We now return to the representation (8.104). Since for $\mathbf{x}$ in $R^2 \backslash \bar{D}$ we have

$$\int_{\partial D} \frac{\partial}{\partial \nu(y)} \log R \, ds(y) = 0 \tag{8.116}$$

we can deduce from the asymptotic behavior of the Hankel function that

$$\begin{aligned} u_s(\mathbf{x}) &= \frac{i}{4} \int_{\partial D} u(\mathbf{y}) \frac{\partial}{\partial \nu(y)} H_0^{(1)}(kR) \, ds(y) \\ &= \tfrac{1}{4} e^{i(kr + \pi/4)} \left(\frac{2}{\pi k r} \right)^{1/2} F(\theta; k) + 0\left(\frac{1}{r^{3/2}} \right) \end{aligned} \tag{8.117}$$

where F is the far field pattern defined by

$$F(\theta; k) = \int_{\partial D} u(\mathbf{y}) \frac{\partial}{\partial \nu(y)} \exp[-ik\rho \cos(\theta - \phi)] \, ds(y) \tag{8.118}$$

and

$$\mathbf{x}=x_1+\mathrm{i}x_2=r\mathrm{e}^{\mathrm{i}\theta}, \qquad \mathbf{y}=y_1+\mathrm{i}y_2=\rho\mathrm{e}^{\mathrm{i}\phi}. \tag{8.119}$$

We are now in a position to begin considering the inverse problem. Assume that the far field pattern is known for $-\pi\leqslant\theta\leqslant\pi$ and small values of the wave number k. If we expand F in a Fourier series

$$F(\theta; k)=\sum_{-\infty}^{\infty} a_n(k)\mathrm{e}^{\mathrm{i}n\theta} \tag{8.120}$$

then from (8.118) and the well-known integral representation of Bessel's function we have that (cf. Erdélyi *et al.* [53])

$$\begin{aligned} a_n(k)&=\frac{1}{2\pi}\int_{-\pi}^{\pi}\int_{\partial D} u(\mathbf{y})\frac{\partial}{\partial\nu(\mathrm{y})}\exp\left[-\mathrm{i}n\theta-\mathrm{i}k\rho\cos(\theta-\phi)\right]\mathrm{d}s(\mathrm{y})\,\mathrm{d}\theta \\ &=\mathrm{i}^{-n}\int_{\partial D} u(\mathbf{y})\frac{\partial}{\partial\nu}\left[J_n(k\rho)\mathrm{e}^{-\mathrm{i}n\phi}\right]\mathrm{d}s \end{aligned} \tag{8.121}$$

where J_n denotes Bessel's function. Since $\rho^n\mathrm{e}^{-\mathrm{i}n\phi}$ is a harmonic function

$$\int_{\partial D}\frac{\partial}{\partial\nu}\left[\rho^n\mathrm{e}^{-\mathrm{i}n\phi}\right]\mathrm{d}s=0, \tag{8.122}$$

and hence, from (8.111), (8.122) and the Taylor series expansion of Bessel's function, we have that for $n\geqslant1$ and k sufficiently small

$$a_n(k)=\frac{\mathrm{i}^{-n+1}k^{n+1}}{2^n n!}\int_{\partial D} u_0(\mathbf{y})\frac{\partial}{\partial\nu}\left[\rho^n\mathrm{e}^{-\mathrm{i}n\phi}\right]\mathrm{d}s+0(|k^{n+2}\log k|). \tag{8.123a}$$

Since $J_n(k\rho)=(-1)^nJ_{-n}(k\rho)$, the same expression (up to conjugation and a factor of plus or minus one) holds for $n\leqslant-1$ and thus no new information can be gathered from the negatively indexed Fourier coefficients. For $n=0$ we have

$$a_0(k)=-\frac{k^2}{4}\int_{\partial D}\frac{\partial}{\partial\nu}\rho^2\,\mathrm{d}s+0(|k^3|). \tag{8.123b}$$

Thus the problem of determining the scattering obstacle D from a knowledge of the far field pattern F has been reduced to the following: given

$$\begin{aligned} \mu_n&=\mathrm{i}^{n-1}\lim_{k\to0}\frac{a_n(k)}{k^{n+1}}; \qquad n\geqslant1 \\ \mu_0&=-\lim_{k\to0}\frac{a_0(k)}{k^2} \end{aligned} \tag{8.124}$$

determine ∂D from the relations

$$\mu_0 = \frac{1}{4}\int_{\partial D} \frac{\partial}{\partial \nu} \rho^2 \, ds \tag{8.125a}$$

$$\mu_n = \frac{1}{2^n n!}\int_{\partial D} u_0(\mathbf{y}) \frac{\partial}{\partial \nu}[\rho^n e^{-in\phi}] \, ds; \qquad n \geq 1 \tag{8.125b}$$

where u_0 is given by (8.115). Note that due to (8.121) the fact that (8.115) is only valid module a constant does not affect our determination of ∂D from the relation (8.125b).

We first consider (8.125a). From Green's identity we have

$$\mu_0 = \frac{1}{4}\iint_D \Delta_2 \rho^2 \, dy_1 \, dy_2$$

$$= \iint_D dy_1 \, dy_2 = \text{area of } D. \tag{8.126}$$

We now want also to express μ_0 in terms of the Laurent coefficients of the inverse conformal mapping f^{-1} taking the exterior of the unit disk onto the exterior of D. To this end we make use of the following theorem (cf. Nehari [135]).

Theorem 8.16 (*Area theorem*) *Let A = area of D and $f^{-1}: R^2\backslash\Omega \to R^2\backslash D$ conformally map the exterior of the unit disk onto the exterior of D. If f^{-1} has the Laurent expansion*

$$f^{-1}(w) = w/a + b_0 + b_1/w + b_2/w^2 + \cdots \tag{8.127}$$

then

$$A = \pi/a^2 - \pi \sum_{n=1}^{\infty} n\,|b_n|^2.$$

Proof The unit tangent vector to ∂D is $(dy_1/ds)\vec{\imath} + (dy_2/ds)\vec{\jmath}$ and hence

$$\vec{\nu} = (dy_2/ds)\vec{\imath} - (dy_1/ds)\vec{\jmath}. \tag{8.128}$$

Therefore, from (8.125a), (8.126)

$$A = \frac{1}{4}\int_{\partial D} \frac{\partial}{\partial \nu}(y_1^2 + y_2^2) \, ds$$

$$= \frac{1}{2}\int_{\partial D} y_1 \, dy_2 - y_2 \, dy_1$$

$$= -\frac{i}{2}\int_{\partial D} (y_1 - iy_2)(dy_1 - i\,dy_2) \tag{8.129}$$

since

$$\int_{\partial D} y_1\,dy_1 = \int_{\partial D} y_2\,dy_2 = 0. \tag{8.130}$$

But, $y_1 + iy_2 = f^{-1}(w)$ for w on $\partial\Omega$ and hence

$$A = -\frac{i}{2}\int_0^{2\pi} \overline{f^{-1}(w)}\frac{\partial f^{-1}(w)}{\partial\theta}\,d\theta \tag{8.131}$$

for $w = e^{i\theta}$. Since ∂D is smooth, the Laurent series (8.127) converges on $\partial\Omega$, and hence from (8.127) and (8.131) we have

$$A = \pi/a^2 - \pi\sum_{n=1}^{\infty} n\,|b_n|^2. \tag{8.132}$$

We now turn to the relation (8.125b). From (8.128) we have

$$(\partial/\partial\nu)[\rho^n e^{-in\phi}]\,ds = (\partial/\partial\nu)[y_1 - iy_2]^n\,ds = in[y_1 - iy_2]^{n-1}[dy_1 - i\,dy_2] \tag{8.133}$$

and hence from (8.115), (8.125b), (8.133) we have that for $n \geq 1$,

$$\begin{aligned}
\bar{\mu}_n &= \frac{-i}{2^n(n-1)!\,a}\int_{\partial D}\left[f(z) + \frac{1}{f(z)}\right]z^{n-1}\,dz \\
&= \frac{-i}{2^n(n-1)!\,a}\oint_{|w|=1}\left[w + \frac{1}{w}\right][f^{-1}(w)]^{n-1}\frac{df^{-1}}{dw}\,dw \\
&= \frac{i}{2^n n!\,a}\oint_{|w|=1}\left[1 - \frac{1}{w^2}\right][f^{-1}(w)]^n\,dw,
\end{aligned} \tag{8.134}$$

where the bar denotes conjugation. We now use (8.134) to compute the coefficients of the Laurent expansion of f^{-1} as given by (8.127). Without loss of generality, we assume that the coefficient b_0 is zero, since this simply represents a translation in the z plane. Making this assumption and computing the residue in (8.134) gives

$$\begin{aligned}
\bar{\mu}_1 &= -\frac{\pi}{a}\left[b_1 - \frac{1}{a}\right] \\
\bar{\mu}_2 &= -\frac{\pi}{2a^2}b_2 \\
\bar{\mu}_3 &= -\frac{\pi}{8a}\left[\frac{b_3}{a^2} - \frac{b_1}{a^2} + \frac{b_1^2}{a}\right] \\
&\;\;\vdots
\end{aligned} \tag{8.135}$$

In general we can write $f^{-1}(w) = w/a + g(w)$ where

$$g(w) = \frac{b_1}{w} + \frac{b_2}{w^2} + \cdots \tag{8.136}$$

and note that

$$[f^{-1}(w)]^n = \left[\frac{w}{a} + g(w)\right]^n = \sum_{k=0}^{n} \binom{n}{k} \left(\frac{w}{a}\right)^{n-k} [g(w)]^k. \tag{8.137}$$

The highest coefficient of f^{-1} appearing in the residue of (8.134) is now seen to come from the $k = 1$ term in (8.137) and this implies

$$\bar{\mu}_n = -\frac{\pi b_n}{a^n 2^{n-1}(n-1)!} + \text{lower order coefficients.} \tag{8.138}$$

Hence (8.134) allows us to compute each coefficient b_n recursively in terms of the measured data μ_k, $k = 1, \ldots, n$, and the (unknown) mapping radius a^{-1}.

We are now presented with a serious computation problem since, in practice, we only know a finite number of the μ_n, say for $n = 0, 1, \ldots, N$. From the information gathered thus far we would then have to solve (8.135) for each of the b_n, $n = 1, \ldots, N$, where $b_n = b_n(a)$, and then substitute back into (8.132) (recalling that $\mu_0 = A$) for the determination of a (truncating the series in (8.132) at $n = N$). This is a highly impractical and inaccurate method for determining a (and hence the coefficients b_n) and, in addition, gives us no idea of the magnitude of the coefficients b_n for $n > N$. A more practical approach is as follows. Send in a second plane wave along the x_2-axis (any other direction not along the x_1-axis would also suffice), i.e. move the transmitter. Then, mathematically, the problem can be formulated as (8.100a)–(8.100d) with $\exp(ikx_1)$ replaced by $\exp(ikx_2)$. All the analysis goes through exactly as before, except that the harmonic function u_0 is now replaced by u_0^* where

$$u_0^*(\mathbf{x}) = a^{-1} \operatorname{Im}[f(z) - 1/f(z)]. \tag{8.139}$$

We note that $f(z) - 1/f(z)$ is purely imaginary on ∂D. The equation corresponding to (8.134) is now

$$\overline{\mu_n^*} = -\frac{1}{2^n n!\, a} \oint_{|w|=1} \left[1 + \frac{1}{w^2}\right] [f^{-1}(w)]^n \, dw; \qquad n \geq 1 \tag{8.140}$$

where

$$\overline{\mu_n^*} = i^{n-1} \lim_{k \to 0} \frac{a_n^*(k)}{k^{n+1}} \tag{8.141}$$

with $a_n^*(k)$ being the nth Fourier coefficient of the far field pattern. From

(8.140) we now have that

$$\overline{\mu_1^*} = -\frac{\pi i}{a}\left[b_1 + \frac{1}{a}\right]. \tag{8.142}$$

Hence from (8.135) and (8.142) we have

$$a = \left(\frac{2\pi}{\bar{\mu}_1 + i\overline{\mu_1^*}}\right)^{1/2}. \tag{8.143}$$

Having obtained a we can now determine b_n, $n = 1, 2, \ldots, N$ from (8.135) and from (8.126), (8.132) we have an estimate on the coefficients b_n for $n > N$ given by

$$\sum_{n=N+1}^{\infty} n\,|b_n|^2 = \pi/a^2 - \sum_{n=1}^{N} n\,|b_n|^2 - \mu_0. \tag{8.144}$$

Finally, we note that the above procedure for determining the coefficients of the conformal mapping f^{-1} is highly unstable for the higher order coefficients. This is due to the fact that in practice the μ_n become increasingly difficult to measure for large n and hence the computation of b_n from (8.138) is subject to a sharp decline in accuracy. This completes our discussion of the inverse scattering problem for a 'hard' cylinder.

We now consider the inverse scattering problem for a cylinder under the assumption that the obstacle is 'soft', i.e. that the boundary condition is of Dirichlet type. Our aim is again to obtain a nonlinear 'moment' problem for the conformal mapping taking the exterior of the unit disk Ω onto the exterior of the obstacle D and then to solve this moment problem. The main problem to be faced is in the first step: that of solving the exterior Dirichlet problem by iteration. The difficulty here is that it is no longer possible to 'shift' the eigenvalues as with the Neumann problem (cf. the discussion preceding Theorem 8.15) and we are forced to modify the kernel of the integral operator itself. Having done this, we can again derive a nonlinear moment problem for the conformal mapping $f^{-1}: R^2\backslash\Omega \to R^2\backslash D$. We shall then solve this moment problem and, by means of the area theorem, obtain error estimates for the case when only a finite number of moments are known (this is, of course, always the case in practice).

To begin with, we again let D be a bounded, simply connected domain with C^2 boundary ∂D and let ν denote the outward unit normal to ∂D. We assume, further, that D contains the origin. We first want to consider two boundary value problems for Laplace's equation which we shall denote by Problem I and Problem II. The existence of a unique solution to these two problems is guaranteed by the results of Section 8.3, and our aim is to obtain these solutions by iteration. Problem I is to find

$u \in C^2(R^2 \backslash \bar{D}) \cap C^1(R^2 \backslash D)$ such that

$$u(\mathbf{x}) = \log \frac{1}{|\mathbf{x}|} + u_s(\mathbf{x}), \qquad \mathbf{x} \in R^2 \backslash D \tag{8.145a}$$

(I) $\Delta_2 u = 0$ in $R^2 \backslash \bar{D}$ (8.145b)

$u = 0$ on ∂D (8.145c)

u_s is bounded as $r = |\mathbf{x}|$ tends to infinity. (8.145d)

Since u_s is bounded as r tends to infinity we can conclude that $\partial u_s/\partial r = 0(1/r^2)$ and

$$\lim_{r \to \infty} u_s(\mathbf{x}) = \alpha \tag{8.146}$$

exists, where α is a constant. Then, from Green's formula, we have (where $R = |\mathbf{x} - \mathbf{y}|$)

$$\frac{1}{2\pi} \int_{\partial D} \left[\log R \frac{\partial u_s}{\partial \nu(y)} - u_s \frac{\partial}{\partial \nu(y)} \log R \right] ds(y)$$
$$= \begin{cases} u_s(\mathbf{x}) - \alpha; & \mathbf{x} \in R^2 \backslash \bar{D} \\ \frac{1}{2} u_s(\mathbf{x}) - \alpha; & \mathbf{x} \in \partial D \end{cases} \tag{8.147}$$

$$\frac{1}{2\pi} \int_{\partial D} \left[\log R \frac{\partial}{\partial \nu(y)} \log |\mathbf{y}| - \log |\mathbf{y}| \frac{\partial}{\partial \nu(y)} \log R \right] ds(y)$$
$$= \begin{cases} \log |\mathbf{x}|; & \mathbf{x} \in R^2 \backslash \bar{D} \\ \frac{1}{2} \log |\mathbf{x}|; & \mathbf{x} \in \partial D \end{cases}$$

and hence, for $\mathbf{x} \in R^2 \backslash D$, we have

$$\frac{1}{2\pi} \int_{\partial D} \log R \frac{\partial u}{\partial \nu} ds(y) = u(\mathbf{x}) - \alpha. \tag{8.148}$$

Since

$$\frac{1}{2\pi} \int \frac{\partial u}{\partial \nu} ds(y) = -1 \tag{8.149}$$

we have that (8.148) is equivalent to

$$\frac{1}{2\pi} \int_{\partial D} [\log R - \tfrac{1}{2} \log |\mathbf{x}|] \frac{\partial u}{\partial \nu} ds(y) = u(\mathbf{x}) - \alpha + \tfrac{1}{2} \log |\mathbf{x}| \tag{8.150}$$

and, using Theorem 8.10, we finally arrive at

$$\frac{\partial u(\mathbf{x})}{\partial \nu} - \frac{1}{\pi} \int_{\partial D} \frac{\partial}{\partial \nu(x)} [\log R - \tfrac{1}{2} \log |\mathbf{x}|] \frac{\partial u}{\partial \nu} ds(y) = -\frac{\partial}{\partial \nu} \log |\mathbf{x}| \tag{8.151}$$

for $\mathbf{x}$ on ∂D. (8.151) is an integral equation for the determination of $\partial u/\partial \nu$ for $\mathbf{x}$ on ∂D and, if this equation can be solved, u and α can be determined from (8.148) and the fact that $u=0$ for $\mathbf{x}$ on ∂D.

We now consider Problem II, which is to find $u \in C^2(R\backslash\bar{D})\cap C^1(R^2\backslash D)$ such that

$$\text{(II)}\quad \begin{cases} u(\mathbf{x})=u_i(\mathbf{x})+u_s(\mathbf{x}), \quad \mathbf{x}\in R^2\backslash\bar{D} & (8.152\text{a}) \\ \Delta_2 u=0 \quad \text{in} \quad R^2\backslash D & (8.152\text{b}) \\ u=0 \quad \text{on} \quad \partial D & (8.152\text{c}) \\ u_s \text{ is bounded as } r=|\mathbf{x}| \text{ tends to infinity,} & (8.152\text{d}) \end{cases}$$

where u_i is a known solution of Laplace's equation in all of R^2. The existence of a unique solution to Problem II is again assured by the results of Section 8.3, and our aim is to reformulate the problem in such a manner that it can be solved by iteration. Following the analysis of (8.147)–(8.151) with $\log 1/|\mathbf{x}|$ replaced by $u_i(\mathbf{x})$ we arrive at

$$u_i(\mathbf{x})+\frac{1}{2\pi}\int_{\partial D}\log R\frac{\partial u}{\partial \nu}\,ds(y)=u(\mathbf{x})-\alpha \tag{8.153}$$

for $\mathbf{x}$ in $R^2\backslash D$ and hence, using the fact that in this case

$$\frac{1}{2\pi}\int_{\partial D}\frac{\partial u}{\partial \nu}\,ds=0, \tag{8.154}$$

we have

$$\frac{\partial u(\mathbf{x})}{\partial \nu}-\frac{1}{\pi}\int_{\partial D}\frac{\partial}{\partial \nu(x)}[\log R-\tfrac{1}{2}\log|\mathbf{x}|]\frac{\partial u}{\partial \nu}\,ds(y)=2\frac{\partial u_i(\mathbf{x})}{\partial \nu} \tag{8.155}$$

for $\mathbf{x}$ on ∂D. (8.155) is an integral equation for the determination of $\partial u/\partial \nu$ for $\mathbf{x}$ on ∂D and if this equation is solvable then u and α can be determined from (8.153) and the fact that $u=0$ on ∂D.

We now examine the solvability of the integral equations (8.151) and (8.155). In particular, we shall show that both of these equations can be solved by iteration and the solution expressed as a Neumann series. It is for this reason that the term $\frac{1}{2}(\partial/\partial \nu)\log|\mathbf{x}|$ was added to the kernel of the integral operator appearing in these equations. We first define

$$\langle u, v\rangle=\int_{\partial D}uv\,ds \tag{8.156}$$

and let $\mathbf{B}$ be the operator of potential theory defined in (8.84). Recall from Theorem 8.14 that $\sigma(\mathbf{B})\subset[-1, 1)$. If we define the operator $\mathbf{L}_0$: $C(\partial D)\to C(\partial D)$ by

$$\mathbf{L}_0[\psi]=\mathbf{B}[\psi]+1/2\pi(\partial/\partial \nu)\log|\mathbf{x}|\,\langle\psi, 1\rangle \tag{8.157}$$

then we can write both (8.151) and (8.155) in the form

$$g(\mathbf{x}) = \psi(\mathbf{x}) + (\mathbf{L}_0\psi)(\mathbf{x}) \tag{8.158}$$

where g is a given continuous function prescribed on ∂D and $\psi = \partial u/\partial \nu$.

Theorem 8.17 $\sigma(\mathbf{L}_0) \subset (-1, 1)$.

Proof Our proof is modelled on an argument due to Kress [106]. Let $\mathbf{A}$ be the adjoint of $\mathbf{B}$ in the dual system $\langle C(\partial D), C(\partial D)\rangle$ and $N(\mathbf{I}+\mathbf{B})$ the null space of $\mathbf{I}+\mathbf{B}$. Then, from Section 8.3, we have that $N(\mathbf{I}+\mathbf{B}) = \text{span}\,(\psi_0)$ for some continuous function ψ_0. We shall first show that $\langle \psi_0, 1\rangle \neq 0$. Assume the contrary, i.e. $\langle \psi_0, 1\rangle = 0$. Then, by the Fredholm alternative, $\phi + \mathbf{A}\phi = 1$ is solvable and the double-layer potential

$$v(\mathbf{x}) = -\int_{\partial D} \phi(\mathbf{y}) \frac{\partial}{\partial \nu(y)} \log R \, ds(y); \qquad \mathbf{x} \in R^2 \backslash \bar{D} \tag{8.159}$$

satisfies $v_+(\mathbf{x}) = 1$ for $\mathbf{x}$ on ∂D. Then from Green's formula

$$\begin{aligned}\int_{R^2\backslash D} |\text{grad}\, v|^2 \, dx &= -\int_{\partial D} v_+(\partial \bar{v}_+/\partial \nu) \, ds \\ &= -\int_{\partial D} (\partial \bar{v}_+/\partial \nu) \, ds \\ &= -\int_{\partial D} (\partial \bar{v}_-/\partial \nu) \, ds \\ &= 0.\end{aligned} \tag{8.160}$$

Hence v is constant in $R^2\backslash D$. But, since $v = 0(1/r)$, we have that $v = 0$ in $R^2\backslash D$, a contradiction. Hence $\langle \psi_0, 1\rangle \neq 0$. Now note that for any ψ in $C(\partial D)$ we have

$$\langle \mathbf{B}\psi, 1\rangle = \langle \psi, \mathbf{A}1\rangle = -\langle \psi, 1\rangle \tag{8.161}$$

and, since

$$\int_{\partial D} (\partial/\partial \nu) \log |\mathbf{x}| \, ds = 2\pi, \tag{8.162}$$

we have

$$\langle \mathbf{L}_0\psi, 1\rangle = \langle \mathbf{B}\psi, 1\rangle + \langle \psi, 1\rangle = 0. \tag{8.163}$$

Since $\mathbf{L}_0$ is compact, its spectrum consists only of a discrete set of eigenvalues. We shall show that if λ is an eigenvalue of $\mathbf{L}_0$ then $\lambda \in (-1, 1)$. Suppose $\mathbf{L}_0\psi = \lambda\psi$ where $\lambda \notin [-1, 1)$. Then

$$\lambda\langle \psi, 1\rangle = \langle \mathbf{L}_0\psi, 1\rangle = 0 \tag{8.164}$$

from (8.163) and hence $\langle\psi, 1\rangle = 0$ and $\mathbf{L}_0\psi = \mathbf{B}\psi$. But $\sigma(\mathbf{B}) \subset [-1, 1)$ and this contradicts the assumption that $\lambda \notin [-1, 1)$. Hence $\lambda \in [-1, 1)$. We now show that $\lambda = -1$ is not an eigenvalue. Suppose $\mathbf{L}_0\psi + \psi = 0$. Then, from (8.164), we again have $\langle\psi, 1\rangle = 0$ and $\mathbf{L}_0\psi = \mathbf{B}\psi$. Hence $\psi = a\psi_0$ for some constant a. But then

$$0 = \langle\psi, 1\rangle = \langle a\psi_0, 1\rangle = a\langle\psi_0, 1\rangle \tag{8.165}$$

and, since $\langle\psi_0, 1\rangle \neq 0$, we have $a = 0$ and hence $\psi = 0$, a contradiction. Therefore -1 is not an eigenvalue and $\sigma(\mathbf{L}_0) \subset (-1, 1)$.

Theorem 8.17 implies that we can solve the integral equation (8.158) by successive approximations. Indeed, since $\sigma(\mathbf{L}_0) \subset (-1, 1)$ the Neumann series for $(\mathbf{I} + \mathbf{L}_0)^{-1}$ converges and hence the solution of (8.158) is given by

$$\psi = (\mathbf{I} + \mathbf{L}_0)^{-1}[g] = \sum_{n=0}^{\infty} (-1)^n \mathbf{L}_0^n[g]. \tag{8.166}$$

We now consider the exterior Dirichlet problem for the Helmholtz equation which we shall call Problem III. This problem is to determine $u \in C^2(R^2 \backslash \bar{D}) \cap C^1(R^2 \backslash D)$ such that

$$u(\mathbf{x}) = u_i(\mathbf{x}) + u_s(\mathbf{x}), \qquad \mathbf{x} \in R^2 \backslash D \tag{8.167a}$$

$$\text{(III)} \qquad \Delta_2 u + k^2 u = 0 \quad \text{in} \quad R^2 \backslash \bar{D} \tag{8.167b}$$

$$u = 0 \quad \text{on} \quad \partial D \tag{8.167c}$$

$$\lim_{r \to \infty} \sqrt{r}(\partial u_s/\partial r - iku_s) = 0 \tag{8.167d}$$

where u_i is a solution of the Helmholtz equation in all of R^2 and the radiation condition (8.167d) is assumed to hold uniformly in all directions. The existence of a unique solution to Problem III for k sufficiently small follows again from the results of Section 8.3, and our aim is to obtain this solution by iteration for k suitably small. We proceed exactly as in Problem II, noting that $u_s = 0(1/\sqrt{r})$, and arrive at

$$u_i(\mathbf{x}) - \frac{i}{4}\int_{\partial D} H_0^{(1)}(kR)\frac{\partial u}{\partial \nu}\, ds(y) = u(\mathbf{x}) \tag{8.168}$$

for $\mathbf{x} \in R^2 \backslash D$ where $H_0^{(1)}$ is Hankel's function defined by (8.94). Let D be contained in a disk of radius a. Then, since $u = 0$ on ∂D, we have from Green's formula and the radiation condition satisfied by u_s that

$$\begin{aligned} \frac{i}{4}\int_{\partial D} H_0^{(1)}(k\,|\mathbf{y}|)\frac{\partial u}{\partial \nu}\, ds &= \frac{i}{4}\int_{|\mathbf{y}|=a} \left[H_0^{(1)}(k\,|\mathbf{y}|)\frac{\partial u}{\partial \nu} - u\frac{\partial}{\partial \nu}H_0^{(1)}(k\,|\mathbf{y}|)\right] ds \\ &= \frac{i}{4}\int_{|\mathbf{y}|=a} \left[H_0^{(1)}(k\,|\mathbf{y}|)\frac{\partial u_i}{\partial \nu} - u_i\frac{\partial}{\partial \nu}H_0^{(1)}(k\,|\mathbf{y}|)\right] ds \\ &= u_i(0). \end{aligned} \tag{8.169}$$

Hence (8.168) is equivalent to

$$\frac{\partial u(\mathbf{x})}{\partial \nu}+\frac{\mathrm{i}}{2}\int_{\partial D}\frac{\partial}{\partial \nu(x)}\left[H_0^{(1)}(kR)+\frac{\pi \mathrm{i}}{4\log k}H_0^{(1)}(k\,|\mathbf{x}|)H_0^{(1)}(k\,|\mathbf{y}|)\right]\frac{\partial u}{\partial \nu}\,\mathrm{d}s(y)$$

$$=2\frac{\partial u_{\mathrm{i}}(\mathbf{x})}{\partial \nu}+\frac{\pi \mathrm{i} u_{\mathrm{i}}(0)}{2\log k}\frac{\partial}{\partial \nu}H_0^{(1)}(k\,|\mathbf{x}|) \tag{8.170}$$

for $\mathbf{x}$ on ∂D.

We now want to show that for k sufficiently small the integral equation (8.170) can be solved by iteration. To this end we define $\mathbf{L}_0$ by (8.157) and let $\mathbf{L}_k$ be the integral operator defined in (8.170). Then from (8.94) we have that for k positive and sufficiently small

$$\|\mathbf{L}_k-\mathbf{L}_0\|=0(1/|\log k|) \tag{8.171}$$

and, since the spectral radius of $\mathbf{L}_0$ is less than one, we can conclude from the proof of Theorem 8.15 that there exists a positive number k_0 such that for $|k|\leqslant k_0$ the spectral radius of $\mathbf{L}_k$ is less than one. Hence the integral equation (8.170) can be solved by iteration and the solution expressed as a Neumann series.

We are now in a position to consider the inverse scattering problem for a soft cylinder, i.e. we consider Problem III when the obstacle D is unknown but u_{i} is given along with the behavior of u_{s} at infinity

$$u_{\mathrm{s}}(\mathbf{x})=\tfrac{1}{4}\mathrm{e}^{i(kr+\pi/4)}\left(\frac{2}{\pi kr}\right)^{1/2}F(\theta;k)+0\left(\frac{1}{r^{3/2}}\right) \tag{8.172}$$

where F is the far field pattern. We assume that F is known for $|k|<k_0$ and $-\pi\leqslant\theta\leqslant\pi$. From (8.168) we have that

$$u_{\mathrm{s}}(\mathbf{x})=-\frac{\mathrm{i}}{4}\int_{\partial D}\frac{\partial u}{\partial \nu}H_0^{(1)}(kR)\,\mathrm{d}s(y) \tag{8.173}$$

and from (8.172), (8.173) and the asymptotic behavior of Hankel's function we see that F is given by

$$F(\theta;k)=-\int_{\partial D}\frac{\partial u}{\partial \nu}\exp\left[-\mathrm{i}k\rho\cos(\theta-\phi)\right]\mathrm{d}s(y) \tag{8.174}$$

where $\mathbf{x}=r\mathrm{e}^{\mathrm{i}\theta}$, $\mathbf{y}=\rho\mathrm{e}^{\mathrm{i}\phi}$. Expanding F in a Fourier series we have

$$F(\theta;k)=\sum_{-\infty}^{\infty}a_n(k)\mathrm{e}^{\mathrm{i}n\theta} \tag{8.175}$$

where

$$a_n(k)=-\left(\frac{1}{2\pi}\right)\int_{-\pi}^{\pi}\int_{\partial D}\frac{\partial u}{\partial \nu}\exp\left[-\mathrm{i}n\theta-\mathrm{i}k\rho\cos(\theta-\phi)\right]\mathrm{d}s(y)\,\mathrm{d}\theta$$

$$=-\mathrm{i}^{-n}\int_{\partial D}\frac{\partial u}{\partial \nu}J_n(k\rho)\mathrm{e}^{-\mathrm{i}n\phi}\,\mathrm{d}s. \tag{8.176}$$

To proceed further we need to obtain a low-frequency approximation to $\partial u/\partial \nu$ evaluated on the (unknown) boundary ∂D. To this end we first assume $u_i(\mathbf{x}) = \exp(ikx_1)$ where $\mathbf{x} = (x_1, x_2)$. Then we have that for $|k| < k_0$, $\mathbf{x}$ on ∂D,

$$\frac{\partial u(\mathbf{x})}{\partial \nu} = \sum_{n=0}^{\infty} (-1)^n \mathbf{L}_k^n \left[\frac{1}{\log k} \frac{\partial}{\partial \nu} \log \frac{1}{|\mathbf{x}|}\right] + 0(|k|)$$

$$= \sum_{n=0}^{\infty} (-1)^n \mathbf{L}_0^n \left[\frac{1}{\log k} \frac{\partial}{\partial \nu} \log \frac{1}{|\mathbf{x}|}\right] + 0\left(\frac{1}{|\log k|^2}\right)$$

$$= \frac{1}{\log k} \frac{\partial u_0(\mathbf{x})}{\partial \nu} + 0\left(\frac{1}{|\log k|^2}\right) \tag{8.177}$$

where, for $\mathbf{x} \in R^2 \backslash D$, we can identify u_0 as the solution of Problem I. If $w = f(z)$, $z = x_1 + \mathrm{i}x_2$, is the (unique) analytic function that conformally maps the exterior of D onto the exterior of the unit disk such that at infinity f has the Laurent expansion (8.114), then we can write

$$u_0(\mathbf{x}) = -\log |f(z)|. \tag{8.178}$$

Hence, from (8.176), (8.178) and the Taylor series expansion of Bessel's function, we have that for $n = 0, 1, 2, \ldots$

$$a_n(k) = \frac{\mathrm{i}^{-n} k^n}{2^n n! \log k} \int_{\partial D} \frac{\partial}{\partial \nu} \log |f(\rho \mathrm{e}^{\mathrm{i}\phi})| \, \rho^n \mathrm{e}^{-\mathrm{i}n\phi} \, \mathrm{d}s + 0\left(\frac{|k|^n}{|\log k|^2}\right). \tag{8.179}$$

Since $J_n(k\rho) = (-1)^n J_{-n}(k\rho)$, the same expression (up to conjugation and a factor of plus or minus one) holds for $n \leqslant -1$ and thus no new information can be gathered from the negatively indexed Fourier coefficients. Thus, if we define (for $n \geqslant 0$)

$$\mu_n = \mathrm{i}^n 2^n n! \lim_{k \to 0} \frac{a_n(k) \log k}{k^n} \tag{8.180}$$

we can reformulate the inverse scattering problem as follows: given μ_n, $n = 0, 1, 2, \ldots$ as defined by (8.180), to determine ∂D from the relation

$$\bar{\mu}_n = \int_{\partial D} \frac{\partial}{\partial \nu} \log |f(\rho \mathrm{e}^{\mathrm{i}\phi})| \, \rho^n \mathrm{e}^{\mathrm{i}n\phi} \, \dot{\mathrm{d}}s \tag{8.181}$$

where f is defined as in (8.114) and we have taken the complex conjugate of (8.179).

We now proceed to determine ∂D from (8.181). More specifically, we shall show that from (8.181) we can compute the Laurent coefficients of $f^{-1}(w)$ module the determination of the mapping radius a^{-1} where a is defined in (8.114). The mapping radius will then be obtained by deriving a second moment problem analogous to (8.181). From the Cauchy–Riemann equations

$$(\partial/\partial\nu)\log|f(\rho e^{i\phi})| = -(\partial/\partial s)\arg f(\rho e^{i\phi}) \tag{8.182}$$

and hence from (8.181) we have

$$\begin{aligned}\bar{\mu}_n &= -\int_{\partial D}\frac{\partial}{\partial s}\arg f(\rho e^{i\phi})\rho^n e^{in\phi}\,ds\\ &= -\int_{|w|=1}\frac{\partial \arg w}{\partial w}[f^{-1}(w)]^n\,dw\\ &= i\int_{|w|=1}\frac{1}{w}[f^{-1}(w)]^n\,dw.\end{aligned} \tag{8.183}$$

From (8.183), (8.127), and the fact that (8.127) is uniformly convergent on $|w|=1$ we now have

$$\begin{aligned}\bar{\mu}_0 &= -2\pi\\ \bar{\mu}_1 &= -2\pi b_0\\ \bar{\mu}_2 &= -2\pi\left[b_0^2+\frac{2b_1}{a}\right]\\ &\;\;\vdots\end{aligned} \tag{8.184}$$

and, in general,

$$\bar{\mu}_n = -2\pi n a^{-n+1}b_{n-1} + \text{lower order coefficients} \tag{8.185}$$

for $n \geqslant 1$, where b_n are the Laurent coefficients of f^{-1} as defined by (8.127). Hence, module the mapping radius a^{-1}, we can determine the Laurent coefficients of f^{-1} recursively in terms of the measured far field data μ_n.

In order to determine the mapping radius we use the extra information gained by measuring the far field pattern arising from transmitting the incident wave $u_i(\mathbf{x}) = \exp(-ikx_1)$ or combined with $u_i(\mathbf{x}) = \exp(ikx_1)$, $u_i(\mathbf{x}) = \sin kx_1$. Then, from our previous results, we have that if u^* is the solution of Problem III for $u_i(\mathbf{x}) = \sin kx_1$, $|k| \leqslant k_0$, then for $\mathbf{x}$ on ∂D we have for k sufficiently small

$$\begin{aligned}\frac{\partial u^*(\mathbf{x})}{\partial\nu} &= \sum_{n=0}^{\infty}(-1)^n\mathbf{L}_k^n\left[2k\frac{\partial x_1}{\partial\nu}\right]+0(|k^3|)\\ &= \sum_{n=0}^{\infty}(-1)^n\mathbf{L}_0^n\left[2k\frac{\partial x_1}{\partial\nu}\right]+0\left(\left|\frac{k}{\log k}\right|\right)\\ &= k\frac{\partial u_0^*(\mathbf{x})}{\partial\nu}+0\left(\left|\frac{k}{\log k}\right|\right)\end{aligned} \tag{8.186}$$

where, for $\mathbf{x}\in R^2\backslash D$, we can identify u_0^* as the solution of Problem II for $u_i(\mathbf{x})=x_1$. Then, in terms of the conformal mapping f, we can write

$$u_0^*(\mathbf{x})=\frac{1}{a}\operatorname{Re}\left[f(z)-\frac{1}{f(z)}\right], \tag{8.187}$$

and, if $a_n^*(k)$ are the Fourier coefficients of the far field pattern, we have from (8.176) that for $n\geqslant 0$

$$\begin{aligned} a_n^*(k) &= -\mathrm{i}^{-n}\int_{\partial D}\frac{\partial u^*}{\partial \nu}J_n(k\rho)\mathrm{e}^{-\mathrm{i}n\phi}\,\mathrm{d}s \\ &= \frac{-\mathrm{i}^{-n}k^{n+1}}{2^n n!}\int_{\partial D}\frac{\partial u_0^*}{\partial \nu}\rho^n\mathrm{e}^{-\mathrm{i}n\phi}\,\mathrm{d}s+0\left(\frac{|k|^{n+1}}{|\log k|}\right). \end{aligned} \tag{8.188}$$

Thus, if we define

$$\mu_n^*=\mathrm{i}^n 2^n n!\lim_{k\to 0}\frac{a_n^*(k)}{k^{n+1}}, \tag{8.189}$$

we have

$$\overline{\mu_n^*}=-\int_{\partial D}\frac{\partial u_0^*}{\partial \nu}\rho^n\mathrm{e}^{\mathrm{i}n\phi}\,\mathrm{d}s=\int_{\partial D}\frac{\partial v_0^*}{\partial \nu}\rho^n\mathrm{e}^{\mathrm{i}n\phi}\,\mathrm{d}s \tag{8.190}$$

where v_0^* is the harmonic conjugate of u_0^* defined by

$$v_0^*(\mathbf{x})=\frac{1}{a}\operatorname{Im}\left[f(z)-\frac{1}{f(z)}\right]. \tag{8.191}$$

Since $u_0^*=0$ on ∂D we now have

$$\begin{aligned} \overline{\mu_n^*} &= \frac{1}{a}\int_{\partial D}\frac{\partial}{\partial s}\left[f(\rho\mathrm{e}^{\mathrm{i}\phi})-\frac{1}{f(\rho\mathrm{e}^{\mathrm{i}\phi})}\right]\rho^n\mathrm{e}^{\mathrm{i}n\phi}\,\mathrm{d}s \\ &= \frac{1}{a}\int_{|w|=1}\frac{\partial}{\partial w}\left[w-\frac{1}{w}\right][f^{-1}(w)]^n\,\mathrm{d}w \\ &= \frac{1}{a}\int_{|w|=1}\left[1+\frac{1}{w^2}\right][f^{-1}(w)]^n\,\mathrm{d}w. \end{aligned} \tag{8.192}$$

From (8.127), (8.192) we compute

$$\overline{\mu_1^*}=\frac{2\pi\mathrm{i}}{a}\left[b_1+\frac{1}{a}\right] \tag{8.193}$$

and hence from (8.184), (8.193) we can determine a in terms of μ_1, $\overline{\mu_1^*}$ and $\bar{\mu}_2$.

Having obtained a we can now determine $b_0, b_1, \ldots$ from (8.184) and hence the conformal mapping f^{-1}. However, in practice only a finite number of the μ_n can be determined, and hence we need an estimate on the error made in approximating f^{-1} by its truncated Laurent series on $|w|=1$. To this end we assume that we know *a priori* that D contains a disk of radius r_0. Then, if A denotes the area of D we have, from the area theorem, that if the coefficients $b_0, b_1, \ldots, b_N$ are assumed known and f_N^{-1} is defined by

$$f_N^{-1}(w)=\frac{w}{a}+b_0+\frac{b_1}{w}+\cdots+\frac{b_N}{w^N}, \tag{8.194}$$

then

$$\begin{aligned}\int_{-\pi}^{\pi}|f^{-1}(e^{i\theta})-f_N^{-1}(e^{i\theta})|^2\,d\theta &= \sum_{n=N+1}^{\infty}|b_n|^2\\ &\leqslant \frac{1}{N+1}\sum_{n=N+1}^{\infty} n\,|b_n|^2\\ &= \frac{1}{N+1}\left[\frac{1}{a^2}-\sum_{n=0}^{N} n\,|b_n|^2-\frac{A}{\pi}\right]\\ &\leqslant \frac{1}{N+1}\left[\frac{1}{a^2}-r_0^2\right].\end{aligned} \tag{8.195}$$

Note that if no *a priori* information is available we must set r_0 equal to zero in (8.195).

Exercises

(1) Prove that (8.3) is valid.

(2) Consider the inverse scattering problem for a 'soft' sphere of radius one contained in a spherically stratified medium of compact support. Mathematically, the 'direct' problem can be formulated as (8.20a)–(8.20c) plus the boundary condition $u(1, \theta)=0$. Show that the determination of $m(r)$ (as given in (8.20b)) can be reduced to a moment problem of the form

$$\mu_n=\int_1^{r_0} m(r)[r^{2n+2}+r^{-2n}-2r]\,dr$$

and show that if it exists, the solution of this moment problem is unique.

(3) Prove Theorem 8.10.

(4) Prove Theorem 8.11.

(5) Use potential theoretic methods to establish the existence of a solution to the interior Dirichlet and Neumann problems for Laplace's equation.
(6) Establish the uniqueness of a solution to the exterior Dirichlet and Neumann problems for the Helmholtz equation in R^2.
(7) Improve the estimate (8.195) by making additional assumptions on D or ∂D.

Bibliography

1 Agmon, S., *Lectures on Elliptic Boundary Value Problems*, Princeton, N.J., Van Nostrand, 1965.

2 Agmon, S., *Unicité et Convexité dans les Problemes Differentiels*, Montreal, University of Montreal Press, 1966.

3 Ahlfors, L. V., *Complex Analysis*, New York, McGraw-Hill, 1953.

4 Ahner, J. F., Low frequency Neumann scattering problems in two dimensions, *AFOSR Scientific Report TR-72-1243*, University of Delaware, Newark, Delaware, 1972.

5 Ahner, J. F., Scattering by an inhomogeneous medium, *J. Inst. Math. Appl.*, **19**, 425–439, 1977.

6 Ahner, J. F. and Hsiao, G. C., On the two dimensional exterior boundary value problems of elasticity, *SIAM J. Appl. Math.*, **31**, 677–685, 1976.

7 Ahner, J. F. and Kleinman, R. E., The exterior Neumann problem for the Helmholtz equation, *Arch. Rat. Mech. Anal.*, **52**, 26–43, 1973.

8 Anger, G., *Posed Problems in Differential Equations*, Berlin, Akademie-Verlag, 1979.

9 Backus, G. and Gilbert, F., Uniqueness in the inversion of inaccurate gross earth data, *Phil. Trans. Roy. Soc. Lond.*, **266**, 123–197, 1970.

10 Bates, R. H. T. and Wall, D. J. N., Null field approach to scalar diffraction, *Phil. Trans. Roy. Soc. Lond.*, **287**, 45–114, 1977.

11 Bauer, F., Garabedian, P. and Korn, D., *Supercritical Wing Sections*, Berlin, Springer-Verlag: Lecture Notes in Economics and Mathematical Systems, Vol. 66, 1972.

12 Bergman, S., *Integral Operators in the Theory of Linear Partial Differential Equations*, Berlin, Springer-Verlag, 1969.

13 Bergman, S., *The Kernel Function and Conformal Mapping*, Providence, R.I., American Mathematical Society, 1970.

14 Bergman, S. and Herriot, J. G., Numerical solution of boundary value problems by the method of integral operators, *Numer. Math.*, **7**, 42–65, 1965.
15 Bergman, S. and Schiffer, M., *Kernel Functions and Elliptic Differential Equations in Mathematical Physics*, New York, Academic Press, 1953.
16 Bers, L., *Theory of Pseudoanalytic Functions*, New York, NYU Lecture Notes, 1953.
17 Bers, L., John, F. and Schecter, M., *Partial Differential Equations*, New York, Interscience Publishers, 1964.
18 Bloom, C. O. and Kazarinoff, N. D., *Short Wave Radiation Problems in Inhomogeneous Media: Asymptotic Solutions*, Berlin, Springer-Verlag: Lecture Notes in Mathematics, Vol. 522, 1976.
19 Browder, F. E., Approximation by solutions of partial differential equations, *Amer. J. Math.*, **84**, 134–160, 1962.
20 Buzbee, B. and Carasso, A., On the numerical computation of parabolic problems for preceding times, *Math. Comp.*, **27**, 237–266, 1973.
21 Cannon, J. R., Some numerical results for the solution of the heat equation backwards in time, in: *Numerical Solutions of Nonlinear Differential Equations*, Donald Greenspan (ed.), New York, John Wiley, pp. 21–54, 1966.
22. Carasso, A. and Stone, A. P., *Improperly Posed Boundary Value Problems*, London, Pitman, 1975.
23. Carroll, R. and Showalter, R. E., *Singular and Degenerate Cauchy Problems*, New York, Academic Press, 1976.
24. Chadan, K. and Sabatier, P. C., *Inverse Problems in Quantum Scattering Theory*, New York, Springer-Verlag, 1977.
25. Chang, Y. F. and Colton, D., The numerical solution of parabolic partial differential equations by the method of integral operators, *Int. J. Computer Math.*, **6**, 229–239, 1977.
26 Coleman, B. D., Duffin, R. J. and Mizel, V. J., Instability, uniqueness and nonexistence theorems for the equation $u_t = u_{xx} - u_{xtx}$ on a strip, *Arch. Rat. Mech. Anal.*, **19**, 100–116, 1965.
27 Colton, D., *Partial Differential Equations in the Complex Domain*, London, Pitman, 1976.
28 Colton, D., *Solution of Boundary Value Problems by the Method of Integral Operators*, London, Pitman, 1976.
29 Colton, D., On the inverse scattering problem for axially symmetric solutions of the Helmholtz equation, *Quart. J. Math.*, **22**, 125–130, 1971.
30 Colton, D., Pseudoparabolic equations in one space variable, *J. Diff. Eqns*, **12**, 559–565, 1972.

31 Colton, D., The inverse Stefan problem for the heat equation in two space variables, *Mathematika*, **21**, 282–286, 1974.

32 Colton, D., Integral operators and reflection principles for parabolic equations in one space variable, *J. Diff. Eqns*, **15**, 551–559, 1974.

33 Colton, D., A reflection principle for solutions to the Helmholtz equation and an application to the inverse scattering problem, *Glasgow Math. J.*, **18**, 125–130, 1977.

34 Colton, D., On reflection principles for parabolic equations in one space variable, *Proc. Edinburgh Math. Soc.*, **21**, 143–147, 1978.

35 Colton, D., The approximation of solutions to the backwards heat equation in a nonhomogeneous medium, *J. Math. Anal. Appl.*, **72**, 418–429, 1979.

36 Colton, D., The inverse scattering problem for a cylinder, *Proc. Roy. Soc. Edinburgh*, **84A,** 135–143, 1979.

37 Colton, D., Schwarz reflection principles for solutions of parabolic equations, to appear.

38 Colton, D. and Kleinman, R. E., The direct and inverse scattering problems for an arbitrary cylinder: Dirichlet boundary conditions, *Proc. Roy. Soc. Edinburgh.*

39 Colton, D. and Kress, R., The construction of solutions to acoustic scattering problems in a spherically stratified medium, *Quart. J. Mech. Appl. Math.*, **31**, 9–17, 1978.

40 Colton, D. and Kress, R., The construction of solutions to acoustic scattering problems in a spherically stratified medium, II, *Quart. J. Mech. Appl. Math.*, **32**, 53–62, 1979.

41 Colton, D. and Watzlawek, W., Complete families of solutions to the heat equation and generalized heat equation in R^n, *J. Diff. Eqns*, **25**, 96–107, 1977.

42 Colton, D. and Wimp, J., Asymptotic behaviour of the fundamental solution to the equation of heat conduction in two temperatures, *J. Math. Anal. Appl.*, **69**, 411–418, 1979.

43 Courant, R. and Hilbert, D., *Methods of Mathematical Physics*, Vol. II, New York, Interscience Publishers, 1961.

44 Davis, P. J., *Interpolation and Approximation*, New York, Dover Publications, 1975.

45 De Alfaro, D. and Regge, T., *Potential Scattering*, Amsterdam, North-Holland, 1965.

46 Diaz, J. B., On an analogue of Euler–Cauchy polygon method for the numerical solution of $u_{xy} = f(x, y, u, u_x, u_y)$, *Arch. Rat. Mech. Anal.*, **1**, 357–390, 1958.

47 Dolph, C. L. The integral equation method in scattering theory, in: *Problems in Analysis*, R. C. Gunning (ed.), Princeton, N.J., Princeton University Press, pp. 201–227, 1970.

48 Dolph, C. L. and Scott, R. A., Recent developments in the use of complex singularities in electromagnetic theory and elastic wave propagation, in: *Electromagnetic Scattering*, P. L. E. Uslenghi (ed.), New York, Academic Press, pp. 503–570, 1978.

49 Egorov, Ju. V., On the solvability of differential equations with simple characteristics, *Russian Math. Surveys*, **26**, 113–130, 1971.

50 Eisenstat, S., On the rate of convergence of the Bergman–Vekua method for the numerical solution of elliptic boundary value problems, *SIAM J. Numer. Anal.*, **11**, 654–680, 1974.

51 Epstein, B., *Partial Differential Equations—An Introduction*, New York, McGraw-Hill, 1962.

52 Erdélyi, A., Singularities of generalized axially symmetric potentials, *Comm. Pure Appl. Math.*, **9,** 403–414, 1956.

53 Erdélyi, A., Magnus, W., Oberhettinger, F. and Tricomi, F. G., *Higher Transcendental Functions*, Vol. II, New York, McGraw-Hill, 1953.

54 Ewing, R., The approximation of certain parabolic equations backward in time by Sobolev equations, *SIAM J. Math. Anal.*, **6**, 283–294, 1975.

55 Filippenko, V., On the reflection of harmonic functions and of solutions of the wave equation, *Pacific J. Math.*, **14**, 883–893, 1964.

56 Fock, V. A., *Electromagnetic Diffraction and Propagation Problems*, Oxford, Pergamon Press, 1965.

57 Friedman, A., *Partial Differential Equations of Parabolic Type*, Englewood Cliffs, N.J., Prentice Hall, 1964.

58 Friedman, A., *Partial Differential Equations*, New York, Holt, Rinehart and Winston, 1969.

59 Friedrichs, K. O., On the differentiability of the solutions of elliptic differential equations, *Comm. Pure Appl. Math.*, **6,** 299–326, 1953.

60 Gajewski, J. and Zacharias, K., Zur Regularisierung einer Klasse nichtkorrekter Probleme bei Evolutionsgleichungen, *J. Math. Anal. Appl.*, **38**, 784–789, 1972.

61 Garabedian, P. R., *Partial Differential Equations*, New York, John Wiley, 1964.

62 Garabedian, P. R., An example of axially symmetric flow with a free surface, *Studies in Mathematics and Mechanics Presented to Richard von Mises*, New York, Academic Press, pp. 149–159, 1954.

63 Garabedian, P. R., An integral equation governing electromagnetic waves, *Quart. Appl. Math.*, **12**, 428–433, 1955.

64 Garabedian, P. R., Partial differential equations with more than two independent variables in the complex domain, *J. Math. Mech.*, **9**, 241–271, 1960.

65 Garabedian, P. R., Applications of the theory of partial differential equations to problems of fluid mechanics, *Modern Mathematics for the Engineer: Second Series*, E. Beckenbach (ed.), New York, McGraw-Hill, pp. 347–372, 1961.
66 Garabedian, P. R., An unsolvable problem, *Proc. Amer. Math. Soc.*, **25**, 207–208, 1970.
67 Garabedian, P. and Lieberstein, H., On the numerical calculation of detached bow shock waves in hypersonic flow, *J. Aero. Science*, **25**, 109–118, 1958.
68 Gelfand, I. M. and Levitan, B. M., On the determination of a differential equation from its spectral function, *Amer. Math. Soc. Trans.*, **2 (1)**, 253–304, 1955.
69 Gilbert, R. P., *Singularities of Three-Dimensional Harmonic Functions*, Thesis, Pittsburgh, Carnegie Inst. Tech., 1958.
70 Gilbert, R. P., *Function Theoretic Methods in Partial Differential Equations*, New York, Academic Press, 1969.
71 Gilbert, R. P., *Constructive Methods for Elliptic Equations*, Berlin, Springer-Verlag: Lecture Notes in Mathematics, Vol. 365, 1974.
72 Gilbert, R. P., *Elliptic Systems in the Plane*, to appear.
73 Gilbert, R. P., On the singularities of generalized axially symmetric potentials, *Arch. Rat. Mech. Anal.*, **6**, 171–176, 1960.
74 Gilbert, R. P., The construction of solutions for boundary value problems by function theoretic methods, *SIAM J. Math. Anal.*, **1**, 96–114, 1970.
75 Gilbert, R. P. and Colton, D., On the numerical treatment of partial differential equations by function theoretic methods in: *Numerical Solution of Partial Differential Equations—II*, B. Hubbard (ed.), New York, Academic Press, pp. 273–326, 1971.
76 Gilbert, R. P. and Linz, P., The numerical solution of some elliptic boundary value problems by integral operator methods, *Constructive and Computational Methods for Solving Differential and Integral Equations*, Berlin, Springer-Verlag: Lecture Notes in Mathematics, Vol. 430, pp. 237–252, 1974.
77 Gilbert R. P. and Weinacht, R. J., *Function Theoretic Methods in Differential Equations*, London, Pitman, 1976.
78 Glasstone, S. and Edlund, M. C., *The Elements of Nuclear Reactor Theory*, Princeton, N.J., Van Nostrand, 1952.
79 Grushin, V. V., A certain example of a differential equation without solutions, *Math. Notes*, **10**, 499–501, 1971.
80 Haack, W. and Wendland, W., *Lectures on Partial and Pfaffian Differential Equations*, Oxford, Pergamon Press, 1972.
81 Hartman, P. and Wilcox, C., On solutions of the Helmholtz equation in exterior domains, *Math. Zeit.*, **75,** 228–255, 1961.

82 Hellwig, G., *Partial Differential Equations*, New York, Blaisdell Publishing, 1964.

83 Henrici, P., On the domain of regularity of generalized axially symmetric potentials, *Proc. Amer. Math. Soc.*, **8**, 29–31, 1957.

84 Hill, C. D., Parabolic equations in one space variable and the non-characteristic Cauchy problem, *Comm. Pure Appl. Math.*, **20**, 619–633, 1967.

85 Hill, C. D., A method for the construction of reflection laws for a parabolic equation, *Trans. Amer. Math. Soc.*, **133**, 357–372, 1968.

86 Hill, R. N., Kleinman, R. and Pfaff, E. W., Convergent long wavelength expansion method for two-dimensional scattering problems, *Canadian J. Physics*, **51**, 1541–1564, 1973.

87 Hochstadt, H., *Integral Equations*, New York, John Wiley, 1973.

88 Hong, S. and Goodrich, R. F., Applications of conformal mapping to scattering and diffraction problems, in: *Electromagnetic Wave Theory, Part II*, J. Brown (ed.), Oxford, Pergamon Press, pp. 907–914, 1967.

89 Hopf, E., Elementare Vemerkungen über die Lösungen partieller Differentialgleichungen zweiter Ordnung vom elliptischen Typus, *Sitzungsber. d. Preuss. Akad. d. Wiss.*, **19**, 147–152, 1927.

90 Hopf, E., Über den Funktionalen, inbesondere den analytischen Charakter der Lösungen elliptischer Differentialgleichungen zweiter Ordnung, *Math. Zeit.*, **34**, 194–233, 1931.

91 Hopf, E. A remark on linear elliptic differential equations of the second order, *Proc. Amer. Math. Soc.*, **3**, 791–793, 1952.

92 Hörmander, L., *Linear Partial Differential Operators*, Berlin, Springer-Verlag, 1963.

93 John, F., *Plane Waves and Spherical Means Applied to Partial Differential Equations*, New York, Interscience Publishers, 1955.

94 John, F., On linear partial differential equations with analytic coefficients, unique continuation of data, *Comm. Pure Appl. Math.*, **2**, 209–253, 1949.

95 John, F., Continuation and reflection of solutions of partial differential equations, *Bull. Amer. Math. Soc.*, **63**, 327–344, 1957.

96 John, F., Continuous dependence on data for solutions of partial differential equations with prescribed bound, *Comm. Pure Appl. Math.*, **13**, 551–585, 1960.

97 Jones, B. F., An approximation theorem of Runge type for the heat equation, *Proc. Amer. Math. Soc.*, **52**, 289–292, 1975.

98 Jones, D. S., *The Theory of Electromagnetism*, Oxford, Pergamon Press, 1964.

99 Jörgens, K., *Lineare Integraloperatoren*, Stuttgart, Teubner-Verlag, 1970.

100 Karp, S., A convergent 'farfield' expansion for two-dimensional radiation functions, *Comm. Pure Appl. Math.*, **14**, 427–434, 1961.
101 Keller, J. B. and Lewis, R. M., *Asymptotic Theory of Wave Propagation and Diffraction*, New York, John Wiley, to appear.
102 Kleinman, R. E. and Roach, G. F., Boundary integral equations for the three-dimensional Helmholtz equation, *SIAM Review*, **16**, 214–236, 1974.
103 Kleinman, R. E. and Wendland, W., On Neumann's method for the exterior Neumann problem for the Helmholtz equation, *J. Math. Anal. Appl.*, **57**, 170–202, 1977.
104 Knops, R. J., *Symposium on Non-well-posed Problems and Logarithmic Convexity*, Berlin, Springer-Verlag: Lecture Notes in Mathematics, Vol. 316, 1973.
105 Korevaar, J., *Mathematical Methods*, Vol. I., New York, Academic Press, 1968.
106 Kress, R., *Integral Equations and Their Application to Boundary Value Problems of Mathematical Physics*, Glasgow, University of Strathclyde: Lecture Notes, 1977.
107 Krzywoblocki, M. Z. v., *Bergman's Linear Integral Operator Method in the Theory of Compressible Fluid Flow*, Berlin, Springer-Verlag, 1960.
108 Lagnese, J., General boundary value problems for differential equations of Sobolev type, *SIAM J. Math. Anal.*, **3**, 105–119, 1972.
109 Lattes, R. and Lions, J. L., *The Method of Quasireversibility*, New York, American Elsevier, 1969.
110 Lavrentiev, M. M., *Some Improperly Posed Problems in Mathematical Physics*, New York, Springer-Verlag, 1967.
111 Lavrentiev, M. M., Romanov, V. G. and Vasiliev, V. G., *Multidimensional Inverse Problems for Differential Equations*, Berlin, Springer-Verlag: Lecture Notes in Mathematics, Vol. 167, 1970.
112 Lax, P. D., A stability theory of abstract differential equations and its application to the study of local behavior of solutions of elliptic equations, *Comm. Pure Appl. Math.*, **9**, 747–766, 1956.
113 Lax, P. D., and Phillips, R. S., *Scattering Theory*, New York, Academic Press, 1967.
114 Leis, R., *Vorlesungen über Partielle Differentialgleichungen Zweiter Ordnung*, Mannheim, BI Mannheim, 1967.
115 Levin, B. Ja., *Distribution of Zeros of Entire Functions*, Providence, R.I., American Mathematical Society, 1964.
116 Lewy, H., Neuer Beweis des analytischen Charakters der Lösungen elliptischer Differentialgleichungen, *Math. Ann.*, **101**, 609–619, 1929.
117 Lewy, H., An example of a smooth linear partial differential equation without solution, *Ann. Math.*, **66**, 155–158, 1957.

118 Lewy, H., On the reflection laws of second order differential equations in two independent variables, *Bull. Amer. Math. Soc.*, **65**, 37–58, 1959.

119 Lewy, H., On the extension of harmonic functions in three variables, *J. Math. Mech.*, **14**, 925–927, 1965.

120 Magenes, E., Sull' equazione del calore: teoremi di unicità e teoremi di completezza connessi col metodo di integrazione di M. Picone I, *Rend. Sem. Mat. Univ. Padova,* **21**, 99–123, 1952.

121 Magenes, E., Sull' equazione del calore: teoremi di unicità e teoremi di completezza connessi col metodo di integrazione di M. Picone II, *Rend. Sem. Mat. Univ. Padova,* **21**, 136–170, 1952.

122 Majda, A., High-frequency asymptotics for the scattering matrix and the inverse problem of acoustic scattering, *Comm. Pure Appl. Math.*, **30**, 165–194, 1977.

123 Malgrange, B., Existence et approximation des solutions des équations aux derivées partielles et des équations de convolutions, *Ann. Inst. Fourier*, **6**, 271–355, 1956.

124 Marsden, J. E., *Basic Complex Analysis*, San Francisco, W. H. Freeman, 1973.

125 Meister, V. E., Weck, N. and Wendland, W., *Function Theoretic Methods for Partial Differential Equations*, Berlin, Springer-Verlag: Lecture Notes in Mathematics, Vol. 561, 1976.

126 Millar, R. F., Singularities of two-dimensional exterior solutions of the Helmholtz equation, *Proc. Camb. Phil. Soc.*, **69,** 175–188, 1971.

127 Millar, R. F., The Rayleigh hypothesis and a related least squares solution to scattering problems for periodic surfaces and other scatterers, *Radio Science*, **8**, 785–796, 1973.

128 Millar, R. F., The singularities of solutions to analytic elliptic boundary value problems, *Function Theoretic Methods for Partial Differential Equations*, Berlin, Springer-Verlag: Lecture Notes in Mathematics, Vol. 561, pp. 73–87, 1976.

129 Miller, K., Least square methods for ill-posed problems with a prescribed bound, *SIAM J. Math. Anal.*, **1**, 52–74, 1970.

130 Miller, K., Stabilized quasireversibility and other nearly best possible methods for non-well-posed problems, in: *Symposium on Non-well-posed Problems and Logarithmic Convexity*, R. J. Knops (ed.), Berlin, Springer-Verlag: Lecture Notes in Mathematics, Vol. 316, pp. 161–176, 1973.

131 Müller, C., *Foundations of the Mathematical Theory of Electromagnetic Waves*, Berlin, Springer-Verlag, 1969.

132 Müller, C., Radiation patterns and radiation fields, *J. Rat. Mech. Anal.*, **4**, 235–246, 1955.

133 Nashed, M. Z., Proceedings of the *Symposium on Ill-Posed Problems: Theory and Practice*, to appear.

134 Nashed, M. Z., Approximate regularized solutions to improperly posed linear integral and operator equations, *Constructive and Computational Methods for Differential and Integral Equations*, Berlin, Springer-Verlag: Lecture Notes in Mathematics, Vol. 430, pp. 289–332, 1974.

135 Nehari, Z., *Conformal Mapping*, New York, Dover Publications, 1975.

136 Newton, R. G., *Scattering Theory of Waves and Particles*, New York, McGraw-Hill, 1966.

137 Nirenberg, L., *Lectures on Linear Partial Differential Equations*, Regional Conference Series in Mathematics, No. 17, Providence, R.I., American Mathematical Society, 1973.

138 Nirenberg, L., A strong maximum principle for parabolic equations, *Comm. Pure Appl. Math.*, **6**, 167–177, 1953.

139 Nirenberg, L. and Trèves, F., On local solvability of linear partial differential equations. I. Necessary conditions, *Comm. Pure Appl. Math.*, **23**, 1–38, 1970.

140 Nirenberg, L. and Trèves, F., On local solvability of linear partial differential equations. II. Sufficient conditions, *Comm. Pure Appl. Math.*, **23**, 459–510, 1970.

141 Oleĭnik, O. A., On properties of some boundary value problems for equations of elliptic type, *Math. Sbornik, N. S.*, **30(72),** 695–702, 1952.

142 Payne, L. E., *Improperly Posed Problems in Partial Differential Equations*, Philadelphia, Society for Industrial and Applied Mathematics, 1975.

143 Payne, L. E. and Stakgold, I., Nonlinear problems in nuclear reactor analysis, *Proc. Conference on Nonlinear Problems in Physical Sciences and Biology*, Berlin, Springer-Verlag: Lecture Notes in Mathematics, Vol. 322, pp. 298–307, 1972.

144 Payne, L. E. and Stakgold, I., On the mean value of the fundamental mode in the fixed membrane problem, *Applicable Anal.*, **3**, 295–303, 1973.

145 Prosser, R. T., Can one see the shape of a surface? *Amer. Math. Monthly*, **84**, 259–270, 1977.

146 Protter, M. and Weinberger, H. F., *Maximum Principles in Differential Equations*, Englewood Cliffs, N.J., Prentice-Hall, 1967.

147 Rellich, F., Über das asymptotische Verhalten der Lösungen von $\Delta u + \lambda u = 0$ in unendlichen Gebieten, *Jahresber. Deutsch. Math. Verein.*, **53,** 57–65, 1943.

148 Rorres, C., Low-energy scattering by an inhomogeneous medium and by a potential, *Arch. Rat. Mech. Anal.*, **39**, 340–357, 1970.

149 Rosenbloom, P. C. and Widder, D. V., Expansions in terms of heat polynomials and associated functions, *Trans. Amer. Math. Soc.*, **92**, 220–266, 1959.

150 Rubinstein, L. I., *The Stefan Problem*, Providence, R.I., American Mathematical Society, 1971.
151 Rundell, W., The solution of initial-boundary value problems for pseudoparabolic partial differential equations, *Proc. Royal Soc. Edinburgh*, **74A**, 311–326, 1974.
152 Rundell, W. and Stecher, M., Remarks concerning the supports of solutions of pseudoparabolic equations, *Proc. Amer. Math. Soc.*, **63**, 77–81, 1977.
153 Rundell, W. and Stecher, M., A method of ascent for parabolic and pseudoparabolic partial differential equations, *SIAM J. Math. Anal.*, **7**, 898–912, 1976.
154 Ruscheweyh, S., *Funktionen theoretische Methoden bei Partiellen Differentialgleichungen*, Bonn, Gessellschaft für Mathematik und Datenverarbeitung, Nr. 77, 1973.
155 Sabatier, P. C., On geophysical inverse problems and constraints, *J. Geophysics*, **43**, 115–137, 1977.
156 Schaefer, P. and Sperb, R., Maximum principles for some functionals associated with the solution of elliptic boundary value problems, *Arch. Rat. Mech. Anal.*, **61**, 65–76, 1976.
157 Schryer, N. L., Constructive approximation of solutions to linear elliptic boundary value problems, *SIAM J. Numer. Anal.*, **9**, 546–572, 1972.
158 Serrin, J. B., On the Harnack inequality for linear elliptic equations, *J. d'Analyse Math.*, **4**, 292–308, 1954/56.
159 Showalter, R. E., *Hilbert Space Methods for Partial Differential Equations*, London, Pitman, 1977.
160 Showalter, R. E., The final value problem for evolution equations, *J. Math. Anal. Appl.*, **47**, 563–572, 1974.
161 Sleeman, B. D., The three-dimensional inverse scattering problem for the Helmholtz equation, *Proc. Camb. Phil. Soc.*, **73,** 477–488, 1973.
162 Stakgold, I., *Green's Functions and Boundary Value Problems*, New York, John Wiley, 1979.
163 Stakgold, I., Global estimates for nonlinear reaction and diffusion, *Proc. Dundee Conf. on Ordinary and Partial Differential Equations*, Berlin, Springer-Verlag: Lecture Notes in Mathematics, Vol. 415, pp. 252–266, 1974.
164 Taylor, A. E., *Introduction to Functional Analysis*, New York, John Wiley, 1958.
165 Tikhonov, A. N. and Arsenin, V. Y., *Solutions of Ill-Posed Problems*, Washington, D.C., V. H. Winston and Sons, 1977.
166 Ting, T. W., A cooling process according to two-temperature theory of heat conduction, *J. Math. Anal. Appl.*, **45**, 23–31, 1974.
167 Titchmarsh, E. C., *Theory of Functions*, London, Oxford University Press, 1939.

168 Tutschke, W., *Partielle komplexe Differentialgleichungen in Einer und in Mehreren Komplexen Variablen*, Berlin, VEB Deutscher Verlag der Wissenschaften, 1977.

169 Ursell, F., On the short-wave asymptotic theory of the wave equation $(\nabla^2 + k^2)\phi = 0$, *Proc. Camb. Phil. Soc.*, **53**, 115–133, 1957.

170 Ursell, F., On the exterior problem of acoustics, *Proc. Cambridge Phil. Soc.*, **74,** 117–125, 1973.

171 Vekua, I. N., *Generalized Analytic Functions*, Oxford, Pergamon Press, 1962.

172 Vekua, I. N., *New Methods for Solving Elliptic Equations*, Amsterdam, North-Holland, 1967.

173 Vekua, I. N., On completeness of a system of metaharmonic functions, *Dokl. Akad. Nauk. SSSR*, **90**, 715–718, 1953.

174 Weinstein, A., Generalized axially symmetric potential theory, *Bull. Amer. Math. Soc.*, **59**, 20–38, 1953.

175 Wendland, W., *Elliptic Systems in the Plane*, London, Pitman, 1979.

176 Wendland, W., Die Fredholmsche Alternative für Operatoren, die bezuglich eines bilinearen Functionals adjungiert sind, *Math. Zeit.*, **101**, 61–64, 1967.

177 Wendland, W., Bemerkungen über die Fredholmschen Sätze, *Meth. Verf. Math. Physik*, **3**, 141–176, 1970.

178 Weston, V. H., Bowman, J. J. and Ar, E., On the inverse electromagnetic scattering problem, *Arch. Rat. Mech. Anal.*, **31**, 199–213, 1968.

179 Widder, D., *The Heat Equation*, New York, Academic Press, 1975.

180 Wilcox, C. H., *Scattering Theory for the D'Alembert Equation in Exterior Domains*, Berlin, Springer-Verlag: Lecture Notes in Mathematics, Vol. 442, 1975.

181 Wilton, D. R. and Mittra, R., A new numerical approach to the calculation of electromagnetic scattering properties of two-dimensional bodies of arbitrary cross-section, *IEEE Trans. Antennas Prop.*, **20**, 310–317, 1972.

182 Yasuura, K., A view of numerical methods in diffraction problems, in: *Radio Waves and Circuits; Radio Electronics*, W. V. Tilston and M. Sauzade (eds), Brussels, URSI, pp. 257–270, 1971.

183 Yu, C. L., Reflection principle for solutions of higher order elliptic equations with analytic coefficients, *SIAM J. Appl. Math.*, **20**, 358–363, 1971.

184 Yu, C. L., Reflection principle for systems of first order elliptic equations with analytic coefficients, *Trans. Amer. Math. Soc.*, **164**, 489–501, 1972.

185 Yu, C. L., Cauchy problem and analytic continuation for systems of first order elliptic equations with analytic coefficients, *Trans. Amer. Math. Soc.*, **185**, 429–443, 1973.

186 Yu, C. L., Integral representation, analytic continuation and the reflection principle under the complementing boundary condition for higher order elliptic equations in the plane, *SIAM J. Math. Anal.*, **5**, 209–223, 1974.

Index